T0329671

Geminos's *Introduction to the Phenomena*

Geminos's *Introduction to the Phenomena*

A TRANSLATION AND STUDY OF
A HELLENISTIC SURVEY OF ASTRONOMY

James Evans and J. Lennart Berggren

PRINCETON UNIVERSITY PRESS

PRINCETON AND OXFORD

Copyright © 2006 by Princeton University Press
Published by Princeton University Press, 41 William Street, Princeton, New Jersey 08540
In the United Kingdom: Princeton University Press, 3 Market Place, Woodstock,
Oxfordshire OX20 1SY

Library of Congress Cataloging-in-Publication Data

Geminus.
[Introduction to astronomy. English]
Geminos's introduction to the phenomena : a translation and study of a Hellenistic survey
of astronomy / [translated and commentary by] James Evans and J. Lennart Berggren.
p. cm.
Includes bibliographical references and index.
ISBN-13: 978-0-691-12339-4 (alk. paper)
ISBN-10: 0-691-12339-X (alk. paper)
1. Astronomy, Greek—Textbooks. 2. Astronomy—Early works to 1800. I. Evans,
James, 1948– II. Berggren, J. L. III. Title.

QB21.G4513 2006
520—dc22   2006049372

British Library Cataloging-in-Publication Data is available

This book has been composed in Sabon

Printed on acid-free paper. ∞

pup.princeton.edu

Printed in the United States of America

10 9 8 7 6 5 4 3 2 1

Contents

FRAGMENTS 1 AND 2, FROM GEMINOS'S OTHER WORKS

REFERENCE MATERIALS

Illustrations

Figures *have been added by the authors to provide historical context and to illustrate astronomical concepts. They occur in our Introduction, as well as in our Commentary to the various chapters in Geminos's* Introduction to the Phenomena.

Diagrams *are photographs (or, in one case, a drawing) of illustrations that occur in the actual medieval manuscripts. They occur in the body of Geminos's chapters i and ii, and in our commentary to his chapters x and xi.*

FIGURES IN OUR INTRODUCTION

FIGURES IN OUR COMMENTARY TO GEMINOS'S
VARIOUS CHAPTERS

DIAGRAMS

Tables

Preface

A famous mathematician of the twentieth century once lamented, "The algebraic topologist has practically ceased to communicate with the point-set topologist!" This remark is characteristic of our time and culture, in which knowledge has become fractured into thousands of specialties and subspecialties, and in which no one science can claim to hold a privileged place. It was not so in Greek Antiquity, when astronomy was the central science, with vital links to nearly every other aspect of the culture.

Astronomy had important relations with other sciences, such as physics (or philosophy of nature) and mathematics. As Aristotle pointed out, the motions of the celestial bodies were the best clues to the physics (or essential natures) of these bodies. But the methods of investigation, as well as of demonstration and application, in astronomy were so thoroughly mathematical that astronomy was often considered to be a branch of applied mathematics. It was partly for this reason that Plato included it in the quadrivium of mathematical arts recommended for the education of the guardians of his ideal state. Astronomy also had links to ancient religion, for the planets were widely held to be divine, and the celestial phenomena commanded the attention of the poets, who from the time of Hesiod had sung of the celestial signs and of the revolving year. Astronomy provided subject matter for craftsmen, who represented the heavens in the form of ingenious globes and mechanisms. And, finally, it was one of the most significant channels of intellectual exchange between ancient civilizations, most notably between the Babylonians and the Greeks. Geminos's *Introduction to the Phenomena* manifests all these cultural affiliations of ancient astronomy. This graceful manual of astronomy, written probably in the first century B.C. by a man who had had some experience of teaching, remains today an engaging introduction to the central natural science of Antiquity.

Much of ancient astronomy requires of the reader an approach over a long and difficult road. This includes Ptolemy's *Almagest* as well as the planetary theories of the Babylonian scribes. And much else is either devoted to special problems (such as Aristarchos's treatise *On the Sizes and Distances of the Sun and Moon*) or consists of repetitive material arranged in theorems and proofs that survived because it was useful for teaching (such as Autolykos's *On Risings and Settings*). Finally, there is a good deal of low-level, nontechnical material written for ancient readers who were not willing to try very hard (such as the astronomical portions of Pliny's *Natural History*), material that cannot really give a modern reader a fair appreciation of the ancient science.

Geminos's *Introduction to the Phenomena* is one of a very small number of ancient astronomical works that can be read with appreciation and understanding by a nonspecialist, but one that offers, nevertheless, a competent and reasonably comprehensive account of its subject. The English translation of the *Introduction to the Phenomena* here presented is the first complete one ever to be published. We hope it will be of interest and use not only to historians of science, but also to students of ancient civilization, as well as to scientists who want to know more about the origins of their art.

The book that the reader now holds had a long gestation. JE encountered Geminos while completing a doctoral dissertation at the University of Washington. Working from Germaine Aujac's relatively recent edition of the Greek text (Aujac 1975), he translated most of Geminos's chapter v, on the circles of the celestial sphere, for his students to read in a course he was teaching on the history of astronomy. He was struck by Geminos's patience and clarity, and charmed by his frequent use of literary examples to illustrate a point of astronomy. Geminos was an excellent writer for students to read—the astronomy was accurate and useful, but the priorities and concerns of the ancient thinker came through loud and clear as well. A student could read Geminos with scientific as well as historical senses open. In 1983–84, JE spent a year in Paris, with the aid of a Fulbright Grant, working at the Centre Koyré under the patronage of the late René Taton, and going regularly to the history of astronomy sessions of the *Équipe Copernic* (Copernicus team) at the Paris Observatory. He spent most of the year working on eighteenth-century physics. But in his spare time, simply for pleasure and as a way of keeping up his Greek, he completed a draft translation of the whole of Geminos's *Introduction to the Phenomena*. Some time later, at the International Congress of the History of Science, held at Berkeley, he had opportunity to meet Germaine Aujac, who responded generously by lending him her microfilms of the most important Geminos manuscripts. JE used his translation of Geminos for many years in teaching a course on the history of astronomy at the University of Puget Sound. A short extract from Geminos's chapter v appeared in his *The History and Practice of Ancient Astronomy* (Evans 1998). While working on other projects, he occasionally took time out for his ongoing commentary on Geminos.

JLB and JE had known each other for a long time before beginning a collaboration on Geminos. JLB works on both medieval Arabic mathematics and ancient Greek mathematics. With R.S.D. Thomas, he had published a translation of and commentary on Euclid's *Phenomena*

(Berggren and Thomas 1996), which well equipped him for further work on Greek phenomena literature. And, with Alexander Jones, he had published an annotated translation of the theoretical chapters of Ptolemy's *Geography* (Berggren and Jones 2000). Through a happy alignment of their stars, JLB and JE had free time, at the same time, to devote to getting Geminos into final form. JLB undertook a complete review and revision of the translation. The two translators consulted regularly on issues raised by the Greek text and its translation, striving not only for accuracy and readable English but also for fidelity to Geminos's style and cadences. JLB also reviewed and corrected the draft commentary, adding to it his own insights. The two authors wrote the introduction and appendices together. Although they were able to do much of their work apart, communicating by telephone, e-mail, and fax, they retain fond memories of working together at the Evans's dining room table in Seattle, at the Berggren's house in Coquitlam, British Columbia, and their mountain retreat in Whistler, as well as in the bar of the Sylvia Hotel in Vancouver. The final push to completion of the manuscript was carried out at the Helen Riaboff Whiteley Center, in Friday Harbor, Washington.

The authors are grateful to friends and colleagues who helped in many different ways in the course of this project. Our greatest scholarly debt is to Germaine Aujac, whose Greek text provided the basis for our translation. Alexander Jones, Liba Taub, and Noel Swerdlow read considerable portions of the manuscript and were generous with comments and suggestions, many of which resulted in improvements or saved us from errors. Marinus Taisbak helped with several translations from the Latin, and Tasoula Berggren proofread the Greek of the glossary and typed the index. JE remains grateful to the late Will Humphreys for a day-long discussion of Proklos's citations of Geminos in the *Commentary on the First Book of Euclid's Elements*. Daryn Lehoux generously lent us his own translation of the Geminos *parapēgma* in advance of the publication of his book on *parapēgmata*. Other scholars took the trouble to respond to questions, among whom we particularly thank Lawrence Bliquez, David Lupher, and A. Mark Smith. We alone, of course, are responsible for any errors or shortcomings in the final product.

Ernst Künzl and Rudolph Schmidt helped obtain photographs for use as illustrations, and Ross Mulhausen aided with photographic work and image processing. The University of Puget Sound provided sabbatical leave that enabled Professor Evans to concentrate on the project, as Simon Fraser University did for Professor Berggren. Both institutions also provided financial support for the payment of fees for the reproduction of some of the images appearing in the book. The Helen Riaboff Whiteley Center, at Friday Harbor on San Juan Island, generously welcomed us for a stay while we were completing the final version of the book. It has also been a privilege and a pleasure to work with the capable staff of Princeton University Press, including Ingrid Gnerlich, editor; Jill Harris, production editor; and Bill Carver, copy editor. We are grateful to the following institutions and their helpful staffs for supplying photographs and for permission to use them in this book: Musée du Louvre, Museo Archeologico Nazionale di Napoli, Museo Archeologico Nazionale di Aquileia, Bibliotheca Nazionale Marciana (Venice), Bibliotheca Apostolica Vaticana, Römisch-Germanisches Zentralmuseum Mainz, Antikensammlung (Berlin), University of Washington Libraries, The British Museum, Trinity College Dublin, and the Musée National d'Histoire et d'Art Luxembourg.

It remains only to say that we owe our deepest thanks to Sharon Evans and Tasoula Berggren for their understanding and support during the years it took us to complete this work.

J.E.
J.L.B.
January 2006

Introduction

Geminos, a Greek scientific writer of wide-ranging interests, has been assigned dates ranging from the first century B.C. to the first century A.D.,
with, we believe, the first century B.C. the more likely. We know nothing
of the circumstances of his life. Of three works he is believed to have
written, only one, the *Introduction to the Phenomena*, has come down
to us. (This work is also frequently referred to as the *Isagoge*, from the
first word of its Greek title, *Eisagōgē eis ta phainomena*.) The translation
of his *Introduction to the Phenomena* here presented is the first complete
English version ever published.

For the modern reader, Geminos provides a vivid impression of an ed-
ucated Greek's view of the cosmos and of astronomy around the beginning of our era. Moreover, he is frequently a graceful and charming
writer, constantly aware of his audience, and his book remains quite
readable today. Indeed, it is one of a very small number of works of ancient astronomy that can be read right through with appreciation and
understanding by a nonspecialist. Because Geminos covers most of the
central topics of ancient Greek astronomy, his text provides an excellent
general survey of those parts of that astronomy not dependent on sophisticated mathematical models. An English translation of the *Introduction to the Phenomena* should thus be useful not only to historians of
astronomy but also to historians of science more generally, to those interested in classical civilization, and to astronomers who would like to
know more about the history of their discipline.

We have furnished our translation with a commentary, printed at the
foot of the page and signaled in the text by superscript numerals. The purpose of the commentary is not to summarize all that is known on the
topics at hand, but to open up Geminos's text, to make it more comprehensible, and to reveal its connections with other ancient sources—
philosophical and literary, as well as scientific. It should serve, as well, to
direct readers to the specialized scholarly literature. Textual notes, signaled in Geminos's text by superscript roman letters, are grouped together
in appendix 1.

1. Significance of Geminos's *Introduction to the Phenomena*

Geminos's *Introduction to the Phenomena*, a competent and engaging introduction to astronomy, was probably written in conjunction with teaching. Geminos discusses all of the important branches of Greek astronomy, except planetary theory. This he promises to take up "elsewhere." Perhaps he did discuss planetary theory in another work, but if so, it has not survived. Topics covered in Geminos's *Introduction* include the zodiac, solar theory, the constellations, the theory of the celestial sphere, the variation in the length of the day, lunisolar cycles, phases of the Moon, eclipses, heliacal risings and settings of the fixed stars, terrestrial zones, and an introduction to Babylonian lunar theory. Because the work was written for beginners, it does not often get into technical detail—except in the discussion of lunisolar cycles, where Geminos does indulge in a bit of arithmetic.

Geminos's book is important to the task of filling gaps in the history of Greek astronomy in several ways. In general terms, Geminos provides an overview of most of astronomy in the period between Hipparchos (second century B.C.) and Ptolemy (second century A.D.), and thereby provides a good deal of insight into what was current and common knowledge in Geminos's own day. One of the more charming aspects of his work, frequently in evidence, is his desire to set straight common misconceptions about astronomical matters. In this way, he offers us valuable information about the beliefs of his own audience.

More specifically, Geminos provides detailed discussions of several topics not very well treated by other ancient sources. (1) His discussion of Babylonian lunar theory is an important piece of the story of the adaptation of Babylonian methods by Greek astronomers. (2) His discussion of the 8- and 19-year lunisolar cycles is the most detailed by any extant Greek source. (3) His discussion of Hipparchos's rendering of the constellations provides information not found in other sources. (4) His refutation of the then-common view that changes in the weather are caused by the heliacal risings and settings of the stars is the most patient and detailed such argument that has come down to us.

In the extant manuscripts, Geminos's book concludes with a *parapēgma* (star calendar) that permits one to know the time of year by observation of the stars. Many scholars believe that this compilation is older than Geminos by a century or more. Whether by Geminos or not, this *parapēgma* is one of our most important sources for the early history of the genre. The Geminos *parapēgma* was based substantially upon three earlier *parapēgmata*—those by Euktēmōn (fifth century B.C.), Eudoxos (early fourth century B.C.), and Kallippos (late fourth century B.C.). Because the Geminos *parapēgma* scrupulously cites its sources, it permits us to trace the stages in the evolution of the *parapēgma* between the time of Euktēmōn and the time of Kallippos. Our book includes a translation of the Geminos *parapēgma*, as well as a synoptic table of its contents (appendix 2), which should be useful in the study of this important historical document.

Although ancient and medieval Greek readers would have recognized Geminos's book as belonging to a class of "phenomena" literature (see sections 3 and 4 below), we cannot be sure that *Introduction to the Phenomena* is the title that Geminos himself gave it. This is a common difficulty with ancient scientific texts, the conventional titles of which are not always authorial. The Greek manuscripts of Geminos's text do provide good evidence for the commonly accepted title, although there are several variants. Indeed, the three best and oldest Greek manuscripts present a bit of a puzzle: one gives as its title *Geminos's Introduction to the Phenomena*; another gives *Geminos's Introduction to the Things on High* (*meteōra*); and still another gives neither title nor author's name, since the copyist never filled in this information. Some later Greek manuscripts simply have *"The Phenomena" of Geminos*.[1] As we shall see below (sec. 14), the Latin and Hebrew translations made in the twelfth and thirteenth centuries (from an Arabic intermediary) also show that there was considerable confusion about the title and author of the text. For the sake of simplicity, we shall always refer to Geminos's book by the title commonly used today, and best supported by the Greek manuscripts, *Introduction to the Phenomena*.

2. GEMINOS'S OTHER WORKS

Geminos was the author of two other works that have not come down to us. One was a mathematical work of considerable length that discussed, among other things, the philosophical foundations of geometry. Fortunately, a large number of passages from this work (whether in quotation or in paraphrase) are preserved by Proklos[2] in his *Commentary on the First Book of Euclid's Elements*. The exact title of Geminos's book is uncertain, but in one passage Proklos remarks, "so much have I selected from the *Philokalia* of Geminos."[3] (*Philokalia* means "Love of the Beautiful.") In one passage of considerable interest, Geminos discussed the

---

[1] For the Greek titles, see the first textual note (appendix 1).

[2] Proklos (c. A.D. 410–485) was a prolific Neoplatonist philosopher, best known for his *Platonic Theology* and his commentaries on Plato. His extant scientific works include a *Commentary on the First Book of Euclid's Elements* and a *Sketch of Astronomical Hypotheses*.

[3] Friedlein 1873, 177; Morrow 1970, 139. The title of Geminos's mathematical work has been disputed. See the introduction to fragment 1 for a discussion of this issue.

branches of mathematical science and their relationships to one another. This is the most detailed such discussion that has come down to us from the Greeks. Moreover, it is clear that Geminos was discussing, not merely abstract divisions of mathematics, but actual genres of mathematical writing. Because several of Geminos's branches of mathematics pertain to astronomy (e.g., *sphairopoiïa*, dioptrics, and gnomonics), his discussion sheds light on the relationship of astronomy to other mathematical endeavors. Because of its interest for the history of astronomy, we have included a translation of this passage from Geminos's *Philokalia* as fragment 1.

Geminos was also the author of a meteorological work, which was perhaps a commentary on, or an abridgement of, a now lost *Meteorology* of Poseidōnios.[4] A fragment of some length is preserved by Simplikios[5] in his *Commentary on Aristotle's Physics*. Apparently, by Simplikios's time, Geminos's meteorological book had been lost, for Simplikios makes it clear that he is quoting Geminos, not from Geminos's own work, but from some work by Alexander of Aphrodisias.[6] In the course of his citation, Simplikios says that Alexander drew these remarks from Geminos's "*Concise Exposition of the Meteorology of Poseidōnios*."[7] The fragment from Geminos preserved by Simplikios is of considerable interest, for it is devoted to the limits of astronomical knowledge. In this passage, Geminos discusses the relationship of astronomy to physics (or natural philosophy), arguing that astronomy is, of itself, unable to decide between competing hypotheses and must rely on physics for guidance about first principles. We include a translation of this passage from Geminos's lost meteorological work as fragment 2.

3. On "The Phenomena" in Greek Astronomy

Geminos's *Introduction to the Phenomena* had its roots in a well-established genre. In order to explain what the writers and readers of this genre considered to be relevant, we must say a little about what Greek

---

[4] Poseidōnios (c. 135 to c. 51 B.C.) was a Stoic philosopher who wrote also on history, geography, and astronomy. No complete works survive, but a large number of fragments have been collected. See Edelstein and Kidd 1989; Kidd 1999. Geminos's possible debt to Poseidōnios will be discussed below.

[5] Simplikios, a Neoplatonist of the sixth century A.D., was the author of commentaries on Aristotle's *Physics* and *On the Heavens* and was one of the philosophers who left Athens after the emperor Justinian closed the pagan schools of philosophy in 529.

[6] Alexander of Aphrodisias, who flourished around A.D. 200, was the author of commentaries on Aristotle, many of which survive.

[7] Diels 1882, 291. See fragment 2, below, for the complete passage.

astronomical writers mean by the *phenomena*. The word "phenomena" is a participle of the passive verb *phainomai*, which carries the meanings of "to come to light, come to sight, be seen, appear." The last two are definitive for the astronomical sense of the word, which is "things that are seen/appear in the heavens."

A late source, Simplikios, quotes Sosigenēs as having attributed to Plato the statement that the task of astronomy was to show how, by a combination of uniform circular motions, one could "save (i.e., account for) the phenomena." The ascription to Plato is controversial (see sec. 10 below), but in any case the word *Phenomena* appears as the title of a work by an associate of Plato, Eudoxos of Knidos (early fourth century B.C.). Eudoxos's work has not survived, but its essence is preserved in a poem of the same name by Aratos (early third century B.C.). The poetic *Phenomena* of Aratos was the subject of a commentary by the great astronomer Hipparchos of Rhodes (second century B.C.), who was able to compare it with the text of Eudoxos and demonstrate that Aratos had indeed relied upon Eudoxos. It appears from these sources that Eudoxos's work was devoted to a detailed description of the placement of the fixed stars and the constellations, relative to some standard reference circles on the celestial sphere. The following passages give a sense of the character of Eudoxos's book, and also an idea of what sort of "phenomena" it was occupied with. We quote directly from Hipparchos's *Commentary*, and in each case Hipparchos has made it clear that he is himself directly reporting on Eudoxos's text:

> There is a certain star that remains always in the same spot; this star is the pole of the universe.[8]

> Between the Bears is the tail of the Dragon, the end-star of which is above the head of the Great Bear.[9]

> Aratos, following Eudoxos, says that it [the Dragon's head] moves on the always-visible circle, using these words: "Its head moves where the limits of rising and setting are confounded."[10]

Because Aratos includes in his poem a discussion of the principal circles of the celestial sphere (ecliptic, equator, tropics, arctic circle, as well as the Milky Way), we may surmise that the same material was treated, in more detail, by Eudoxos. So, by the early fourth century, the basic theory of the celestial sphere had been established, and a detailed descrip-

---

[8] Hipparchos, *Commentary on the Phenomena of Eudoxos and Aratos* i 4.1. Hipparchos denounces this as erroneous, pointing out that the place of the celestial north pole was at that time not occupied by a star.

[9] Hipparchos, *Commentary* i 2.3.

[10] Hipparchos, *Commentary*, i 4.7. Quotation from Aratos: *Phenomena* 61–62.

tion of the constellations given. Such were the phenomena of Eudoxos.[11]

The oldest extant work named *The Phenomena* is that of Euclid (c. 300 B.C.).[12] Unlike the work of Eudoxos, Euclid's book has no place for uranography. Rather, a short (and possibly spurious) preface introduces the north celestial pole[13] and the principal circles on the celestial sphere (including the parallel circles, the ecliptic, the horizon, and the Milky Way). The author also introduces the arctic and antarctic circles relative to a given locality and the consequent division of stars into those that never rise, those that rise and set, and those that never set. Thus Eudoxos's descriptions of the constellations have been eliminated in favor of a geometrical exploration of the sphere.

After this beginning, Euclid's treatise proceeds by a series of propositions with proofs and accompanying diagrams, in the style of his more famous *Elements*. These begin with proposition 1 on the central position of the Earth in the cosmos, and then progress through three propositions on the risings and settings of stars. Propositions 8–13 deal with the risings and settings of arcs of the ecliptic, particularly the zodiacal signs, and the work concludes with five propositions on how long it takes equal arcs of the ecliptic to cross the visible and invisible hemispheres. The very format of the work illustrates what had become a commonplace among Greek thinkers, namely that celestial phenomena can be explained rationally.

Other extant early Greek texts for which the celestial phenomena form the subject matter include two works of Euclid's contemporary, Autolykos of Pitanē, both of them written in the theorem-proof style one finds in Euclid's book. In *On the Moving Sphere*, Autolykos treats some of the phenomena arising from the uniform rotation of a sphere around its axis relative to a horizon that separates the visible from the invisible portions of the sphere. It is striking that in *On the Moving Sphere*, the descriptions of all circles other than the horizon are as abstract and geometrical as possible, and there is no explicit mention of the astronomical applications of the theorems. As an example we quote proposition 8: Great circles tangent to the same [parallel circles] to which the horizon is tangent will, as the sphere rotates, fit exactly onto the horizon. The abstract character of many of these propositions illustrates how far the Greek geometrization of astronomy had been carried by the time of Euclid and Autolykos. Many of the propositions are hard to prove, but are easy to illustrate on a celestial globe.

[11] Aristotle (*On the Heavens* ii 13), who was Eudoxos's younger contemporary, also uses the word "phenomena" in its astronomical sense.

[12] For an English translation and commentary, see Berggren and Thomas, 1996.

[13] Here, as in Eudoxos's *Phenomena*, also claimed to be occupied by a star.

Autolykos's other book, *On Risings and Settings*, is devoted to heliacal risings and settings—the annual cycle of appearances and disappearances of the fixed stars. This had been a part of Greek popular astronomy from the earliest days, as illustrated by Hesiod's use of the heliacal risings and settings of the Pleiades, Arcturus, and Sirius to tell the time of year in his poem, *Works and Days* (c. 650 B.C.). Clearly, the sidereal events in the annual cycle were a part of what the Greeks considered "phenomena." Autolykos's goal in *On Risings and Settings* is to provide a mathematical foundation, in the form of theorems, for a field that had previously been in the domain of popular lore. Geminos devotes chapter xiii of his *Introduction to the Phenomena* to the same subject. Indeed, Geminos's heading for chapter xviii is the same as the title of Autolykos's book. As we point out in our commentary on that chapter, Geminos follows Autolykos in all significant details, but eliminates the proofs.

The other major writer on the phenomena was Theodosios of Bithynia (c. 100 B.C.), whose *On Habitations* and *On Days and Nights* are the earliest extant works devoted to a discussion of how the phenomena change from one locality to another: as an observer moves north or south, the stars that are visible will become different and the lengths of the day and night may change. An example of a proposition from the first of these is:

For those living under the north pole[14] the same hemisphere of the cosmos is always visible and the same hemisphere of the cosmos is always invisible, and none of the stars either sets or rises for them, but those in the visible hemisphere are always visible and those in the invisible [hemisphere] are always invisible.[15]

Geminos's use of Theodosios is quite clear, for the Greek heading of Geminos's chapter xvi is the same as that of Theodosios's *On Habitations*,[16] and the heading of chapter vi is only trivially different (singular nouns instead of plurals) from that of Theodosios's *On Days and Nights*.

Many of the founding works on the phenomena, such as those by Euclid, Autolykos, and Theodosios, survived because they were short enough and elementary enough for use in teaching. They became staples of the curriculum for mathematics and astronomy, and so survived through late Antiquity and into the Middle Ages, in both the Arabic and Latin worlds.

The motions of the Sun, Moon, and planets around the zodiac are also part of what the Greeks considered "phenomena." Several features of

---

[14] Recall that for the Greeks the north pole was a point on the celestial sphere.

[15] Berggren and Eggert-Strand, forthcoming.

[16] But in our translation we have chosen the more descriptive rendering, "On Geographical Regions," for the chapter title.

planetary motion posed challenges for explanation: the Sun appears to move more slowly at some times of year, and more rapidly at others. The planets are even more puzzling, since they occasionally stop and reverse direction in what is known as retrograde motion. Most scholars believe that the earliest Greek effort to explain the complex motions of the planets was the book *On Speeds* by Eudoxos. It is lost, but we have two rather lengthy discussions of it, one by Aristotle, who was a contemporary of Eudoxos, and one by Simplikios, who lived 900 years later, and whose account must therefore be used with caution. Probably by the time of Apollōnios of Pergē (late third century B.C.) and certainly by the time of Hipparchos, Eudoxos's approach of modeling the planetary phenomena by the gyrations of nested, homocentric spheres had given way to eccentric circles and epicycles lying in a plane. But this was daunting material to address in an elementary work.[17]

4. The Greek Genre of Astronomical Surveys

In the Hellenistic period, there emerged a demand for popular surveys—works that would take students through the celestial phenomena without forcing them through theorems and proofs. The poetic *Phenomena* of Aratos can be considered one of the first such popularizations. The new popular surveys eschewed the austere geometrical demonstrations of Euclid, Autolykos, and Theodosios tended simply to summarize mathematical results in plain language. They also tended to include a greater variety of subjects of interest to the broad public—phases of the Moon, eclipses, and elements of astronomical geography, such as the theory of terrestrial zones. Of course, all of these topics had deep roots in the history of Greek science. What was new was the attempt to produce comprehensive astronomy textbooks written at an elementary level.

The popular surveys of astronomy could be read for their own sake, but some were clearly intended to form part of the curriculum of studies expected of a well-born student. The geographical writer Strabo (c. 64 B.C. to c. A.D. 25) mentions that students can learn in the elementary mathematics courses all the astronomy they will need for the study of geography. He mentions as an example of the standard astronomical curriculum the theory of the celestial sphere—tropics, equator, zodiac, arctic circle, and horizon.[18] The sort of elementary astronomy course that

---

[17] Of all the elementary writers on astronomy, only Theōn of Smyrna does a good job with planetary phenomena. Geminos (chapter i) gives only an explanation of the eccentric-circle theory of the Sun's motion, a vague reference to the *sphairopoiïa* for each planet, and a brief mention of the basic planetary phenomena.

[18] Strabo, *Geography* i 1.21.

Strabo had in mind is well represented by Geminos's *Introduction to the Phenomena*. Diogenēs Laertios tells us that instruction in basic astronomy was part of the curriculum of Stoic teachers.[19] And, of course, astronomy had long been part of the quadrivium of mathematical studies in the Platonist school.[20] Whether for the sake of popular reading, or for liberal education, or as part of the preparation for more advanced studies, introductions to the astronomical phenomena permeated Greek culture from about 200 B.C. to the end of Antiquity.

It is quite appropriate, then, that Geminos's work is named *Introduction to the Phenomena*, for *eisagōgē* ("introduction") carries two meanings. On one hand, this is a regular word for an elementary treatise on a subject; on the other, it can denote a conduit, or channel, into a harbor. Thus an *eisagōgē* could serve either as a liberal arts survey of astronomy, complete in itself, or as the preparatory course for higher studies in the subject.

Geminos occasionally employs demonstrative mathematical arguments (e.g., in his treatment of lunisolar cycles in chapter viii), and he did not write his book for those who were afraid of numbers or geometry. However, his motto seems to have been "mathematics if necessary, but not necessarily mathematics"—and in any case he makes no use of formal mathematical proofs. Nor does Geminos's work smell of the mathematics classroom. There is none of the graded progression from the easy to the complicated that one finds in, for example, Euclid's *Phenomena*. Had Geminos intended to write a textbook of mathematics he would surely have put chapters iv (the axis and the poles) and v (circles on the sphere) at the beginning, and in any case before chapter i (on the zodiac). A third feature of his work is its blending of the topics of the two earlier genres of phenomena literature (the descriptive uranography of Eudoxos and the mathematical topics of Euclid and his successors) with topics outside of these traditions, namely those he treats in chapters viii–xii, xvii, and xviii. Geminos even stretches the definition of the phenomena to include the astrological aspects of the zodiac signs, in chapter i. In summary, Geminos, in his account of the celestial phenomena, extended the tradition of topics treated to include virtually anything having to do with the fixed stars, the Sun, and the Moon. And he did so in a way that was not simply systematic or mathematical, but discursive and, in a broad sense of the word, scientific.

Geminos's *Introduction to the Phenomena* is but one of several Greek elementary textbooks of astronomy that survive from Antiquity. The two most nearly comparable examples are Theōn of Smyrna's *Mathematical*

---

[19] Diogenēs Laertios, *Lives and Opinions* vii 132.
[20] Plato, *Republic* vii 527d.

Knowledge Useful for Reading Plato[21] (second century A.D.) and Kleomēdēs' *Meteōra*[22] (probably early third to mid-fourth century A.D.). These three surveys have a fair amount of overlap—for example, they all discuss the eccentric-circle theory of the motion of the Sun. But each of the three also treats subjects not covered by the other two. For example, Theōn of Smyrna gives an introduction to the deferent-and-epicycle theory of planetary motion, a subject avoided by Kleomēdēs and Geminos. Kleomēdēs, for his part, is our most detailed source for the famous measurement of the Earth by Eratosthenēs. And Geminos gives a detailed discussion of lunisolar cycles, a subject avoided by Theōn and Kleomēdēs.

These three textbooks of astronomy also differ markedly in tone. While Theōn's book is pervaded by Platonism, Kleomēdēs' book is steeped in Stoic physics and concludes with a savage attack on the Epicureans. Theōn and Kleomēdēs, then, give us nice examples of how an introduction to astronomy could be incorporated into a general course in philosophy—and we have examples in two flavors, Platonist and Stoic. By contrast, Geminos's *Introduction to the Phenomena* is remarkable for its comparative freedom from philosophy, for he is very much a straightforward astronomer. Geminos does, however, display a certain literary bent, and is fond of quoting poets, such as Aratos or Homer, in illustration of astronomical points. His *Introduction to the Phenomena* is also considerably earlier than the textbooks of Theōn and Kleomēdēs, and sheds light on the Greeks' reactions to Babylonian astronomy and astrology, which, in Geminos's day, were in the process of being absorbed and adapted.

An earlier, though shorter and much less polished, survey of astronomy is the *Celestial Teaching* (*Ouranios Didascalea*) of Leptinēs.[23] See fig. I.1. This famous papyrus, conserved in the Louvre, is the oldest existing Greek astronomical document with illustrations. It was composed in the decades before 165 B.C. by a certain Leptinēs as an introduction to astronomy for members of the Ptolemaic court. (So it seems that, despite

---

[21] For a French translation of Theōn of Smyrna, see Dupuis 1892.

[22] For Kleomēdēs, see Todd 1990 (text) and Bowen and Todd 2004 (translation). The original title of Kleomēdēs' work is uncertain, and a number of different titles have been used by editors and translators. On the title issue, see Goulet 1980, 35; Todd 1985; and Bowen and Todd 2002, 1n1. The dating of Kleomēdēs is also difficult. Kleomēdēs says that Antares and Aldebaran are diametrically opposite in the zodiac, the first at Scorpio 15° and the second at Taurus 15°. Using this datum, Neugebauer (1975, 960) arrived at a date for Kleomēdēs around A.D. 370. Bowen and Todd situate Kleomēdēs around A.D. 200, because his work reflects the Stoic polemics against the Peripatetics that began to fade after that period, and because works of Stoic pedagogy become rare after the second century.

[23] Earlier writers call this P. Parisinus 1, but it is now known in the Department of Egyptian Antiquities at the Louvre as N 2325. For the text, see Blass 1887. There is a French translation in Tannery 1893, 283–94. On the history of this papyrus see Thompson 1988, 252–65.

Fig. I.1. A portion of the *Celestial Teaching* of Leptinēs on a papyrus, written shortly before 165 B.C. The left column treats the circles of the celestial sphere and the celestial poles. The right column explains that the stars are called fixed because the constellations always retain their forms and their relationships to one another. Département des Antiquités Egyptiennes, Inv. N. 2325, Musée du Louvre. Photo: Maurice and Pierre Chuzeville.

what Euclid is supposed to have said about geometry, there *was* a royal road to astronomy.) Modern writers sometimes refer to this tract as the "Art of Eudoxos," a name that comes from an acrostic poem on the verso of the papyrus, in which the initial letters of the twelve lines of verse spell out *Eudoxou Techne*. But the colophon on the recto clearly gives the title as the *Ouranios Didascalea of Leptinēs*. In any case, the contents of the treatise are certainly not by Eudoxos. Rather, the tract is a brief and rather choppy account of standard astronomical matters. The text includes a short *parapēgma*, an account of the progress of the Sun and Moon around the zodiac, descriptions of the circles on the celestial sphere, a discussion of eclipses, and values for the lengths of the four seasons according to various authorities. This fare overlaps considerably with the material treated more gracefully by Geminos in the next century.

Finally, numerous commentaries on Aratos's poem *Phenomena* often served as introductions to astronomy. One of the most complete is that of Achilleus (often called Achilles Tatius, probably third century A.D.), whose *Introduction to the "Phenomena" of Aratos* formed a part of his On the All (*Peri tou Pantos*).[24] In our commentary on Geminos, we shall occasionally make comparisons to these other works, which can be thought of as constituting a genre of elementary astronomy textbooks.

5. Geminos's Sources for His *Introduction*

Appendix 4 lists the writers that Geminos cites in his *Introduction to the Phenomena*. He enjoys quoting the poets Homer, Hesiod, and Aratos in illustration of scientific points. This reflects not only his own tastes but also his concession to the literary training of his students and readers. He is not, however, one to ascribe too much scientific knowledge to Homer, and feels that critics such as Kratēs have sometimes gone overboard in this regard. (The occasional use of poetry occurs in other elementary surveys as well, e.g., those of Kleomēdēs, Theōn of Smyrna, and Leptinēs.)

Of the astronomical writers, Geminos names Euktēmōn, Kallippos, Philippos, Eratosthenēs, and Hipparchos, though he may not have known the works of all these people firsthand. Geminos was quite well-informed about lunisolar cycles, but we cannot tell from his remarks on those matters whose works he really had access to. He seems to have used some work of Hipparchos on the constellations that was different from Hipparchos's *Commentary on The Phenomena of Eudoxos and Aratos*. For, in chapter iii, he mentions three decisions of Hipparchos regarding the constellations that have no counterpart in the *Commentary*.

[24] See Maass 1898, 25–85 for what remains of Achilleus's *Commentary* on Aratos. On Achilleus, see Mansfield and Runia 1997, 299–305. Hipparchos's extant *Commentary on the Phenomena of Eudoxos and Aratos* is not a part of this genre, since it is highly technical and numerical in its content.

The clearest and most significant of these is the attribution of the constellation Equuleus (*Protomē hippou*) to Hipparchos. Geminos's is the first mention of this constellation in the Greek tradition. Perhaps it comes from Hipparchos's star catalogue. In any case, Ptolemy adopted this constellation name in the *Almagest*. Among writers on such geographical questions as mountain heights, the extent of Ocean, and the arrangement and habitability of the zones, Geminos cites Dikaiarchos, Pytheas, Kleanthēs, and Polybios.

Geminos was clearly influenced by the Stoic Poseidōnios in his philosophical musings and in his work on meteorology. (See fragment 2.) In sec. 7 we address the controversial question of whether Geminos, in writing the *Introduction to the Phenomena*, might have used a lost textbook of Stoic astronomy and physics written by Poseidōnios. Here, it suffices to point out that he does not mention Poseidōnios a single time in the *Introduction to the Phenomena*. The material of Geminos's *Introduction* consists largely of notions that were the common property of all astronomers. His contribution was in the selection and shaping of material, in his graceful prose, and in the tasteful incorporation of literary examples.[25] He would have needed no help from Poseidōnios for this.

But Geminos does leave some of his most important sources unnamed. For as we have seen, and though he does not cite them by name, Geminos clearly knows the material in Euclid's *Phenomena*, Autolykos's *On the Moving Sphere* and *On Risings and Settings*, and Theodosios's *On Habitations* and *On Days and Nights*. We shall see below that he probably knew also Hypsiklēs of Alexandria's *Anaphorikos*. Geminos's merit as a teacher is to absorb all this rather dry mathematical material and transform it into graceful prose—though often at the expense of the original mathematical rigor.

Highly significant are Geminos's citations of the "Chaldeans," by which he means Babylonian astronomers. We should say a few words about this term. The Chaldeans were a group of tribes who moved into southern Mesopotamia by about 1000 B.C. They assumed a growing importance, and in the eighth century succeeded in putting a king on the throne of Babylonia. Within a few decades, the Chaldean kings lost control to the Assyrian kings, who intervened repeatedly in Babylonian affairs. But under Nabopolassar a new Chaldean dynasty was established, which ruled Babylonia from 625 B.C. until the Persian conquest in 539.[26] Ancient Greek writers often used the term "Chaldeans" (*Chaldaioi*) simply to mean Babylonians. But because Babylon had a reputation for arcane knowledge, "Chaldean" also came to mean *an astronomer or*

---

[25] Compare Aujac 1975, lxxxviii, n1.
[26] On the Chaldeans, see Oates (1986), 111–14.

astrologer of Babylon. Here are a few examples that span the range of meanings from "Babylonian" to "astronomer of Babylon" to "astrologer or magus": In the *Almagest*, Ptolemy refers to the "Chaldean" (i.e., Babylonian) calendar. Vitruvius says that Berossus came from the "Chaldean city or nation" to spread the learning of this people. Theōn of Smyrna says that the Chaldeans save the phenomena by using arithmetic procedures. For Herodotos, the Chaldeans are priests of Bel (i.e., Marduk). This is quite reasonable, since astronomy and astrology were concentrated in the temples, and many of the practitioners were priestly scribes. In *Daniel* 2.2–4, the Chaldeans are interpreters of dreams and are associated with magicians and sorcerers. For Sextus Empiricus, Chaldeans are astrologers.[27]

By about 300 B.C. the Babylonians had developed very successful theories for the motions of the planets, Sun, and Moon. These theories were based upon arithmetic rules, rather than on the geometrical models that characterized the Greek approach. When the Greeks began to deal quantitatively with planetary theory, they were able to base their geometrical models on numerical parameters borrowed from the Babylonians. This process was well under way in the second century B.C. In the *Almagest* (second century A.D.), Ptolemy begins with planetary periods that he ascribes to Hipparchos (second century B.C).[28] But in fact these parameters were of Babylonian origin and turn up on cuneiform tablets. In his discussion of the Moon's mean motions, Ptolemy again starts with Hipparchos's values, but in this case says explicitly that Hipparchos had made use of Chaldean observations.[29] Hipparchos's works on lunar and planetary theory have not come down to us, so we do not know exactly how he came into contact with the Babylonian parameters.

In the period between Hipparchos and Ptolemy, the Greek geometrical planetary theories had not yet reached maturity, and were not capable of yielding accurate numerical values for planet positions. But the rise of astrology (which entered the Greek world from Babylonia in the second or first century B.C.) imposed a need for quick, reliable methods of calculating planetary phenomena. Greek astronomers and astrologers adopted the Babylonian planetary theories with enthusiasm. Astronomical papyri from Egypt show Greeks of the first century A.D. using Babylonian planetary theories with complete facility. Ptolemy's publication of his planetary theories and tables in the *Almagest* and the *Handy Tables*

---

[27] Ptolemy, *Almagest* ix 7 and xi 7. Vitruvius, *On Architecture* ix 2.1. Theōn of Smyrna, *Mathematical Knowledge Useful for Reading Plato* iii 30. Herodotos, *Histories* i 181–84. Sextus Empiricus, *Against the Professors* v 2–3.

[28] Ptolemy, *Almagest* ix 3. For a discussion, see Neugebauer 1975, 150–52.

[29] Ptolemy, *Almagest* iv 2. See Neugebauer 1975, 69–71, 309–10.

produced a major change in the way practical astronomy was done. But calculating methods based on Babylonian procedures still existed side by side with methods based on Ptolemy's tables in the fourth century A.D.

In chapter ii of the *Introduction to the Phenomena*, Geminos shows that he is familiar with some features of Chaldean astrology, though he mentions only a few doctrines in passing, and does not seem intensely interested in the subject. In any case, nothing about the level of his familiarity with Chaldean astrology is surprising for a writer of his time. Far more detailed and more historically significant is Geminos's discussion of the Babylonian lunar theory in chapter xviii. His discussion there is important because his is the oldest extant classical text to display familiarity with the technical details of a Babylonian planetary theory based on an arithmetic progression. In particular, Geminos explains a scheme for the motion of the Moon, according to which the daily displacement increases by equal intervals from day to day, until it reaches a maximum, then falls by equal increments from one day to the next. The numerical parameters of Geminos's theory are in exact agreement with cuneiform sources. Geminos's treatment of the Babylonian lunar theory is discussed below in sec. 13, below, where we also address the question of the form that his source for the Babylonian lunar theory might have taken. In chapter xi, Geminos mentions that eclipses of the Moon take place in an eclipse zone (*ekleiptikon*) that is 2 degrees wide. Though he does not mention the Chaldeans in this passage, the 2-degree eclipse zone also comes from Babylonian astronomy. In total, Geminos's remarks provide important information about the adoption and adaptation of Babylonian knowledge by the Greeks of his time. By contrast, Geminos cites the "Egyptians" simply for the general structure of the Egyptian calendar and the circumstances of a festival of Isis.

6. Geminos's Country and Date

Modern scholars sometimes refer to our astronomer as "Geminos of Rhodes,"[30] but there is no ancient mention of his native land or city. The few ancient writers who cite him refer to him simply as Geminos, or as "Geminos the mathematician." The evidence for placing him in Rhodes is suggestive, but not conclusive. In several passages in the *Introduction to the Phenomena*, Geminos uses Rhodes as an example in making some astronomical point—involving the length of the longest day, or the portion of the summer tropic cut off above the horizon, or the meridian alti-

---

[30] For example, Sarton 1970, vol. 2, 305; Bowen and Todd 2004, 194n2.

tude of the star Canopus, or the date of the morning rising of the Dog Star.[31] But although he does use Rhodes most frequently for such examples, he also gives examples for Alexandria, Greece, Rome, and the Propontis.[32] Does his proclivity for using Rhodes suggest a fondness for his native city, or merely reflect Rhodes's usefulness in astronomical examples, owing to its roughly central location in the Greek world? Geminos remarks (xiv 12) that celestial globes and armillary spheres were commonly constructed for this *klima*, or band of latitude. And it is noteworthy that in the second century A.D., Ptolemy, who lived at Alexandria, still found it natural to construct examples for the parallel through Rhodes, "where the elevation of the pole is 36 degrees and the longest day 14½ hours."[33] Or perhaps, as Dicks suggests,[34] Geminos's use of the *klima* of Rhodes reflects examples he found in his sources, which may have included the geographical or astronomical works of Hipparchos. Blass makes an interesting point about Geminos's use of two geographical examples. Geminos (xvii 3) refers to Mt. Kyllēnē, and immediately specifies that it is "the highest mountain in the Peloponnesos"; but in the very next sentence he refers to Mt. Atabyrion without making any similar specification that it is on the island of Rhodes.[35] Does this suggest that he expected his readers to be familiar with Rhodian geography?

Finally, we know that Geminos wrote some sort of abridgment of, or commentary on, the *Meteorology* of Poseidōnios, whose native land was Rhodes. And a likely dating of Geminos's *Introduction to the Phenomena* would make Geminos a younger contemporary of Poseidōnios, and thus potentially his student. (Geminos's possible debts to Poseidōnios will be discussed below.) Near the end of her own discussion of this issue, Aujac concludes, "Let us allow, then, since no other better hypothesis presents itself, that Geminos was born at Rhodes and that he there received his first instruction."[36] This is not an unreasonable position to take, since no convincing evidence exists for placing him elsewhere.[37]

[31] *Introduction to the Phenomena* i 10, 12; iii 15; v 25; vi 8; xvii 40.

[32] *Introduction to the Phenomena* iii 15; v 23; vi 8.

[33] Ptolemy, *Almagest* ii 2, trans. Toomer 1984, 76. Ptolemy's example continues, using the same latitude, in ii 3.

[34] Dicks 1972.

[35] Blass 1883, 5. The mss. actually call the mountain *Satabyrion*, which was corrected by Petau (followed by Aujac). See also Dicks 1960, 30n3.

[36] Aujac 1975, xv.

[37] However, Schmidt (1884b) argues that Geminos wrote the *Introduction to the Phenomena* at Rome for a Roman audience. Against this, Tannery (1887, 37) points out that Geminos is not mentioned by Pliny, who would certainly have included him in his long list of authorities, if he had ever heard of him. Manitius (1898, 247) believes that Geminos wrote at Rhodes, but that the work we have is due to a much later excerptor, who lived in the *klima* of 15ʰ, and most likely at Constantinople.

But we simply do not know. In any case, as Tannery has pointed out,[38] all the writers who cited Geminos were associated with Alexandria or with Athens, which suggests that his works circulated mainly in the Greek world of the eastern Mediterranean.

Whether *Geminos* is a Greek name or a Hellenization of a Latin name (Geminus) has been the subject of dispute. As Aujac remarks, "Petau made it a Latin name, Manitius a Greek name, Tittel again a Latin name (!)"[39] A crucial point in the argument is the length of the central vowel— a long vowel favoring the Greek. Whatever the origin of his name, Geminos was thoroughly Greek in education, intellectual interests, and manner of expression.

But when did Geminos write? There are two ways to narrow the possibilities. Appendix 4 lists the writers that Geminos mentions. The latest datable writers cited in the *Introduction to the Phenomena* are Hipparchos, Polybios, and Kratēs of Mallos, who all flourished in the middle of the second century B.C. Conversely, Geminos was quoted by Alexander of Aphrodisias, the Aristotelian commentator, who flourished at the end of the second century A.D. Thus we may place Geminos between 150 B.C. and A.D. 200.

It appears possible to date Geminos more closely by his remark (viii 20–22) concerning the wandering year of the Egyptians:

> . . . most of the Greeks suppose the winter solstice according to Eudoxos to be at the same time as the feasts of Isis [reckoned] according to the Egyptians, which is completely false. For the feasts of Isis miss the winter solstice by an entire month. . . . 120 years ago the feasts of Isis happened to be celebrated at the winter solstice itself. But in 4 years a shift of one day arose; this of course did not involve a perceptible difference with respect to the seasons of the year. . . . But now, when the difference is a month in 120 years, those who take the winter solstice according to Eudoxos to be during the feasts of Isis [reckoned] according to the Egyptians are not lacking an excess of ignorance.

The feasts of Isis (*ta Isia*) were celebrated at a fixed date in the Egyptian year. But as the Egyptian year consists of 365 days (with no leap day), the feast days shift with respect to the solstice by 1 day every 4 years. Because the Egyptian year is too short, the feast days gradually fall earlier and earlier in the natural, or solar, year. If we knew the Egyptian calendar date on which this Isis festival was observed, it would be easy

---

[38] Tannery 1887, 37.
[39] Aujac 1975, xiv, n2. See Manitius 1898, 251; and Tittel 1910, col. 1027. Heath (1921, vol. 2, 222) provides a summary of the debate up to his time. Dicks (1972) makes the name Latin, but stresses that Geminos's "works and manner are patently Greek."

to calculate the year in which the festival coincided with the winter solstice. We would then place Geminos 120 years after that year.

Most writers on the subject have tried to date Geminos by the use of a remark by Plutarch (late first to early second century A.D.):

> . . . they say that the disappearance of Osiris occurred in the month of Athyr. . . . Then, among the gloomy rites which the priests perform, they shroud the gilded image of a cow with a black linen vestment, and display her as a sign of mourning for the goddess, inasmuch as they regard both the cow and the Earth as the image of Isis; and this is kept up for four days consecutively, beginning with the seventeenth of the month.[40]

Denis Petau, in his *Uranologion* of 1630,[41] used Plutarch's remark to date Geminos's composition, with the following result:

> year 4537 of the Julian period,
> fourth year of Olympiad 175,
> year 677 after the founding of Rome,

or, as we would say, 77 B.C. Petau was followed by most later writers on the subject, with only minor adjustments. Thus, most writers who have accepted this evidence put Geminos's composition of the *Introduction to the Phenomena* in the 60s or 70s B.C.[42] But as we shall see, the margin of error should be taken quite a bit wider.

The reasoning is straightforward. Let us work with 19 Athyr, the 3rd day of the 4-day festival. Athyr is the 3rd month of the Egyptian calendar, so 19 Athyr is the 79th day of the Egyptian year.[43] In Table I.1, the first column lists years of the Julian calendar. In the second column, we have written the date of 1 Thoth, the 1st day of the Egyptian year that began in the course of the given Julian calendar year.[44] Thus, in −200, a new Egyptian year began on 12 October.[45] To obtain column 3, we add 78 days to the dates in column 2. In this way, we move from the 1st day of the Egyptian year (1 Thoth) to the 79th day (19 Athyr). Thus, in the Julian year −200, the 19th of Athyr fell on 29 December. The 4th column

[40] Plutarch, *Isis and Osiris* 39, 366D–E, F.C. Babbitt, trans.

[41] Petavius 1630, 410–11.

[42] For summaries of multiple dating attempts by various writers, see Manitius 1898, 238, and Jones 1999a, 256n2.

[43] For an introduction to the Egyptian calendar, see Evans 1998.

[44] For Julian calendar equivalents of 1Thoth over the centuries, see Bickerman 1980, 110–11.

[45] We use "astronomical reckoning." The years A.D. are written as positive numbers. The B.C. years are shifted by one, in order to introduce a year zero, and written as negative numbers. Thus, +1 = A.D. 1, 0 = 1 B.C., −1 = 2 B.C., etc.

TABLE I.1
Comparing 19 Athyr and Winter Solstice

Year	1 Thoth	19 Athyr	Winter solstice
−250	25 Oct	11 Jan	25 Dec
−200	12 Oct	29 Dec	24 Dec
−150	30 Sep	17 Dec	24 Dec
−100	17 Sep	4 Dec	23 Dec
−50	5 Sep	22 Nov	23 Dec
0	23 Aug	9 Nov	22 Dec
+50	11 Aug	28 Oct	22 Dec

gives the date of the winter solstice[46] for each of the given Julian calendar years. Comparing the 3rd and 4th columns, we see that the winter solstice fell on 19 Athyr sometime between −200 and −150. Interpolation gives −179. Geminos wrote 120 years later, or around the year −59.

An error of 3 days in the date of the solstice could shift the date by ±12 years.[47] Again, Geminos speaks in rough fashion of a whole month, as the difference by which the Isis festival missed the solstice in his own day. He might have spoken in this same way if the actual difference were, say, as small as 28 days or as great as 32, which introduces another ±8 years of uncertainty. Finally, the festival itself stretched over a period of 4 days, which gives us 4 more years of uncertainty after 60 B.C. and 8 more years before. Putting all this together, we find the period 88–36 B.C. as the most likely for the composition of the *Introduction to the Phenomena*, or, to speak in round numbers, 90–35 B.C.

In 1975, Otto Neugebauer proposed a date for Geminos about a century later, around A.D. 50.[48] Although Neugebauer's dating was influential for a while, we shall see that it can no longer be sustained. The argument that follows will be somewhat intricate. But at the end we shall not abandon the dating of 90–35 B.C. that we have just explained. Thus, readers with little enthusiasm for details of ancient chronology should feel no guilt in skipping ahead to the next section.

The key question that Neugebauer posed is whether Petau's argument, based on Plutarch's remark, involved a confusion between the Egyptian

---

[46] The dates of the winter solstices have been calculated from the solar tables in Newcomb 1898. The reason why the date of the solstice slowly shifts is the incorrect length of the Julian calendar year (by approximately 3 days every 400 years), which was corrected by the Gregorian reform of 1582.

[47] Three days is not too large an error. Geminos refers to the winter solstice "according to Eudoxos." In the Geminos *parapēgma*, the winter solstice "according to Eudoxos" is separated by 3 days from the winter solstice "according to Euktēmon."

[48] Neugebauer 1975, 579–81.

and the Alexandrian calendars. After Egypt became a province of the Roman empire, Augustus reformed the Egyptian calendar by introducing a leap day once every four years, the first such day being inserted at the end of the Egyptian year 23/22 B.C. In the reformed calendar, now usually called "Alexandrian" to distinguish it from the original Egyptian calendar, three years of 365 days were followed by a year of 366 days. The reformed calendar thus was very similar to the Julian calendar, which had been used at Rome since 45 B.C. Of course, the Alexandrian calendar continued to use the old Egyptian months of 30 days each, as well as the original Egyptian month names. For dates near 23 B.C., a given day has nearly the same date in both the Egyptian and the Alexandrian calendars. But gradually, at the rate of 1 day in 4 years, the calendars diverge. Moreover, the two calendars continued to be used side by side. For example, Ptolemy, in the *Almagest*, used the old calendar for astronomical calculation, because of its simpler structure, nearly two centuries after it had been abandoned for civil use. In his *parapēgma*, however, Ptolemy adopted the Alexandrian calendar, because the heliacal risings and settings of a given star have more nearly fixed dates in this calendar. When an ancient writer, writing after Augustus's reform, says "the 17th of Athyr," it is not immediately clear whether he is expressing the date in terms of the Egyptian calendar or the Alexandrian calendar. One must examine the context carefully.

Neugebauer was troubled by a second reference in Plutarch to what was apparently the same Isis festival:

> . . . then Osiris got into [the chest] and lay down, and those who were in the plot ran to it and slammed down the lid, which they fastened by nails from the outside and also by using molten lead. They say also that the date on which this deed was done was the seventeenth of Athyr, when the Sun passes through Scorpio. . . . [49]

We have again the date Athyr 17, but now with the added information that the Sun passes through Scorpio during the month of Athyr. As Neugebauer pointed out, this was true in the Alexandrian, but not in the Egyptian calendar, for Plutarch's time. The Alexandrian month of Athyr runs from 28 October to 26 November (Julian), which corresponds rather closely to the sign of Scorpio. In Plutarch's time, say A.D. 118, the Egyptian and Alexandrian calendars were out of phase by 35 days: 1 Athyr (Egyptian) then fell on September 23 (Julian), corresponding to the Sun's entry into Libra, not Scorpio. Neugebauer concluded that Plutarch was using the Alexandrian, and not the Egyptian, calendar. Moreover, he surmised that Plutarch (or his source) took the original

---

[49] Plutarch, *Isis and Osiris* 13 356C, F.C. Babbitt, trans.

date of the Isis festival, as expressed in the Egyptian calendar, and converted it to an Alexandrian equivalent. The Alexandrian calendar was, after all, the one in official use, and the one more likely to be understood by Plutarch's readers in the wider Roman world.

Neugebauer found the Egyptian date for what he took to be the same festival in a hieroglyphic text in the East Osiris Chapel on the roof of the Temple of Hathor in Dendera.[50] The text describes the rituals of an Osiris festival that lasted from 12 to 30 Choiak. The text is not later than 30 B.C. and thus predates the reform of the calendar. Moreover, as Neugebauer also pointed out, the papyrus Hibeh 27 (c. 300 B.C.) mentions an Osiris festival on 26 Choiak.[51] Now in Plutarch's time (A.D. 118), the date 26 Choiak (Egyptian) = 21 Athyr (Alexandrian), which appeared to confirm Plutarch's use of the Alexandrian calendar when he placed the rites on 17–20 Athyr. Neugebauer then computed the year when the winter solstice fell on 15 Choiak (Egyptian). (This date is within the span of rituals mentioned by the text in the East Osiris Chapel.) The answer is the year −70; Geminos wrote 120 years later, or around A.D. 50, according to Neugebauer. The new dating by Neugebauer, pushing Geminos forward into the first century A.D., was gradually adopted by historians of ancient astronomy.

Alexander Jones reexamined the question in 1999.[52] As Jones points out, Neugebauer deserved credit for being the first to use papyrological evidence for the date of the Isis festival. The advantage of such evidence is that it comes from a time when the Isis festival was a living custom, that it comes directly from Egypt without having passed through the hands of other writers, and that some of it comes from a date before the reform of the Egyptian calendar, thus removing any possibility of confusion between the calendars. But there was much more such evidence (in both the Greek and Egyptian languages) available than Neugebauer had realized.

Jones adduces a good deal of evidence showing that Neugebauer had wrongly taken the Osiris festival of 12 to 30 Choiak (Egyptian) to be the same festival as the *Isia* that Plutarch mentions. Jones also points out that Geminos refers to the festival simply as *ta Isia*, without any further specification. This implies that the festival was so well known that Geminos had no fear that it would be confused by his readers with other festivals associated with Isis or Osiris. Now, as Jones points out, there are at least nineteen references in Greek papyri to a festival called the *Isia* (also spelled *Iseia* or *Isieia*). Only a few of these provide calendrical information. But enough do that it is possible to confirm Plutarch's dates of

[50] For a bibliography pertaining to this text, see Porter and Moss 1927, vol. 6, 97.
[51] Grenfell and Hunt 1906, 144, 148.
[52] Jones 1999a.

17–20 Athyr, and to be sure that these dates indeed apply to the old (Egyptian) calendar. For example, several papyri from before the calendar reform are private letters or records, with dates in Athyr, concerned with ordering or issuing supplies (logs and lamp oil) for the *Isia*. One papyrus gives the dates of the *Isia* in terms of the Macedonian calendar. These dates can be converted to the Egyptian calendar (with an uncertainty of 1 day) and indeed correspond to 17–20 Athyr. Slightly altering the chronological assumptions and broadening the error bars, Jones concludes that it is very probable that Geminos wrote his *Introduction to the Phenomena* "between 90 and 25 B.C., and definitely not during the first century of our era."[53] There is an irony in the fact, confirmed by the papyri, that after the calendar reform, the *Isia* continued to be celebrated on days called 17–20 Athyr, *but in the new calendar.* Thus, Plutarch's dates turn out to refer to the reformed calendar after all! (But they should not be converted back into the old calendar to obtain the dates that Geminos would have been familiar with.)

One minor problem with dating Geminos to the first century B.C. involves his mention of Hero of Alexandria in fragment 1. The dating of Hero has been controversial, with suggested dates from the middle of the second century B.C. to the middle of the third century A.D.[54] In *Dioptra* 35, however, Hero mentions a lunar eclipse observed simultaneously in Alexandria and Rome. Although Hero does not mention the year of the eclipse, he is detailed about its other circumstances: 10 days before the vernal equinox, 5th seasonal hour of the night at Alexandria. Neugebauer[55] has shown that these circumstances were satisfied by only one lunar eclipse between about −200 and +300, namely that of March 13, A.D. 62. If Hero used an eclipse of recent memory, we must place him in the second half of the first century A.D. Thus, if the dating of Geminos to the first century B.C. is correct, we must suppose that Proklos or a later copyist interpolated the name of Hero in fragment 1.

Finally, we note that Geminos writes about Babylonian astronomy and astrology as if they were still new to his Greek readers. This well suits a dating to the first century B.C., when this material was still being absorbed and adapted by the Greeks.[56]

---

[53] Jones 1999a, 266.

[54] For a summary of the older estimates, see Heath 1921, vol. 2, 298–307.

[55] Neugebauer 1938; with results summarized in Neugebauer 1975, 846.

[56] Some scholars have attempted to identify the author of the *Introduction to the Phenomena* with other men named Geminos. Aujac (1975, xxii) suggests a certain Cnaeios Pompeios Geminos, active around A.D. 15. Reinhardt (1921, 178–83) prefers to identify the author of the *Introduction* with an earlier Geminos; Tannery (1887, 37) with a later one. Aujac (1975, xx–xxii) is one of the few who discount the usual dating argument based on the feasts of Isis, and seeks to explain this paragraph (viii 20–22) in a completely different way.

7. Geminos and the Stoics

Geminos is usually considered to have been a Stoic, and is often said to have been a disciple of Poseidōnios (c. 130–50 B.C.).[57] Key evidence for this involves two circumstances. First, Simplikios tells us that Geminos wrote an abridgement of, or perhaps a commentary on, a *Meteorology* by Poseidōnios. (See fragment 2.) Moreover, Simplikios says that Geminos's discussion of the relationship between astronomy and physics was based on Poseidōnios's.[58] And, second, following Petau's analysis, most scholars assigned a date of about 70 B.C. to the composition of Geminos's *Introduction to the Phenomena*. According to this dating, Geminos was a younger contemporary of Poseidōnios, and could therefore be visualized sitting at the feet of the master of Stoic philosophy in Rhodes. The fact that Geminos wrote a *Meteorology* based on Poseidōnios's work does seem to imply, at the very least, an interest in Stoic physics. And Geminos's dates *may* overlap Poseidōnios's well enough to permit a student-teacher relationship. But because of the margin of error in the date for the *Introduction to the Phenomena*, this is not certain.

Some have gone so far as to make Geminos's *Introduction to the Phenomena* itself a virtual paraphrase of Poseidōnios's lost *Peri Meteōrōn*. This view was maintained by Blass,[59] who pointed to a number of passages in which Geminos's language closely resembles that of corresponding passages of Kleomēdēs' *Meteōra*, a work that obviously *is* dependent on Poseidōnios and cites him many times. Blass's approach is an example of the *Quellenforschung* ("source research") that characterized much late-nineteenth-century German philology. The idea was to mine the surviving works of ancient scientists or philosophers, in an attempt to reconstruct the lost works of their ancient precursors. Blass's thesis regarding the *Introduction to the Phenomena* was refuted by Tannery, who carefully demonstrated its inconsistencies with the evidence.[60] Blass's theory, in its strong form, has little, if any, support among contemporary scholars. But the parallels between Geminos and Kleomēdēs constitute the best evidence that both made use of the same earlier work,

---

[57] Writers who identify Geminos as a student of Poseidōnios include Sarton (1959, 305) and Kouremonos (1994, 437).

[58] We also have the testimony of Diogenēs Laertios (*Lives and Opinions* vii 132) that the relationship between mathematics (i.e., astronomy) and physics was, indeed, a topic discussed by Stoic writers.

[59] Blass 1883. Blass even proposed that the original title of Geminos's book was *Geminos's exegesis of the phenomena from the Meteorologica of Poseidōnios*. Since Geminos modifies many of Poseidōnios's doctrines, Geminos is therefore, according to Blass, a fraud, who could not have been a student of Poseidōnios, and would not, in any case, have dared to publish such a work in Poseidōnios's lifetime.

[60] Tannery 1887, 29–36.

for which a likely candidate is the *Peri Meteōrōn* of Poseidōnios. This is the third major argument for Geminos's dependence on Poseidōnios. Against this, as Tannery pointed out, there are other ways of explaining the parallel passages. Both Geminos and Kleomēdēs drew upon material that was the common property of astronomers from long before Poseidōnios's time. It is possible, too, that Kleomēdēs made use of other elementary surveys of astronomy, including Geminos's, in writing his own.[61]

At the other end of the opinion scale, Reinhardt studied the uses of "sympathy" in Geminos and in the fragments of Poseidōnios, and concluded that Geminos could not possibly have been a student of Poseidōnios.[62] And Neugebauer, who was dismissive of Poseidōnios's competence in astronomy, remarked that Geminos "had nothing to learn from Posidonius."[63] Indeed, the view that Kleomēdēs gives us of Poseidōnios's contribution to astronomy is not very impressive. For example, Poseidōnios's methods of determining the sizes and distances of the Sun and Moon, which Neugebauer described as "naive," represent a methodological step backward from Aristarchos and Hipparchos. Tannery, too, judged Geminos to be a much better writer than Kleomēdēs, and argued that their books have no more in common than one would expect to see in two elementary textbooks of our own day. Tannery pointed also to the *maladresse* of Kleomēdēs' book ii, chapter 1, almost certainly based on Poseidōnios, in which the philosopher tediously piles up arguments against the Epicureans' doctrine of the immediate validity of sense-evidence, and in particular their view that the Sun is the size it appears to be, namely 1 foot in diameter. One could hardly imagine Geminos wasting his time on such a matter.[64] Tannery concludes, "If, in his *Introduction to the Phenomena*, Geminos never cites Poseidōnios, one must not at all conclude that he is seeking to disguise a plagiarism, but rather that he did not estimate the cosmographical works of the philosopher of Apamea [Poseidōnios] highly enough to rely on their authority."[65]

Teachers of Stoic philosophy commonly offered a complete curriculum of logic, ethics, and physics. Diogenēs Laertios gives an extensive sum-

---

[61] Many of the parallels between Geminos and Kleomēdēs are pointed out in the notes to Todd's [1990] edition of Kleomēdēs. Todd also identifies Kleomēdēs' many parallels with other elementary writers on astronomy.

[62] Reinhardt 1921, 178–183; recapitulated in Reinhardt 1926, 51–53. Reinhardt held that the *Introduction to the Phenomena* belongs to the Stoa of the period before Poseidōnios, and that its author was not the same Geminos who wrote sophisticated mathematical works (the *Philokalia*).

[63] Neugebauer 1975, 579.

[64] For a more positive view of Kleomēdēs' work, see Bowen and Todd 2004.

[65] Tannery 1887, 35.

mary of this curriculum.[66] Although, as he tells us himself, Diogenēs shortens his discussion by merging a number of Stoic writers, he does at times distinguish one Stoic from another, in instances where there were substantive disagreements over doctrine. It is also clear that Diogenēs still had access to a number of Stoic treatises and textbooks that are now lost. Astronomy was treated under the physical part of the curriculum, and Diogenēs gives a short summary of the teachings of the Stoics in astronomy.[67] The astronomical "teachings" are elementary facts that were the common property of all astronomers since the fourth century B.C. Diogenēs' list includes: a description of the parallel circles of the sphere, and of the oblique zodiac, the fact that the Sun is spherical and is larger than the Earth, that the Moon gets its light from the Sun, the causes of lunar and solar eclipses, that the Moon's path is oblique to the zodiac, that the Earth is in the center of the cosmos, and that it has five zones. The astronomy is commonplace, and there is nothing particularly Stoic about it.

The Stoic flavor is lent to the astronomical discussion by the incorporation of Stoic physical doctrines.[68] Among these Diogenēs mentions the following. There are two principles in the world, the active and the passive. The passive principle is substance without quality, that is, matter, while the active principle is the reason inherent in substance, that is, God. God is one and the same with reason, fate, and Zeus. The world is ordered by reason and providence, and harbors no empty space, but is held together by sympathy and tension. The material world is finite, but beyond it lies the infinite void. The Sun is pure fire, a doctrine that Diogenēs attributes specifically to Poseidōnios, "in the seventh book of his *Peri Meteōrōn.*" The blend of Stoic physics and elementary astronomy that Diogenēs describes corresponds well with Kleomēdēs' *Meteōra,* but not so well with Geminos's *Introduction to the Phenomena,* in which the Stoic physics is completely lacking.

As Alan Bowen has aptly said, Stoicism was, in Geminos's day, "like a color." We would not deny a Stoic coloration to Geminos's thought, as is apparent in fragment 2, and perhaps in Geminos's matter-of-fact acceptance of "sympathies" between people born in certain zodiacal relationships to one another (*Introduction to the Phenomena* ii). But it must be said that, if Geminos were truly a Stoic, he wore his Stoicism lightly. Geminos's *Introduction* is remarkably free of philosophical interpolations, unlike those of Kleomēdēs and Theōn of Smyrna, which show us

---

[66] Diogenēs Laertios, *Lives and Opinions* vii 38–160.

[67] Diogenēs Laertios, *Lives and Opinions.* The summary of physical doctrine begins at vii 132, and the specifically astronomical and cosmographical parts are at vii 144–46 and 155–56.

[68] On Stoic physics and cosmology, see Sambusky 1959, Todd 1976, Todd 1982, and Todd 1989.

what philosophically oriented surveys of astronomy look like. Theōn's text is marked by frequent, specific references to the works of Plato. Kleomēdēs begins his book with an account of Stoic physics and cosmology: the *pneuma*, the impossibility of a void place inside the cosmos, the necessity for an infinite void place outside the cosmos, etc. Moreover, Kleomēdēs frequently cites Poseidōnios, while Geminos never mentions Poseidōnios at all.

Geminos does cite three Stoic writers, Kleanthēs of Assos, Kratēs of Mallos, and Boēthos—but in the first two cases, simply to take issue with them. At xvi 21, he mentions Kleanthēs' view that Ocean spreads all over the tropical zone, only to refute it as "quite mistaken" and "alien to both mathematical and physical thought." In several passages (vi 10, 16; xvi 27), Geminos refers to Kratēs' efforts to attribute sound astronomical and geographical knowledge to Homer. Geminos admires Homer and enjoys making use of Homeric verses with astronomical references, but he faults Kratēs for reading too much into Homer: "Kratēs, speaking in marvels, takes things said by Homer for his own purposes and in archaic fashion, and transfers them to the spherical system that accords with reality." Geminos (xvii 48) endorses Boēthos's use of natural signs in forecasting the weather, rather than attributing the weather to the influence of the stars. In this, says Geminos, Boēthos is in accord with Aratos, Eudoxos, and Aristotle.

Moreover, there are a number of technical reasons that make Geminos a poor candidate for paraphraser of Poseidōnios in particular, or dedicated Stoic in general: (1) Geminos (chapter xv) adopts a scheme for the terrestrial zones different from that of Poseidōnios. (2) Geminos (vi 38) adopts a pattern for the variation in the length of the day different from that of Kleomēdēs. (See note 162 to the Introduction, p. 74.) (3) Geminos (xvi 6) follows Eratosthenēs rather than Poseidōnios on the size of the Earth. (4) Geminos denies (xvii 2) that the exhalations from the Earth reach as far as the stars, which is contrary to a view popular among the Stoics, including Poseidōnios, who claimed that the stars were nourished by these exhalations. (5) Geminos devotes an entire chapter (xvii) to refuting the common belief that changes in the weather are caused by the heliacal risings and settings of the stars. Again, this puts him at odds with the Stoics, who were sympathetic to the doctrine of astral influences. (6) Geminos (xvii 15) states that the stars have no sympathy with things occurring on the Earth—a decidedly non-Stoic view. (7) In the same passage, Geminos is simply indifferent to whether the stars are made of fire or *aithēr*, which shows little regard for a question that Poseidōnios thought important.

Though it is perfectly possible that Geminos borrowed from an elementary survey of astronomy by Poseidōnios or someone else, particular chapters of the *Introduction to the Phenomena* seem to depend rather on earlier technical writers, such as Autolykos for chapter xiii. And it is clear that Geminos has made some use of a work by Hipparchos on the fixed stars in chapter iii. It is also likely that Geminos has put in plenty of his own. The long chapter (viii) on lunisolar cycles is probably his own; no other extant Greek text treats the subject in so much detail, and there is no evidence that Poseidōnios wrote on this subject. Similarly, the final chapter (xviii), on the *exeligmos*, is without parallel in early Greek astronomy; we do not find another such discussion until Ptolemy's *Almagest*. Finally, Geminos's incorporation of elements of Babylonian astronomy has no parallel in other Greek works of the period, and for that reason is particularly precious.

8. Geminos on Astronomical Instruments and Models

In the *Introduction to the Phenomena*, Geminos mentions several kinds of astronomical instruments and models and assumes that his reader is familiar with them: the sundial, the *dioptra*, and the celestial sphere, of which there are two kinds, solid and ringed. Today, we refer to these last two as the celestial globe and the armillary sphere. Geminos uses all these pieces of apparatus more as tools of instruction than as instruments of observation. His use of instruments is thus pedagogical—exactly what we would expect in an introductory textbook.

Celestial Globe and Armillary Sphere

Geminos mentions both the "solid sphere," *sphaira sterea* (xvi 12), and the "ringed sphere," *sphaira krikōtē* (xvi 10, 12). By his day, celestial globes were common enough that a writer on astronomy or geography could assume his readers to be familiar with them and could therefore appeal to these instruments in illustration or argument. Strabo devotes a chapter of his *Geography* to the question of what sort of astronomical knowledge students of geography will require. He assures his readers that they need not be experts in astronomy, but warns that they should not be so simple or lazy as never to have seen a globe and the circles inscribed upon it, or to have examined the positions of the tropics, equator, and zodiac.[69] Celestial globes were not only useful tools of instruction, but also potent symbols that could carry philosophical, religious, and (increasingly in the Roman period) political meanings. For this reason they often figure on coins and murals, as well as in sculpture.[70]

---

[69] Strabo, *Geography* i 1.21.

Readers who had not had the opportunity to manipulate a celestial globe would at least have seen globes in artistic representations.

Three complete celestial globes are extant from Antiquity. The oldest and best known is the large marble sphere (fig. I.2) supported by a statue of Atlas, now in the Museo Archeologico Nazionale in Naples. This statue, transferred to its present location from the Farnese Palace in Rome, is commonly called the Farnese Atlas. The Farnese globe, about 65 cm in diameter, is a product of the early Roman empire, but was probably modeled on a Hellenistic original. The positions of the constellations place the original globe most likely in the second century B.C. But since the artist included the constellation of the "Throne of Caesar," the Farnese globe must be placed in the reign of Augustus or later.[71] The Farnese globe is a magnificent display piece, but hardly a suitable tool for classroom instruction.

A small bronze Roman globe (fig. I.3) in Mainz is probably more typical of the globes that astronomy students might have encountered in Geminos's day. Fig. I.3 shows the globe itself, and fig. I.4 shows a modern replica, on which details are more easily seen. Ernst Künzl has dated the globe, on the basis of the engraving technique, to A.D. 150–220. This globe, 11 cm in diameter, is figured with constellations, the principal celestial circles, and the Milky Way, as well as individual stars.[72] A globe of this size could easily have been held in the hand for study of the constellations, as well as the celestial circles that Strabo deemed so important for astronomy instruction. This sphere has a small square hole in its top and a larger round hole in its bottom, which suggest that the globe fit a spike that served as a stand. Or perhaps the globe fit over the tip of a gnomon, as a decorative feature for a large sundial.

A third ancient celestial globe surfaced recently in the Paris antiquities market, only to disappear again after its sale to a private collector. It is a small (6 cm in diameter) and handsome silver object, but its engraver

[70] On the symbolism of the sphere, see Arnaud 1984.

[71] On the date of the copy, see Künzl 2000, 535, and Künzl 2005, 63–66. On the Throne of Caesar, see Pliny, *Natural History* ii 178, and Künzl 2000, 535. For a precession dating of the supposed Hellenistic original, see Schaefer 2005. The general forms of the constellations agree with the tradition established by Eudoxos and Aratos, but the artist made a number of modifications that point to the influence of later sources as well. Schaefer argues that the contours of some constellations were adjusted to accord with corrections that Hipparchos introduced in his *Commentary on the Phenomena of Eudoxos and Aratos*. Schaefer's association of the Farnese globe with Hipparchos has been refuted by Duke 2006. See also Valerio 1987.

[72] See Künzl 2000 for a wonderfully detailed study of this object. One oddity of this globe is that its maker must have wished to show the 48 constellations made canonical by Ptolemy, but he missed two, and simply stuck in two round assemblages of stars in the southern hemisphere to make up the difference. One of these can be seen in the lower right of fig. I.3.

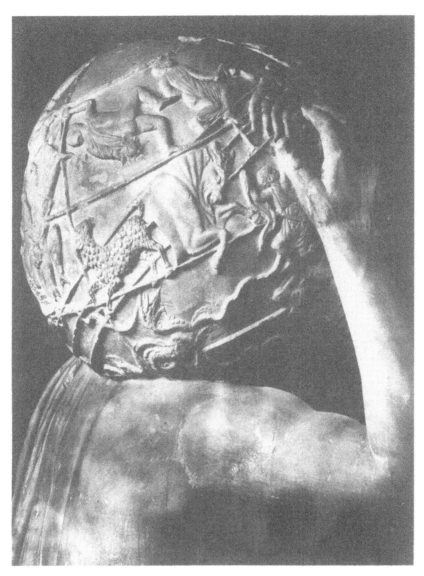

Fig. I.2. The Farnese globe. Roman copy of the first century A.D. or late first
century B.C., after a Greek original of the second century B.C. The statue of Atlas
was heavily restored in the Renaissance, and the arms and head are not original.
Museo Archeologico Nazionale, Naples. The zodiac slants upward from left to
right. The three broadly spaced parallel circles are, from top to bottom, the
tropic of Cancer, the celestial equator, and the tropic of Capricorn.

Fig. I.3. A small, bronze Roman celestial globe from about A.D. 150–220. This view is centered on Ophiuchus. Römisch-Germanisches Zentralmuseum, Mainz (RGZM O. 41339). Photo by Volker Iserhardt.

made a number of errors in the constellations. It was most likely from Asia Minor, and was offered for sale with two other metal objects that could be dated confidently to the second century A.D.[73]

None of the extant globes are of the kind most suitable for teaching— i.e., none is free to turn on an axis that can make an adjustable angle with a fixed horizon. Two ancient writers, however, give directions for

[73] Cuvigny 2004 offers a detailed description of this globe. The same author gives a shorter account, along with excellent color photographs, in Kugel 2002.

Fig. I.4. A galvanized plastic copy of the Mainz globe. On the lower half of the sphere may be seen the zodiac constellations Capricorn and Aquarius. The irregular chain of small dots crossing the upper part of the sphere is the Milky Way. Römisch-Germanisches Zentralmuseum, Mainz. Photo by Ernst Künzl.

building a globe, and both make these features plain. Leontios the mechanic, for example, says that the horizon ring is of the same size as the base of the globe.[74] And Ptolemy describes the meridian and horizon rings quite clearly.[75]

[74] Leontios, *Construction of the Sphere of Aratos*, in Halma 1821 or Maass 1898.

[75] Ptolemy, *Almagest* viii 3. Ptolemy's design is more complicated, since his globe is adjustable for precession.

Fig. I.5. A mosaic of an armillary sphere in the Casa di Leda, at Solunto, near Palermo. Photo courtesy of Rudolf Schmidt.

Similar to the celestial globe, but easier to construct, is the armillary sphere. In the armillary sphere, the sky is represented not as a solid sphere, but by a network of rings. (The Latin word *armilla* signifies an armband or bracelet.) These rings embody the most important circles of the sphere: ecliptic, equator, tropics, arctic and antarctic circles, and the colures. None of these delicate constructions has survived from Antiquity, but a number of illustrations of armillary spheres are preserved in ancient art. A ceiling painting in Stabiae (near Pompeii) displays what is almost certainly meant to be an armillary sphere, but only a portion of the painting is extant and, moreover, the original design seems to have been imperfect in its depiction of the sphere.[76] More impressive is the mosaic in Solunto (near Palermo), shown in Fig. I.5, which depicts very plainly all the features of an armillary sphere, including what is most

[76] For pictures of the Stabiae sphere, see Arnaud 1984, 73; or Carmado and Ferrrara 1989, 67–68.

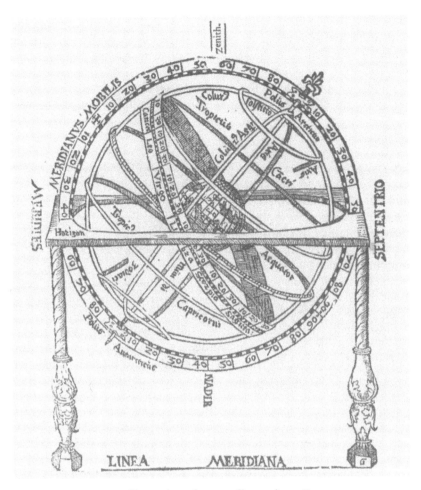

Fig. I.6. A Renaissance illustration of an armillary sphere. From
Cosmographia . . . Petri Apiani & Gemmae Frisii (Antwerp: 1584).
Photo: Special Collections, University of Washington Libraries, Negative
UW18183.

likely intended to be the horizon ring.[77] In fig. I.6 we see a Renaissance
illustration of an armillary sphere that preserves all the essential features
of the ancient prototype that Geminos would have known.

In his teaching, Geminos would have used a celestial globe to illustrate
the circles of the sphere (chapter v), the risings and settings of the zodiac
signs (vii), and the variation of the phenomena with locality (xvi). At xvi
12 he mentions a limitation of the globes that were commonly available.

[77] Von Boeselager 1983, pp. 56–60, and Tafel XV.

According to Geminos, the commercially available globes were inscribed with arctic and antarctic circles for the single *klima* corresponding to latitude 36°. Of course, if the globe were fitted with a meridian ring that was free to turn inside its holding slots in the horizon, one could adjust the globe for any latitude. But an arctic circle drawn or engraved on the surface of the globe could correspond only to a single latitude.

Sundial

For "sundial" Geminos uses three words more or less synonymously: *hōroskopeion* (ii 35 and xvi 13), *hōrologion* (ii 38, 45; vi 33, 46; viii 23; and xvi 18), and *skiothēron* (vi 32). Geminos invokes the sundial, not for genuine observation, but as an appeal to experience with which his reader is familiar, for most of the arguments he makes from the sundial would require the observer to make observations separated by several months in time or by several hundred stades in latitude. Clearly, Geminos never intended his reader to make these observations. Rather, he appeals to sense evidence that is *in principle* easily obtained, and the reader is supposed to grant him the argument.

Greek dialers were very inventive, and many different kinds are preserved among the 300 or so ancient dials extant today.[78] From the theoretical point of view, the simplest is the spherical dial or *skaphē*. The shadow-receiving surface is the interior of a hemisphere, and the tip of the gnomon lies at the center of the spherical surface. Since the dial is merely an inverted image of the celestial sphere, the theory governing the placement of the hour lines, as well as the equator and tropics, is very simple. Moreover, because the Sun cannot be found at just any place on the celestial sphere, but must remain between the tropics, an entire hemisphere of stone is not required. The lower part of the south face of the dial, corresponding to the part of the sky above the tropic of Cancer, can be cut away. Fig. I.7 shows an idealized view of this kind of dial.

Eleven hour curves indicate *seasonal hours*. The period from sunset to sunrise consists always of 12 hours, all equal to one another. Similarly, the night is divided into 12 equal hours. In the summer the day hour is long and the night hour is short, whereas in winter the opposite is true. The *equinoctial hour* that we use today (one twenty-fourth of the whole diurnal period) is a seasonal hour evaluated on the day of equinox. Although Greek astronomers did use the equinoctial hour

---

[78] For a catalogue of 256 stone dials, see Gibbs 1976. For the portable dials and stone dials published since Gibbs's survey, see Arnauldi and Schaldach 1997; Catamo et al. 2000; Field and Wright 1984; Locher 1989; Pattenden 1981; Price 1969; Rohr 1980; Savoie and Lehoucq 2001; Schaldach 2004; and Evans and Marée (forthcoming).

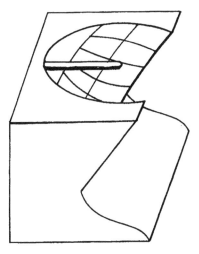

Fig. I.7. The principle of the spherical sundial with cutaway south face. Eleven hour curves mark the seasonal hours. Three parallel circles represent (from top to bottom) the tropic of Capricorn, the celestial equator, and the tropic of Cancer. An iron or bronze gnomon casts a shadow.

when they needed a uniform unit of time for precise calculation, the seasonal hour was the only one used in everyday life. (All surviving Greek and Roman sundials are in fact marked in seasonal hours.) The dial in fig. I.7 is also furnished with three *day curves*, indicating the track of the shadow's tip on (from top to bottom) winter solstice, equinox, and summer solstice.

To judge by the numbers preserved, the most common dial was of the *conical* type. In a conical dial, the shadow-receiving surface is the inner surface of a cone. Typically, the conical surface was cut into a roughly rectangular slab of stone, as with the dial shown in fig. I.8. The stoneworking involved in making a conical dial was easier than that required for a spherical dial. But by compensation, the theory was more complicated: it was necessary to project the celestial sphere onto a conical surface. The eleven more or less vertically oriented curves are the hour lines. On the dial shown in fig. I.8, the hour lines are labeled with Greek numerals, which is a very rare feature.[79]

Large numbers of *plane* dials have also been preserved. Although plane dials are simple from a stonecutter's perspective, they are much more complicated mathematically, for the celestial sphere must be projected onto a plane surface. In the dial of fig. I.9 we see the typical form of a horizontal plane sundial. The upper curve is the shadow track for

---

[79] This is sundial no. 3086 in Gibbs 1976.

Fig. I.8. A conical sundial found near Alexandria at the base of Cleopatra's needle. 40 cm high × 43 cm wide. The Greek numerals labeling the hour lines are an unusual feature and are probably Byzantine additions to the original Ptolemaic dial. British Museum no. 1936 3-9.1. Photo by permission of the Trustees of the British Museum.

summer solstice; the long straight line, the shadow track for equinox; and the lower curve, that for winter solstice. The broken stub of an iron gnomon remains embedded in the lead that fills the hole above the summer solstitial curve. The eleven more or less vertical lines represent the hours. This dial is decorated with the names of eight winds, in Latin, in the circle around the outside of the dial. An unusual feature of this dial is the signature of the dialmaker, a certain M. Antistius Euporus.[80]

[80] No. 4002 in Gibbs 1976.

Fig. I.9. A horizontal-plane sundial found in Aquileia in the north of Italy. The outer diameter of the circle of winds is 66 cm. Photo: Museo Archeologico Nazionale, Aquileia.

Geminos's readers were certainly familiar with dials of these basic types. At ii 35, Geminos appeals to the evidence of sundials to prove that the sign of Cancer is in syzygy with the sign of Gemini, saying that when the Sun is in Cancer, the tip of the gnomon's shadow follows the same curve as it does when the Sun is in Gemini. (See also ii 38 and 45, and vi 46.)

At vi 32–33, Geminos uses the sundial to show that the rate of the Sun's progress in declination (i.e., its motion toward the north or south) varies in the course of the year. Thus, around the time of the solstice, the tip of the shadow follows the same curve, as far as sense is concerned, for about 40 days. By contrast, at the time of the equinox, the tip of the shadow makes a perceptible departure from the equinoctial track in the course of a single day.

Geminos also mentions that around the solstices the lengths of the days and nights scarcely change for 40 days. The length of the day, however, was not something that most of Geminos's readers could have measured directly. The most direct method of measurement, using a sundial calibrated in hours, was not of much help, since the dials were all divided into seasonal hours: every day was 12 hours long.[81] And the water clocks mentioned by scientific writers were neither common nor very reliable keepers of time. Geminos's argument based on the motion of the sundial's shadow therefore provided a bolstering element of objectivity. This was something you could *see*.

Similarly, at viii 23, in criticizing the common belief that the Egyptians celebrate the feasts of Isis around the winter solstice, Geminos remarks that this can be refuted either by noting the lengths of the days or by looking at the track of the shadow tip on a sundial. In Geminos's time, the feast of Isis differed from the solstice by a whole month.

Finally, at xvi 13, 18, Geminos appeals to the sundial to illustrate the meaning of a geographic *klima*. The *klima* remains the same everywhere on the same parallel of latitude, and thus the lines engraved on the sundials remain the same. If we travel north or south, the *klima* changes, but the change is imperceptible for displacements of less than 400 stades. With larger displacements, the lines engraved on the sundials must be different.

Dioptra

The *dioptra* (i 4, v 11, and xii 4) was originally a simple sighting tube, or a rod equipped with two sights. The word itself indicates something that one "looks through." For compactness of expression, we shall refer simply to the *sighting tube*, although a rod fitted with two sights is the more likely form. In its simplest version, the *dioptra* could be aimed at a star and clamped in position. A teacher could use it to point out a particular star to students. If one then waited for 10 or 20 minutes, the star would have moved out of the sighting tube: in this way the motion of the celestial sphere could be demonstrated.

A pedagogical use of the *dioptra* is mentioned by Euclid in his *Phenomena*.[82] Euclid wants to demonstrate that the Earth is at the center of the celestial sphere. Suppose, as in fig. I.10, that *ABC* is the circle of "the horizon in the cosmos," i.e., the intersection of the horizon plane with the celestial sphere. The Earth is at *D*. Look through a *dioptra* at Cancer when it is rising at *C*. If you turn around and look through the other end of the *dioptra* you will see Capricorn setting at *A*, so *A*, *D*, and *C* are three points on a straight line. But Capricorn and Cancer are diametrically opposite one another in the zodiac, so *ADC* is a diameter of zodiac.

---

[81] Several ancient dials do call attention to the variation in the length of the day throughout the year, by comparing the length of the winter solstitial day to the other days of the year: see Gibbs 1976, nos. 1044, 1068, 3046, and 4001. For a discussion of the latter, see Evans 1998, 130–31.

[82] Euclid, *Phenomena*, prop. 1. Berggren and Thomas 1996, 52–53.

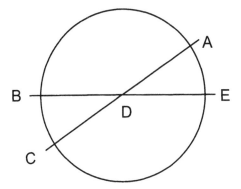

Fig. I.10. Euclid's argument, using a *dioptra*, to show that the Earth is at the center of the cosmos.

In the same way, aim the *dioptra* at *B* when Leo is rising there. If you then look through the other end of the *dioptra* you will see Aquarius setting at *E*. Therefore *EDB* is also a diameter of the zodiac. The two diameters of the zodiac circle intersect at *D*, which is therefore the center of that circle and of the celestial sphere. Since one can make the same argument at any place on the Earth, the Earth is a geometric point—the center of the cosmos. Now this is clearly only a pedagogical argument, and not a sequence of genuine observations. The conventional character of the argument is apparent in Euclid's use of Leo and Cancer as if they were points rather than extended constellations. The appeal to an instrument nevertheless lends the argument an air of authority.

The *dioptra* was eventually equipped with a protractor, which made it suitable for measuring angles. The *dioptra* in this form was used in surveying and in measuring the heights of mountains. It played a role similar to that of the modern surveyor's transit or theodolite. The most detailed extant discussion of these surveying instruments is Hero of Alexandria's *Dioptra* (first century A.D.).[83] No ancient *dioptra* has survived. Most *dioptra*s must have been much simpler than the elaborate instrument described by Hero, and conjecturally reconstructed in fig. I.11.

Yet another sort of *dioptra*, described by Ptolemy, was used for measuring the angular sizes of the Sun and the Moon.[84] (See fig. I.12.) The observer looks though a small hole in a plaque. A movable cylinder is then slid along a rod until the cylinder just barely covers the Sun or Moon. Ptolemy makes it clear that Hipparchos had used a similar instrument in

[83] See Lewis (2001), which includes an English translation of Hero's description of the instrument, pp. 259–62.

[84] Ptolemy, *Almagest* v 4. The most detailed description of this instrument is given by Proklos, *Sketch of Astronomical Hypotheses* iv 87–99.

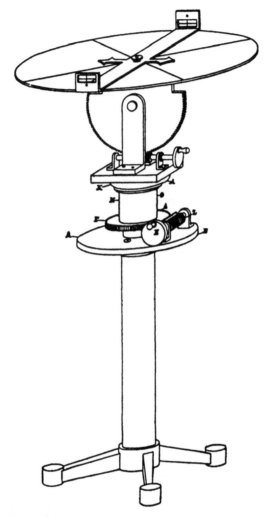

Fig. I.11. A conjectural reconstruction of Hero's *dioptra*. From H. Schöne 1903. The reconstruction is based on Hero's own description, but also reflects nineteenth-century German engineering!

the second century B.C. Moreover, from a brief remark in his *Sand Reckoner*, it seems that Archimēdēs used essentially the same instrument in the third century B.C.[85]

Thus the word *dioptra* covers a broad class of instruments, from quite simple to relatively sophisticated. To gain an idea of what sorts of *diop-*

[85] Dijksterhuis 1987, 364–65.

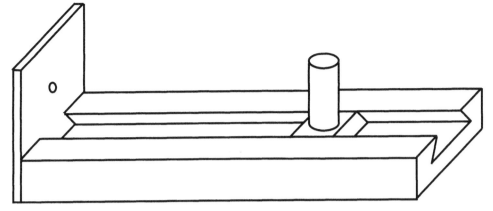

Fig. I.12. A *dioptra* for measuring the angular diameters of the Sun and Moon. The figure shows only a short length of the instrument, which would have extended much farther to the right. Hipparchos used a similar instrument that was 4 cubits long.

*tra*s Geminos had in mind, let us see what uses he makes of them. At i 4, he says that it is possible to divide the zodiac into twelve equal parts by means of the *dioptra*. Here, he is making a point about the conventional character of the zodiac signs, i.e., that they are artificial 30° segments of the zodiac. An instrument rather like Hero's would suffice for this purpose. The table (the large round disk carrying the sights at the top of fig. I.11) must be adjusted to lie in the plane of the ecliptic. One could then, in principle, turn the sights to mark off 30° segments. But it must be stressed that Geminos is appealing to a process that could be *imagined*, and that his readers would therefore grant. This process would have no real astronomical utility.

At v 11 Geminos says that it is possible to trace out the celestial circles (including the tropic and arctic circles) by means of the *dioptra*. At xii 4 he says that "all the stars observed through the *dioptra*s are seen to be making a circular motion during the whole rotation of the *dioptra*s." Both of these appeals require a more astronomically specialized instrument than the surveying instrument described by Hero. It must be possible to fix the sighting tube at a given angle to the axis of the cosmos. And it must be possible to rotate the whole instrument about the axis. This is quite an interesting appeal on Geminos's part, since, as far as we know, no other ancient writer mentions a *dioptra* of precisely this sort. The general principle of this instrument, which we might call the equatorial *dioptra*, is suggested in fig. I.13. The table is adjusted to lie in the plane of the celestial equator. Thus the table must face toward the north and must make an angle with the horizon that is equal to the

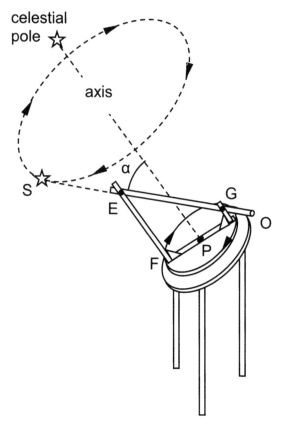

Fig. I.13. A conjectural reconstruction of Geminos's equatorial *dioptra*.

observer's co-latitude. With this orientation achieved, the table lies at right angles to the axis of the cosmos. On top of the table is a disk that is free to turn about a pivot *P*. To this disk is attached the supporting mechanism for a sighting tube *OE*. The sighting tube has a pivot at *G* and a movable clamp at *E*. Thus *E* may be slid up and down arm *EF* and clamped for any desired polar angle α. The observer looks through the tube at *O*, aims at a star *S* that is north of the celestial equator, and clamps the tube into position. Then, in the course of the night, the observer can keep *S* in sight simply by turning the sighting apparatus about pivot *P*. The polar angle or the declination of the star could be read from appropriate markings along *EF*. To sight a star that is south of the celestial equator, the observer simply looks through the other end, *E*, of the sighting tube.

Pure mathematics (concerned with mental objects only)
    Arithmetic (study of odds, evens, primes, squares, etc.)
    Geometry
        Plane geometry
        Solid geometry
Applied mathematics (concerned with perceptible things)
    Logistic (practical calculation, which is analogous to arithmetic)
    Geodesy (practical mensuration, which is analogous to geometry)
    Canonics (theory of musical scales, an offspring of arithmetic)
    Optics (an offspring of geometry)
        Optics proper (straight rays, shadows, etc.)
        Catoptrics (mirrors, etc.)
        Scenography (perspective)
    Mechanics
        Military engineering (engines of war)
        Wonder-working (pneumatics applied to automata)
        Equilibrium and centers of gravity (theory of the lever, etc.)
        Sphere-making (*sphairopoiïa* = models of the heavens)
    Astronomy
        Gnomonics (sundials)
        Meteoroscopy (armillary spheres)
        Dioptrics (the dioptra and related instruments)

Fig. I.14. Geminos's branches of the mathematical arts.

9. Geminos on Mathematical Genres

Geminos offers testimony to the importance of genres in mathematical writing, in a passage of his *Philokalia* that discusses the subdivisions of pure and applied mathematics. Although Geminos's book has not come down to us, his discussion of the organization of the mathematical sciences is preserved by Proklos in his *Commentary on the First Book of Euclid's Elements*. This passage is translated below as fragment 1. Geminos's organization of the mathematical sciences is illustrated in fig. I.14.

Mathematics is divided first of all into what we might call the pure and the applied. But Geminos's own distinction is Platonic—that one branch is concerned "with mental things only," in abstraction from material forms, while the other deals with "perceptible things." Of sciences concerned with mental objects he gives two examples: arithmetic and geometry. It is important to note that, for the Greeks, arithmetic (*arithmētikē*) was not the elementary computation taught in the schools.

This distinction was made at an early date, for Plato distinguishes the art of calculation from the contemplation of pure number.[86] What the Greeks called arithmetic we might call number theory. Thus arithmetic is concerned with the classes and properties of whole numbers and their relationships to one another (for example, odd, even, prime, and perfect numbers, as well as figured numbers such as the triangular, square, and pentagonal). Greek arithmetic originated with the Pythagorean tradition, but books vii–ix of Euclid's *Elements* constitute the oldest surviving treatment of the subject. The first part of Theōn of Smyrna's *Mathematical Knowledge Useful in Reading Plato*, is a survey of arithmetic, without proofs, intended for beginners.[87]

Of the sciences concerned with perceptible things, Geminos gives six examples: logistic, geodesy, canonics, optics, mechanics, and astronomy. Logistic (*logistikē*) is the art of practical computation. As a subject of instruction in the schools, logistic was concerned with addition, subtraction, multiplication, and division, including the handling of fractions. Examples of school problems are provided by the arithmetical epigrams of the *Greek Anthology*.[88] And Heath argued that, despite its title, the *Arithmetica* of Diophantos is a work of logistic.[89]

Geodesy (*geōdesia*) is the science of measuring surfaces and volumes, including surveying. Extant works by Hero of Alexandria provide good examples of this genre. In *Dioptra*, Hero describes the construction of a surveying instrument that may be used for leveling and for measuring angles. The application of the instrument to various surveying problems is treated in some detail. Another of Hero's works, *Stereometrica*, is devoted to mensuration. Hero begins with textbook problems, involving spheres, cones, and pyramids, but progresses to more practical problems, including the calculation of the seating capacity of a theater from the lengths of its highest and lowest rows.

According to Geminos, calculation and geodesy are only "analogous" to arithmetic and geometry, since they do not make propositions about mental numbers or figures, but rather about perceptible ones. Thus, geodesy does not deal with ideal cylinders or cones, but rather with heaps as cones, and with pits as cylinders. And it does not use mental straight lines, but rather lines made perceptible. These perceptible lines can be rather precise representations of ideal lines (e.g., rays of sunlight) or rather crude ones (ropes). Similarly, the calculator (*logistikos*) does not consider the properties of numbers in themselves, but rather as present in perceptible things.

---

[86] Plato, *Republic* vii, 525a–26c.
[87] See also Heath 1921, vol. 1, 65–117.
[88] *Greek Anthology* xiv, in Paton 1918, vol. 5.
[89] Heath 1921, vol. 1, 15–16.

In a similar but somewhat different way, Geminos considers canonics and optics to be "offspring" of arithmetic and geometry, respectively. Canonics (*kanonikē*) is the mathematical theory of the musical scale. A canon (*kanōn*) was originally a rule or line, as used by carpenters and masons. It is also the usual term for the geometer's straightedge. Then, by metaphor, it refers to any norm or standard. The stretched string, or monochord, was the *kanōn*, or basis, of the theory of musical intervals. Thus canonics is the theory of the division of the octave. Canonics is an offspring of arithmetic because it is concerned with ratio—but ratio made perceptible, either to the eye in the form of the lengths of strings, or to the ear in the aspect of harmony. The oldest surviving treatise is the *Division of the Canon*, which is attributed to Euclid, although this attribution has been challenged.[90] The second part of Theōn of Smyrna's *Mathematical Knowledge Useful for Reading Plato*, is an introduction to canonics.

Optics is an offspring of geometry, because it is concerned with lines made perceptible. Under optics, Geminos ranges three subdivisions. Optics proper is concerned with effects that may be explained in terms of the straight-line propagation of the visual ray. The Euclidean *Optics* (third century B.C.) is the prototype of this genre. Among its propositions we find: parallel lines seen from a distance appear to converge (prop. 6), and a circle viewed obliquely appears flattened (prop. 36). The *Optics* has come down to us in two forms. Heiberg conjectured that one is the original, Euclidean version and the other a revised version due to Theōn of Alexandria (fourth century A.D.).[91] Thus it appears that this branch of applied science remained a part of the mathematics curriculum for seven centuries.

The second subdivision of optics is catoptrics (*katoptrikē*), the theory of mirrors (*katoptron* = "mirror"). The prototype is the pseudo-Euclidean *Catoptrics*, which treats thirty propositions involving plane, convex, and concave mirrors. Other important writers on catoptrics include Hero of Alexandria[92] and Ptolemy,[93] who also investigated refraction.

The final subdivision of optics is scenography or "scene-painting." *Skēnographikē* is perspective. It is of use in the production of theater sets of pleasing and convincing proportion. No Greek treatise on scenogra-

---

[90] For a recent study, see Bowen, "Euclid's Sectio canonis and the History of Pythagoreanism," in Bowen 1991. For Ptolemy's treatment of the subject, see Barker 2000.

[91] Both versions are found in Heiberg's text and in Ver Eecke's translation. More recently, scholars have parted with Heiberg over which version is the older: see Jones 1994.

[92] The *Catoptrics* often attributed to Hero survives only in a medieval Latin translation, in which the work is wrongly ascribed to Ptolemy. Jones 2001b provides an edition and English translation and discusses the evidence for and against the attribution to Hero.

[93] For an English translation of Ptolemy's optical treatise, see Smith 1996.

phy is extant, but both Anaxogoras and Dēmokritos are said to have written on this subject.[94]

Hero of Alexandria's *Mechanics* is a good example of a treatise on general mechanics. The Greek text has not come down to us, but the work survives in a medieval Arabic translation. Hero's treatise deals with gear ratios, the parallelogram of forces, centers of gravity, as well as with the basic machines—windlass, lever, pulley, wedge, and screw, both alone and in combination. Book viii of Pappos's *Mathematical Collection* is a substantial treatise on Greek mechanics.[95] And Vitruvius[96] gives a brief overview of mechanics, based largely on Greek sources, including applications to hoisting machines and waterwheels.[97]

Geminos points to military engineering as a subdivision of mechanics. This art may be illustrated by Hero of Alexandria's treatise on catapults, *Belopoiïkē* (missile-making).[98] Vitruvius[99] is another good source on military applications of mechanics.

Geminos's second subdivision of mechanics is wonder-working. *Thaumatopoiïkē*, the "wonder-making" (art or science), was the craft of devising automata and other gadgets, often operated by means of fluid or air pressure. Examples: mechanical blackbirds that sing by means of waterworks; a pneumatically operated toy temple whose doors open when a fire is lit on the altar; a sacrificial vessel from which liquid flows only when money is introduced. Many such contrivances are described by Hero in his *Pneumatics*, which may be considered typical of this genre. Few, if any, of these inventions were of practical use. They were small-scale toys intended to amuse and amaze. Although the theoretical discussion is often defective, the *Pneumatics* nevertheless demonstrates an impressive mastery of hydrostatics in concrete applications.

The earliest surviving work of the science of equilibrium and centers of gravity is that of Archimēdēs (third century B.C.). In his treatise *On*

---

[94] Vitruvius, *On Architecture* vii, introduction, 11. Heath (1921, vol. 1, 178) lists among Dēmokritos's works one titled *Ekpetasmata*, a word which, he notes (p. 181), Ptolemy glosses in his *Geography* as referring to the projection of the armillary sphere on the plane. This is, of course, just a special case of the representation of three-dimensional bodies on a two-dimensional surface that scene painting would require, so Dēmokritos's *Ekpetasmata* may be the work to which Vitruvius refers.

[95] The Arabic text of Pappos's Book viii, the only one that seems to have been translated into Arabic, contains material that dropped out in the course of transmission of the Greek text. Unhappily, no translation of the Arabic version has yet been published.

[96] Vitruvius, *On Architecture* x 1–9.

[97] On ancient mechanics, see Landels 1978 or, more briefly, Lloyd 1973, 91–112.

[98] This treatise was omitted from the Teubner edition of Hero's works, but is available in English translation in Marsden 1971.

[99] Vitruvius, *On Architecture* x 10–16.

the Equilibrium of Planes,[100] Archimēdēs demonstrates the locations of the centers of gravity of various plane figures, such as triangles, parallelograms, and a parabolic sector, and attempts a proof of the law of the lever. Two other works of Archimēdēs on this science are lost: *On Levers* and *On Centers of Gravity*.[101]

Sphere-making (*sphairopoiïa*) is the science of constructing models of the heavens, including celestial globes, armillary spheres, and orreries. A reputation for brilliance in this craft attached to the name of Archimēdēs. Pappos of Alexandria says that Archimēdēs composed a treatise on this subject, which appears, however, to have been lost already by Pappos's time.[102] When the Romans captured Syracuse, Marcellus took two of Archimēdēs' devices back to Rome. One of these seems to have been a celestial globe, which Marcellus placed in the temple of Vesta, where it remained long enough to be seen by Ovid.[103] From the account by Cicero, the other instrument was an orrery that represented the motions of the Sun, Moon, and planets.[104] This orrery of Archimēdēs must have been quite a marvel, for Cicero expresses disapproval of some "who think more highly of the achievement of Archimēdēs in making a model of the revolutions of the firmament than that of nature in creating them, although the perfection of the original shows a craftsmanship many times as great as does the counterfeit."[105]

Under astronomy, Geminos ranges three subdivisions, which are all concerned with instruments of some kind. Gnomonics (*gnōmonikē*) is the science of designing and constructing sundials. The earliest treatises on this science are lost. Vitruvius[106] gives a list of dial types, together with the names of their inventors, at least some of whom presumably wrote short treatises on their inventions. Vitruvius also sets out the

[100] Available in translation in Heath 1912 and Dijksterhuis 1987. Berggren 1976 argues that the extant version of this treatise contains introductory material (including an unsuccessful proof of the law of the lever) that was inserted by a later editor who wished to turn a research treatise of Archimēdēs into a text suitable for instruction.

[101] See Heath 1912, xxxvii.

[102] Pappos, *Mathematical Collection* viii, 3; Hultsch, p. 1026; Ver Eecke, p. 813. Plutarch (*Marcellus* xvii 3–4), promoting his own philosophical line, says that Archimēdēs did not deign to leave behind any treatise on the mechanic arts, judging them ignoble and vulgar. Apparently, he made an exception in the case of *sphairopoiïa*, perhaps because of the nobility the subjects it seeks to represent.

[103] Ovid, *Fasti* vi 277–80.

[104] Cicero, *De republica* i 14. The more elaborate of Archimēdēs' two constructions was probably a gearwork mechanism. There are extant two ancient gearwork mechanisms for reproducing celestial motions: see Price (1974) and Field and Wright (1984). Both of these are, however, from well after Archimēdēs' time.

[105] Cicero, *On the Nature of the Gods* ii 35.

[106] Vitruvius, *On Architecture* ix 8.1.

figure of the analemma, which is a folding down of the celestial sphere onto the plane, and which permits the construction of the hour lines of a sundial by techniques of projective geometry. The analemma described by Vitruvius is certainly not of his own invention, but likely goes back at least to the third century B.C., when Greek sundials began to appear in substantial numbers. The other important surviving ancient work on gnomonics is Ptolemy's *On the Analemma*.[107]

The branch of astronomy called meteoroscopy (*meteōroskopikē*) should not be confused with meteorology (*meteōrologia*). The word *meteoroscopy* means "looking at things on high," and it almost certainly refers to the art of making and using a specialized instrument of observation rather similar to an armillary sphere. Note that Geminos's other two branches of astronomy are also devoted to instruments—sundials and *dioptra*. In the *Geography*, Ptolemy[108] uses *meteoroskopion* for an instrument for taking celestial observations, and says that he has given its description. Proklos[109] uses *meteoroskopeion* for what is some sort of improvement on Ptolemy's armillary astrolabe, described in *Almagest* v 1. And Pappos provides more details for the construction of the meteoroscope than Ptolemy gave in his description of the astrolabe instrument in the *Almagest*.[110] From all this it seems clear that Ptolemy had written a short specialized treatise on the construction of a *meteoroskopion*. That Geminos already considered meteoroscopy a genre of writing shows that the spherical instruments go back at least to his time.

Finally, dioptrics (*dioptrikē*) is the science of using the *dioptra*, a sighting and measuring instrument, several forms of which are described above.

That Geminos's branches of mathematics represent, not mere abstract categories, but actual genres is clear from the fact that we possess, or at least have knowledge of, actual Greek mathematical works that fit every one of them. Naturally, each genre of mathematical writing had its own customs and conventions.[111] Many of these genres had long lives. For example, we have works on optics by Euclid and by Ptolemy, who were separated by nearly 500 years. But, although all the categories listed by Geminos are true genres of mathematical writing, it is clear that he has not listed every genre. There is, for example, nothing to correspond to mathematical planetary theory (although the *Introduction to the Phenomena* makes it clear that he knew of such theories), for Geminos's branches of astronomy are all concerned with instruments of some sort.

[107] Ptolemy, *Opera* ii, 187–223. See Neugebauer 1975, 839–56.
[108] Ptolemy, *Geography* i 3.
[109] Proklos, *Sketch of Astronomical Hypotheses* vi 2.
[110] For a full discussion, see Rome 1927. For the text, see Rome 1931, 1–16.
[111] See Knorr 1986 and Mansfeld, 1998.

10. Reality and Representation in Greek Astronomy

Hypotheses and Phenomena

In chapter i of the *Introduction to the Phenomena*, Geminos makes some remarks about the role of hypotheses in accounting for the phenomena, for which we would like to provide a context. At i 19, he says that "the hypothesis that underlies the whole of astronomy is that the Sun, the Moon, and the five planets move circularly and at constant speed in the direction opposite to that of the cosmos." Moreover, he attributes to the Pythagoreans the principle that uniform circular motion is proper to celestial things. At i 21, still speaking of the Pythagoreans, Geminos says, "they put forward the question: how would the phenomena be accounted for by means of uniform and circular motions?"

On the other hand, Simplikios (sixth century A.D.) attributed to Plato the celebrated homework assignment for the astronomers to "save the phenomena" in terms of uniform circular motions.[112] But there is nothing in Plato's work to confirm that he enunciated a principle of uniform circular motion, and much that is inconsistent with it. Indeed, Knorr argued that the attribution of the principle to Plato was a mistake made by the late Greek commentators.[113] In any case, we find the first clear, surviving statement of the principle in Aristotle.[114] By Geminos's time, the principle that uniform circular motion is appropriate to heavenly bodies had become a commonplace of philosophy. Other astronomical writers who endorse the principle include Theōn of Smyrna, Ptolemy, and Proklos.[115]

One important reason for clearly stating "hypotheses," or assumptions, is that astronomy was commonly regarded as a branch of mathematics, and particularly of geometry. The standard form of a geometrical demonstration involved the clear statement of hypotheses. An excellent astronomical example comes from Aristarchos of Samos's treatise *On the Sizes and Distances of the Sun and Moon*, in which all the necessary astronomical hypotheses are clearly stated before the demonstration begins. The debt of astronomy to mathematical conventions does not necessarily mean that Greek astronomers thought of their hypotheses as

---

[112] Simplikios, *Commentary on Aristotle's* On the Heavens, comments on ii 12. Heiberg 1894, 488.18–24 and 492.15–493.5. The two short passages from Simplikios have been quoted and discussed by many, including Duhem (1908, 3), Samburksy (1956, 59) and Vlastos (1975, 59–61 and 110–11). For an English translation of Simplikios's commentary, see Mueller 2005.

[113] Knorr 1989.

[114] Aristotle, *On the Heavens* i 2 and ii 3, 12.

[115] Theōn of Smyrna, *Mathematical Knowledge* iii 22. Ptolemy, *Almagest* ix 2. Proklos, *Sketch of Astronomical Hypotheses* i 1.7–9.

arbitrary, or as free choices, but it does raise the question of what truth value or epistemic status they attached to them.

When Geminos asks how "the phenomena would be accounted for" (*apodotheiē ta phainomena*), this expression carries the same sense as the more famous prescription "to save the phenomena" (*sōzein ta phainomena*). But Geminos's version of the saying may well be the older. Although there is no reason to take seriously Simplikios's attribution of this program to Plato, the expression "save the phenomena" does crop up in late writers with Platonist affiliations, such as Thēon of Smyrna and Proklos. But according to Goldstein,[116] the oldest extant text in which the expression "save the phenomena" occurs is only of the first century A.D., namely Plutarch's *On the Face in the Orb of the Moon*.[117] (We should point out that Geminos twice uses the expression "save the phenomena in" fragment 2, but we cannot exclude the possibility that this is an interpolation by Simplikios, who is quoting him.) Writing in the fourth century B.C., Aristotle used *apodōsein* ("to give back," or "to render an account of") when he pondered how the spheres of Eudoxos could best be modified "to account for the phenomena."[118]

The program of saving, or accounting for, the phenomena has given rise to a large modern literature of interpretation, which bears strongly on the question of whether Greek astronomy should be thought of as a realist or an instrumentalist endeavor. By scientific *realism* we mean the view that a successful scientific theory contains at least some elements of the true nature of the world. A realist believes that one goal of science is to discover what the world really is, and that at least some features of a good theory correspond to things that truly exist in nature. Scientific realism can be contrasted with *instrumentalism*. In an instrumentalist approach to science, the practitioner may renounce (perhaps as an impossibility) the goal of ever discovering the real nature of things, and may be content to have a theory that "works," i.e., that successfully accounts for known phenomena and allows the prediction of phenomena yet to be investigated. The instrumentalist would not claim that the individual features of a successful theory necessarily correspond to things really existing in nature.

At the beginning of the twentieth century, Pierre Duhem constructed an instrumentalist interpretation of Greek astronomy that has been very influential.[119] In Duhem's account, the testimony of Geminos (in fragment 2) and of Proklos played key roles. Dreyer[120] and Sambursky,[121] too, represented Geminos as an instrumentalist, who was concerned

---

[116] Goldstein 1997, 7.

[117] Plutarch, *On the Face in the Orb of the Moon* 923A.

[118] Aristole, *Metaphysics* 1073b38 and 1074a1.

[119] Duhem 1908.

[120] Dreyer 1906, 132.

[121] Sambursky 1962, 135–37.

only with saving the phenomena, and who renounced any attempt to find the true arrangement of the cosmos. More recently, Lloyd has argued that Duhem misinterpreted some key evidence, and that all the Greek astronomers about whom we know enough to make a judgment (including Geminos) were in fact realists.[122] In any case, Duhem's picture of Greek astronomy as a seamlessly developing science, with a single set of goals that lasted from the time of Plato to the close of Antiquity, is in need of strong correction. For the phenomena judged to be in need of saving changed with time as Greek astronomy matured. Concern with the broad features of planetary phenomena long preceded concern with quantitative details of planetary positions.[123]

Moreover, Greek astronomy often exhibits a split personality, because of its opportunism, and many scholars have overestimated its consistency and allegiance to philosophical principles. Greek writers steeped in the philosophical tradition wrote accounts of the planets based on the circular motions of epicycles and deferents. A good example is Theōn of Smyrna's *Mathematical Knowledge Useful for Reading Plato*. However, in Theōn's day (early second century A.D.), the Greek geometrical theories were not capable of yielding accurate planet positions. Greek astrologers, who needed convenient and reasonably accurate methods of calculating planet positions, relied instead on arithmetic methods borrowed from the Babylonians,[124] and these methods had *nonuniform* motion around the zodiac built into them. It was only with Ptolemy's publication of his planetary tables in the *Almagest* that the Greek geometrical theories became capable of producing accurate planet positions. The new feature that made Ptolemy's planetary models quantitatively accurate was his equant point, which effectively incorporated nonuniform motion.[125] Geminos exhibits the happy coexistence of philosophical principle and arithmetic convenience that was characteristic of Greek astronomy in his day. Thus, in spite of his endorsement in chapter i of the principle of uniform circular motion, in chapter xviii he discusses the Babylonian lunar theory, which involves nonuniform motion modeled by an arithmetic progression.

Sphairopoiïa as World and Representation

Nevertheless, in the *Introduction to the Phenomena*, as well as in fragment 2, Geminos provides us with strong evidence of the position he

[122] Lloyd 1978, with additions in Lloyd 1991, 248–77.

[123] On these issues, see also Mittelstrass 1962, Goldstein 1980, Aiton 1981, Smith 1981, Smith 1982, Lloyd 1987, 285–336, Goldstein 1997, Evans 1998, 216–19, Bowen 2001, and Bowen 2003.

[124] Jones 1999b.

[125] See Evans 1984.

would have taken in a debate about realism and instrumentalism (which are not, of course, ancient terms) in astronomy. To begin with, we can learn a good deal from his discussions of *sphairopoiïa*, the art of modeling the cosmos.

Sphairopoiïa is a word that exhibits an interesting range of meanings. As we have seen, Geminos uses it in fragment 2 to indicate (1) *a branch of mechanics*—the art of constructing working models of the heavens. Moreover, *sphairopoiïa* was a *genre of mathematical writing* that went back at least to Archimēdēs. And 600 years later, Pappos of Alexandria still treated it as a genre.

But *sphairopoiïa* can also mean (2) *a particular mechanical model*. For example, Thēon of Smyrna[126] says that he actually made a *sphairopoiïan* of the nested spindle whorls described by Plato in his mystic vision of the planetary spheres.[127]

In the *Introduction to the Phenomena*, Geminos uses *sphairopoiïa* to mean (3) *a spherical theory of the world*, which can be said to be in agreement with nature. At xvi 28–29, Geminos remarks that "Homer and the ancient poets" believed the Earth to be flat, extending all the way to the sphere of the cosmos. Consequently, according to these poets, the *Aithiopians* near the rising as well as those near the setting are burned by the Sun. Geminos says, "This notion is consistent with their proposed arrangement, but alien to the spherical construction (*sphairopoiïa*) in accord with nature."

Most interestingly, Geminos also uses *sphairopoiïa* to mean (4) *a spherical system that actually exists in nature*. At xii 23, Geminos says that "there is a certain spherical construction proper for each [planet]" that accounts for the motions. Although Geminos provides no details about the *sphairopoiïa* appropriate to the planets, he is probably thinking of epicycles and eccentrics embedded in three-dimensional spheres.[128] That he really means this as a construction existing in nature is confirmed by some of his other uses of the same expression. At xiv 9, he invokes the spherical construction (*sphairopoiïa*) of the cosmos to explain why the patterns of risings and settings are not the same for all stars (e.g., the classes of stars called *doubly visible* and *night-escapers*). At xvi 19, Geminos invokes the spherical construction (*sphairopoiïa*) to show that there exists another temperate zone in the southern hemisphere of

---

[126] Thēon of Smyrna, *Mathematical Knowledge* iii 16.

[127] Plato, *Republic* x 614–17.

[128] Geminos's demonstration of the importance of genre in mathematical writing and his remark about *sphairopoiïa* that actually exist in nature bear on the history of three-dimensional representations of deferent-and-epicycle theory. Evans 2003 argues that deferent-and-epicycle theory was introduced in three-dimensional form at its very beginning.

the Earth. In these two passages, there is clearly no question of a "model" to save the phenomena: Geminos is speaking of the *sphairopoïia* of the world itself.

Thus *sphairopoïia* can mean (1) a branch of mechanics; (2) a particular mechanical model; (3) a theory of the world, which can be said to be after, or according to, nature; or (4) the spherical arrangement or system of the world itself (including the spherical constructions that produce the motions of the planets). These four uses are closely related and reveal a gradation in meaning. And this is a sign of the essential realism of Greek astronomy. No better demonstration could be wished of Geminos's own position as a realist. For Geminos, astronomical models are assertions about the nature of the world: at xvi 27, he speaks of the "spherical construction that accords with truth [or reality]."

Geminos's Realism

Geminos's realist stance is cogently developed in a passage of his *Concise Exposition of the Meteorology of Poseidōnios*, printed below as fragment 2. In this passage, Geminos provides a discussion of the relationship between astronomy and physics that has much to teach us about the goals and limitations of astronomy as the Greeks perceived them. Starting from a remark by Aristotle,[129] Geminos contrasts the methods of the astronomer and the physicist. Physics is concerned with the very natures of the stars, their essential qualities, as well as their origin and destruction. Astronomy, by contrast, is concerned with such things as the size and shape of the Earth, Sun, and Moon, eclipses and conjunctions, and, generally, the motions of the heavenly bodies. Astronomy therefore must rely upon arithmetic and geometry for its demonstrations, rather than on arguments from the nature of things.

As Geminos points out, the astronomer and the physicist may address the same question, but will proceed by different methods, and he gives two good examples of this—proving that the Sun is large, or that the Earth is spherical. That the Sun is large was a common proposition in both Greek natural philosophy and Greek astronomy. Among the pre-Socratic philosophers, Anaximandros, Empedoklês, and Anaxagoras all asserted that the Sun is large (Anaximandros and Empedoklês that the Sun is as large as the Earth, and Anaxagoras that it is larger than the Peloponnesus). An astronomical (or geometrical) method of determining the size of the Sun was first given by Aristarchos of Samos (third century B.C.). All the later Greek efforts, e.g., by Hipparchos and Ptolemy, were modifications of Aristarchos's procedure. Aristarchos's calculations led

[129] Aristotle, *Physics* 193b22.

to the conclusion that the ratio of the diameter of the Sun to the diameter of the Earth is greater than 19/3 but less than 43/6: a large Sun, though much too small by modern standards.[130]

Aristotle gives several arguments for the sphericity of the Earth that nicely illustrate the differing approaches of the physicist and the astronomer.[131] Aristotle begins with a physical or philosophical argument, which he deems to be first in force and importance. The Earth is spherical because of the center-seeking nature of earth as an element: particles of earth, striving to reach the center of the universe, and coming toward that center from all directions, naturally produce a spherical shape. Aristotle does not, however, disdain to add three arguments based on sense-evidence. The first is based on the shape of the Earth's shadow as seen during a lunar eclipse. The second involves the change in the visibility of certain stars as one moves from Egypt or Cyprus to more northerly lands. The third argument from sense-evidence is based on elephants, which may be found at both extremes of the known world—in India and in the region about the Pillars of Hercules. The two astronomical arguments are sound ones and are clearly of a different sort than the physical argument based on Aristotle's own theory of natural motions.

Moreover, according to Geminos, it is not for the astronomer to know the causes of things, or to decide the true nature of the world. Rather, the astronomer must take his first principles from the physicist, "that the motions of the stars are simple, uniform, and orderly." But we cannot know from astronomical observation whether the Earth goes around the Sun or the Sun goes around the Earth, for the phenomena can be saved (or accounted for) under either assumption. In answer to the question of why the Sun, Moon, and planets appear to move irregularly, the astronomer can only say that, if we assume their circles are eccentric, or that the stars go around on epicycle, the apparent irregularity will be accounted for. Here Geminos is referring to the fact that there were two versions of the Greeks' geometrical solar theory.

Fig. I.15 shows the eccentric-circle solar theory that Geminos presents in chapter i of his *Introduction*. In the figure, we look down on the plane of the ecliptic from above its north pole. The Sun S moves at uniform speed on a circle that is centered at a point C somewhat eccentric to the Earth O. Thus angle α increases uniformly with time. In the solar theory of Hipparchos, who was apparently the first to specify numerical values for the parameters of this theory, the eccentricity OC is 4.15% of the radius of the circle. Although the Sun moves uniformly on its circle, to an

---

[130] On Aristarchos, see Heath 1913. Van Helden 1985 gives a good brief account of several attempts by Greek astronomers to determine the size and distance of the Sun.

[131] Aristotle, *On the Heavens* 297a8–98a15.

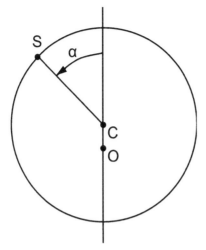

Fig. I.15. An eccentric-circle model for the motion of the Sun. The Sun S moves uniformly around C in the course of the year. But because C is eccentric to the Earth O, the motion appears nonuniform to us.

observer at O it appears to move more slowly on some parts of the circle and more quickly on others, because of the varying distance.

The second version of the solar theory is shown in fig. I.16. The Sun S moves clockwise around a small epicycle, while the center K of the epicycle moves counterclockwise around a deferent circle that is concentric with the Earth O. Both motions are uniform and are completed in exactly 1 year. Thus we always have $\beta = \alpha$.

It happens that *the two theories are mathematically equivalent.* See fig. I.17, which examines the epicycle-plus-concentric theory more closely. Since $\beta = \alpha$, line segment KS remains parallel to OC as K moves around the deferent. Thus the path actually described by the Sun is the circle shown in dashed line. If the radius KS of the epicycle is chosen to be equal to the eccentricity OC in the eccentric-circle model, the two theories are geometrically indistinguishable. Apollōnios of Pergē probably proved this equivalence, around 200 B.C.[132] Both Theōn of Smyrna and Ptolemy give their own proofs.[133]

Although the two forms of the solar theory are mathematically equivalent, nevertheless, a controversy arose over which corresponded to reality. According to Theōn of Smyrna,[134] Hipparchos expressed a preference for the epicycle theory, saying that it was probable that the

[132] See Neugebauer 1975, 262–65.
[133] Theōn of Smyrna, *Mathematical Knowledge* iii 26. Ptolemy, *Almagest* iii 3.
[134] Theōn of Smyrna, *Mathematical Knowledge* iii 34.

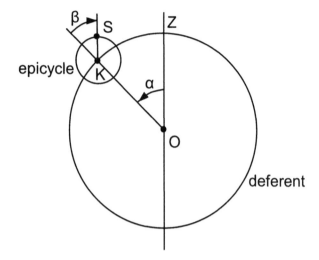

Fig. I.16. A concentric-deferent-and-epicycle model for the motion of the Sun. The Sun *S* moves on an epicycle, while the center *K* of the epicycle moves around a deferent circle centered on the Earth *O*. Both motions are completed in 1 year, and angles β and α are always equal.

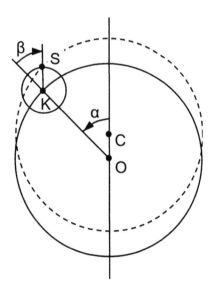

Fig. I.17. Equivalence of the concentric-plus-epicycle model to the eccentric-circle model.

heavenly bodies are placed uniformly with respect to the center of the world. Ptolemy, however, was later to endorse the eccentric-circle form of the theory, saying that it was simpler, since it involved one motion rather than two.[135] Since the two models are mathematically equivalent, the choices of Hipparchos and Ptolemy for one or the other were clearly based on criteria other than agreement with the phenomena—criteria that we may describe as physical or philosophical.

Geminos goes on to say, "It will be necessary to fully examine in how many ways it is possible for these phenomena to be brought about, so that the treatment of the planets may fit the causal explanation (*aitiologia*) that is in accord with acceptable method." The prescription to seek multiple geometrical explanations echoes a remark of Hipparchos, who said that research into the explanation of the same phenomena by hypotheses that are quite different is a task worthy of the attention of the mathematician.[136] According to Geminos, the reason for developing all possible geometrical explanations is to be sure that we will find one that accords with accepted physical principles, by providing a causal explanation of the phenomena.

The term *aitiologia*, "causal explanation" or "investigation into causes," comes from Greek natural philosophy, and the English word "etiology" descends from it. Diogenēs Laertios describes Aristotle as "ready at giving the cause," or "inquiring into causes," but Diogenēs does so in the superlative degree of the adjective (*aitiologikōtatos*), thus meaning that Aristotle was the *most* ready of all philosophers to take up causes.[137] Also according to Diogenēs, *aitiologikos* was one of three parts of Stoic physics and was concerned, in part, with the explanation of things such as vision, and with giving the causes of images in mirrors, clouds, thunder, rainbows, and so on.[138] *Aitiologia* and related words turn up several times in the fragments and testimonia of Poseidōnios.[139] One of the more charming and illuminating occurrences of the word is in a passage of Strabo[140] in which the latter criticizes Poseidōnios for trying to introduce *too much* of this seeking after causes into geography: "For in him there is much inquiry into causes, that is, 'Aristotleizing,' a thing that our School [i.e., the Stoics] avoids because of the concealment of the causes." From all this it is clear that *aitiologia* signifies an inquiry into the causes of natural phenomena in the way pioneered by Aristotle and continued by some of the Stoics, notably Poseidōnios. Geminos uses

[135] Ptolemy, *Almagest* iii 4.
[136] Theōn of Smyrna, *Mathematical Knowledge* iii 26.3.
[137] Diogenēs Laertios, *Lives and Opinions* v 32.
[138] Diogenēs Laertios, *Lives and Opinions* vii 132.
[139] Edelstein and Kidd 1989, fragments 18, 176, and 223.
[140] Strabo, *Geography* ii 3.8.

the word here to insist that it is not enough simply to save the phenomena in terms of some combination of circular motions. One must also have a causal explanation, that is, an explanation based on the nature of things.

Theōn of Smyrna gives another good example of the importance to be placed on physical explanation. Theōn remarks that there is more than one approach to explaining the phenomena. The Greeks, of course, used geometrical methods. The Babylonians, using arithmetic methods, succeeded in confirming the observed facts and in predicting future phenomena. But according to Theōn, the Babylonian methods are nevertheless imperfect, because they were not based on a sufficient understanding of nature, and "one must also examine these matters physically."[141]

Geminos's position on explanation has frequently been misstated. But as we have seen, he was a thoughtful realist. He saw clearly that there were limits to what was astronomically demonstrable, but by no means renounced the effort to determine the true nature of the world.

11. HELIACAL RISINGS AND SETTINGS

Sections 11, 12, and 13 are the most technical parts of our Introduction. Sections 12 and 13 are also somewhat mathematical (although only arithmetic and algebra are used). The present section (sec. 11) is devoted to heliacal risings and settings or, as they are sometimes called, phases of the fixed stars. Geminos treats this subject in chapters xiii and xiv of the *Introduction to the Phenomena*. And, of course, an understanding of the theory of star phases will help the reader appreciate the Geminos *parapēgma*, as well as the analysis in appendix 2. Sec. 12 illustrates the uses of arithmetic progressions in ancient Greek astronomy. Before the development of trigonometry, arithmetic methods allowed for practical solution of problems in astronomy that could not be solved in any other way. And even after the development of trigonometry, the arithmetic methods continued to be used because they were flexible and easy to apply. In several passages of chapters vi, vii, and xviii, Geminos invokes the properties of arithmetic progressions. Sec. 13 is devoted to lunar and lunisolar cycles, topics that Geminos addresses in his chapters viii and xviii. Fixed star phases, arithmetic progressions, and lunisolar cycles were important parts of ancient Greek astronomy, but have long since dropped out of the astronomical curriculum. Because these subjects will not be familiar to most readers, we thought it important to provide reasonably complete discussions. But a reader who wishes to skip these

---

[141] Theōn of Smyrna, *Mathematical Knowledge* iii 30.

three sections on a first reading, and move ahead to sec. 14, may do so with a clear conscience, and return to them when they become necessary for understanding parts of Geminos's text.

Star Phases and Star Calendars in Early Greek Astronomy

The annual cycle of appearances and disappearances of the fixed stars was a central part of early Greek astronomy.[142] Indeed, one of the oldest surviving works of Greek literature is concerned with this lore. Hesiod's *Works and Days* (seventh century B.C.) is a didactic poem in 828 lines. A substantial part of the poem is a sort of farmer's and sailor's calendar, which prescribes the labors to be performed in each season of the year. This section of the *Works and Days* begins with the famous lines:

> When the Pleiades, daughters of Atlas, are rising,
> begin the harvest; the plowing, when they set.[143]

The *rising* of the Pleiades here means their *morning*, or *heliacal rising*, which occurred in late spring, the usual time for reaping what we today call winter wheat. The *setting* of the Pleiades means their *morning setting*. If the farmer saw the Pleiades set in the west in the early morning, just before the Sun came up in the east, he knew that the time had arrived for plowing the ground and sowing the wheat (late fall). Hesiod's theme was very popular in Antiquity, and was continued by poets, encyclopedists, and agricultural writers down to late Roman times.

Star phases were also an early major concern of the Greek scientific writers. The Greek *parapēgma* of later years was a more complete and systematic version of Hesiod's star calendar.[144] It listed in order through the year the heliacal risings and settings of important stars or constellations, usually with associated weather predictions. We know that *parapēgmata* were compiled by Metōn, Euktēmōn, and Dēmokritos—all figures of the late fifth century B.C. And all of the most important of the later Greek astronomers devoted attention to this subject, including Eudoxos, Kallippos, Hipparchos, and Ptolemy.[145]

---

[142] For a more detailed introduction to star phases, see Evans 1998, 190–98, or Schmidt 1952.

[143] Hesiod, *Works and Days* 383–84.

[144] Systematic *parapēgmata* were compiled earlier in Babylonia. MUL.APIN ("Plow Star") is the title of a Babylonian astronomical and astrological compilation, the oldest copies of which date to the seventh century B.C. MUL.APIN includes a chronological list of the dates of the heliacal risings of various constellations. See Hunger and Pingree 1989.

[145] For a study of all known Greek and Latin *parapēgmata*, see Lehoux 2000 and Lehoux (forthcoming). For a recent work on Euktēmōn's *parapēgma*, see Hannah 2002. For a study of the *parapēgma* in the transition between orality and literacy, see Hannah 2001.

Fig. I.18. Fragment of a stone *parapēgma* from Miletus (the "first Miletus *parapēgma*"), late second century B.C. About 35 × 44 cm. Antikensammlung, Staatliche Museen zu Berlin—Preussischer Kulturbesitz (SK 1606 B).

In Greek cities, stone *parapēgmata* were sometimes set up in public places. The *parapēgma* fragment shown in fig. I.18, found at Miletus, has been dated to the second century B.C.[146] For each day of the year, a small hole was drilled in the stone, and next to the hole was inscribed a notice of the heliacal rising or setting of some star or constellation. Someone had the job of moving a wooden peg from one hole to the next each day. (Alternatively, it is possible that numbered pegs for a whole month were inserted all at once.) The holes explain the Greek name for these star calendars: the verb *parapēgnumi* means "to fix beside." This *parapēgma* is arranged according to zodiac signs, just as is the Geminos *parapēgma*. Near the top of the center column is the notice "30," indicating that the Sun is in the sign of Aquarius for 30 days. The next few lines of text read:

> ○The Sun in Aquarius.
> ○[Leo] begins setting in the morning and Lyra sets.
> ○ ○

[146] Diels and Rhem 1904. For a recent, very careful republication of the inscriptions, see Lehoux 2005.

○The Bird begins setting in the evening.

○ ○ ○ ○ ○ ○ ○ ○ ○

○Andromeda begins to rise in the morning.

Holes with no adjacent writing represent days on which there are no celestial phenomena to note. An unusual feature of this *parapēgma* is that the text is confined to star phases and changes of seasons, for example the beginning of the season of the west wind in late winter. It does not attempt to make day-by-day weather predictions, which is one of the main purposes of most *parapēgmata.*[147]

Most Greek *parapēgmata* were probably not inscriptions on stone, but documents on papyrus or other portable media. The *parapēgma* shown in fig. I.19 is the oldest known document of this type. It was written in Greek Egypt around 300 B.C. and provides a nice example of the cultural flexibility of the format. This *parapēgma* is arranged according to months of the Egyptian calendar. The fragment shown in the figure begins:[148]

<Choiak 1> . . . the night is 13 4/45 hours, the day 10 41/45.

16, Arcturus rises in the evening. The night is 12 34/45 hours, the day 11 11/45.

26, the Crown rises in the evening, and the north winds blow, which bring the birds. The night is 12 8/15 hours and the day 11 7/15. Osiris circumnavigates and the golden boat is brought out.

Tybi <5>, the Sun enters Aries.

20, spring equinox. The night is 12 hours and the day 12 hours. Feast of Phitorois.

This *parapēgma* includes notices of important events in the Egyptian religious cycle. It also includes notices of the length of the day according to an old Egyptian scheme, whereby the day length changes by the constant increment of 1/45 hour from one day to the next.

One could, of course, compile a list of heliacal risings and settings simply by observations made at dawn and dusk over the course of a year. There is no need for any sort of theory. In this sense, the *parapēgma* may be considered prescientific. But understanding the annual cycle of star phases was also an important early goal of Greek scientific astronomy. Indeed, one of the oldest surviving works of Greek mathematical astron-

---

[147] This is the so-called first Miletus *parapēgma*. Found on the same site in Miletus were fragments of a second *parapēgma* that does include weather predictions. See Diels and Rhem 1904. Lehoux 2005 makes an important redivision of the four fragments among the two *parapēgmata*. On ancient meteorology and its connections with the *parapēgma* tradition, see Taub 2003.

[148] Translation adapted from Grenfell and Hunt 1906, 152.

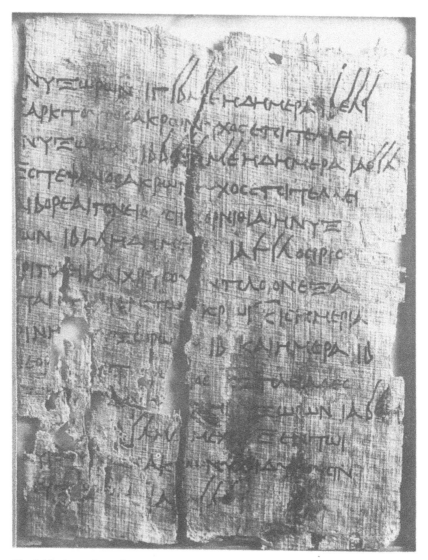

Fig. I.19. A portion of a papyrus *parapēgma*, written about 300 B.C. By permission of The Board of Trinity College Dublin (P. Hibeh i 27. TCD Pap. F 100 r). The diagonal strokes mark fractions.

omy is devoted to this subject. This is Autolykos of Pitanē's *On Risings and Settings* of about 320 B.C.[149] Autolykos defines the various kinds of heliacal risings and settings, then states and proves theorems concerning

[149] The Greek text and a French translation of the works of Autolykos are available in Aujac 1979. There is an English translation (Bruin and Vondijis 1971) based on the Greek

their sequence in time, and the way the sequence depends upon the star's celestial latitude. No individual star is mentioned by name. Autolykos's goal is to provide a theory for understanding the phenomena, and his style is that of Euclid. Nevertheless, Autloykos's treatment established once and for all the Greeks' approach to star phases, as well as the technical vocabulary of the field. Geminos's treatment of the subject is conventional and agrees, in both substance and terminology, with Autolykos.

True Star Phases

Autolykos distinguished between *true* and *visible* star phases. An example of a true star phase is the *true morning rising* (TMR), which occurs when the star rises at the same moment as the Sun. At such a time the star is, of course, invisible, owing to the brightness of the sky. The *visible morning rising* (VMR) occurs some weeks later, after the Sun has moved away from the star. The visible phases are the observable events of interest to farmers, sailors, and poets. However, the true phases are more easily analyzed and described. There are four true phases:

TMR   True morning rising   (Star rises at sunrise)
TMS   True morning setting   (Star sets at sunrise)
TER   True evening rising   (Star rises at sunset)
TES   True evening setting   (Star sets at sunset)

For any star, the TMR and the TER occur half a year apart. Similarly, the TMS and the TES occur half a year apart. These propositions are easily proved. Refer to fig. I.20. Let a star S be rising in the east while the Sun is rising at A. The star is thus making its TMR. The TER will occur when the star is rising at S and the Sun is setting at B. The ecliptic is bisected by the horizon; thus there are six zodiac signs between A and B. If we suppose that the Sun moves uniformly on the ecliptic, it will take the Sun half a year to go from A to B. Thus the TMR and the TER occur six signs (about six months) apart in the year. The same sort of proof is easily made for the TMS and the TES.

The stars have their true phases in different orders, according to whether they are south of the ecliptic, on the ecliptic, or north of the ecliptic.

Ecliptic Stars: If a star is exactly on the ecliptic, its TMR and TES will occur on the same day. Refer to fig. I.21. Let the star S be at ecliptic point A; then S and A rise together in the east, thus producing the star's TMR. In the evening, S and A will set together in the west, thus produc-

---

text in Mogenet 1950. But users of this translation should be forewarned that it often confuses the invisibility of a star phase with the invisibility of the star itself.

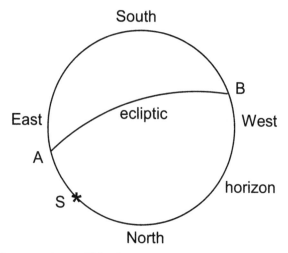

Fig. I.20. True star phases. If the Sun rises at *A* at the same time as the star rises at *S*, the star makes its *true morning rising*.

ing the star's TES. (We assume that the Sun stays at the same ecliptic point for the whole day.) In the same way, one may show that for ecliptic stars, the TER and TMS occur on the same day.

Northern Stars: If a star is north of the ecliptic, the TMR will precede the TES. Let the northern star *S* be making its TMR, rising simultane-

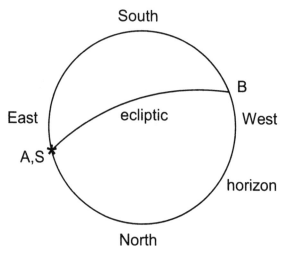

Fig. I.21. For a star on the ecliptic, the true morning rising and the true evening setting occur on the same day.

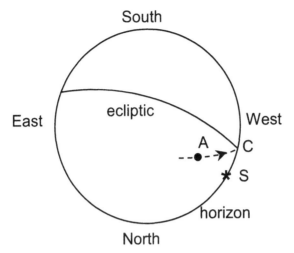

Fig. I.22. The situation shown in fig. I.20, after some hours have gone by and star S is setting in the west. A is below the horizon, so S will not have its true evening setting until the Sun has moved from A to C on the ecliptic.

ously with ecliptic point A, as in fig. I.20. Now, of any two points on the celestial sphere that rise simultaneously, the one that is farther north will stay up longer and set later.[150] (In this and all that follows, we assume the observer is in the northern hemisphere.) S and A rise together. But A will set first. Thus, when S sets, the situation will resemble fig. I.22. S is on the western horizon. A, located farther south on the sphere, will already have set and will be below the horizon. The TES of star S occurs when the Sun is at C. Thus we must wait a few days or weeks for the Sun to advance eastward on the ecliptic from A to C. The TES therefore follows the TMR. The proofs given here are more concise than Autolykos's proofs of the same propositions, but follow his method.

Southern Stars: If a star is south of the ecliptic, the TMR will follow the TES. The proof may be made in the same way. The orders of the true phases for northern, southern, and ecliptic stars are summarized in table I.2.

Visible Star Phases

The visible phases occur in the early dawn (shortly before sunrise), or in the late evening (shortly after sunset). In each case, the star is exactly on

---

[150] Autolykos of Pitanē, *On the Moving Sphere*, Prop. 9.

TABLE I.2
Order of True Star Phases in the Year

Star South of Ecliptic	Star on Ecliptic	Star North of Ecliptic
TMR	TMR and TES	TMR
TMS		TES
TER	TER and TMS	TER
TES		TMS

the horizon; but the Sun is just far enough below the horizon to permit the star to be seen. There are four visible phases:

VMR Visible morning rising (Before sunrise, star is seen rising for the first time)

VMS Visible morning setting (Before sunrise, star is seen setting for the first time)

VER Visible evening rising (After sunset, star is seen rising for the last time)

VES Visible evening setting (After sunset, star is seen setting for the last time)

The visible morning phases come later than the true ones. But the visible evening phases precede the true ones. These propositions follow simply from the fact that the Sun's motion on the ecliptic is from west to east. Let star S be rising while the Sun is rising at A, as in fig. I.23. This is therefore the time of the star's TMR. The star's rising will be invisible. But some weeks later, when the Sun has advanced from A to D on the ecliptic, the star's rising will be visible for the first time. Then the Sun will be far enough below the horizon at the rising of S for the star to be seen. Thus, when the Sun is at D, star S will make its VMR.

Note that the same argument may be applied to the star setting at T. This star makes its TMS when the Sun rises at A. The setting will be invisible, however. The first setting of T to be visible will occur when the Sun has advanced to D. Thus the visible morning phases (whether risings or settings) follow the true ones. Also, the morning phases are the first events to be visible. That is, the VMR is the first visible rising of the star in the annual cycle; the VMS is the first visible setting.

In the same way it may be proved that visible evening phases precede the true ones. Also, the evening phases are the last events to be visible: the VES is the last visible setting of the star in the annual cycle. Similarly, the VER is the last visible rising in the course of the year.

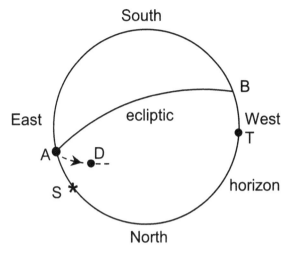

Fig. I.23. The visible morning phases follow the true morning phases in time. Star S makes its true morning rising when the Sun is at A, and its visible morning rising when the sun is at D.

Autolykos's 15° Visibility Rule

The length of time that separates a star's visible rising or setting from its corresponding true one depends on many factors: the brightness of the star, the steepness with which the ecliptic meets the horizon when the star is rising or setting, the observer's latitude, etc. Autolykos dispensed with all these complications by means of one simplifying assumption: any star's rising or setting will be visible if the Sun is below the horizon by at least *half a zodiac sign measured along the ecliptic*. In fig. I.24, let star S rise simultaneously with point X of the ecliptic. Then when the Sun is at X, star S will have its true morning rising. According to Autolykos, S will have its visible morning rising when the Sun reaches Y, which is half a zodiacal sign (15°) from X.

According to modern astronomers, astronomical twilight extends from the Sun's setting until it reaches a position 18° vertically below the horizon, when the sky becomes dark enough to permit observation of even the faintest stars. The brighter stars, however, can be seen when the Sun is only 10° or 12° below the horizon, and these are precisely the stars that play the most prominent roles in the ancient literature on star phases. Autolykos's use of 15° measured obliquely to the horizon is then a fairly good approximation to 12° measured vertically.

Since the Sun moves roughly 1° per day on the ecliptic, 15° corresponds roughly to 15 days. Thus, the VMR of a star follows the TMR by roughly 15 days. Similarly, the VMS follows the TMS by roughly 15

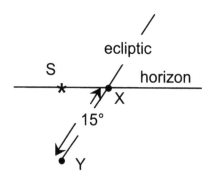

Fig. I.24. Autolykos's 15° visibility rule.

days. But the VER *precedes* the TER by 15 days, and the VES precedes the TES by 15 days. Geminos does not mention Autolykos's 15° visibility rule. We mention it here because it affords an easy way to estimate the dates of the visible phases after one has found the dates of the true phases by (for example) consulting a celestial globe.

Order of the Visible Phases

The visible phases can come in three different orders in the year, depending on where the star is located on the celestial sphere. If the star lies far enough north of the ecliptic, the 15-day correction will not change the order from that of the true phases. Similarly, if the star is far enough south of the ecliptic, the visible phases will come in the same order as the true ones. But for stars near the ecliptic, some of the true phases occur so close together in the year that application of the 15-day visibility rule causes a reversal of order. It turns out that all stars near enough to the ecliptic (whether a little north or south of the ecliptic) have their visible phases in the same order: VMR, VER, VMS, VES. These properties are summarized in table I.3. The orders of the phases given here presuppose that the observer is in the Earth's northern hemisphere, and at a moderate latitude (i.e., not in the arctic zone). The same star may have its visible phases in different orders at different latitudes. Stars near enough to the ecliptic are always "dock-pathed"; but what is "near enough to the ecliptic" varies with the latitude of observation.

In his work on star phases, the *Phaseis*, Ptolemy introduced some convenient nomenclature for the three star classes. Stars sufficiently far south of the ecliptic are called "night-pathed" (*nuktidiexodos*) because of their behavior during the interval between the VMS and the VER.[151] The

[151] Ptolemy, *Phaseis* 6. *Opera*, vol. 2, p. 9.

TABLE I.3
Order of Visible Star Phases in the Year

Star Well South of Ecliptic ("night-pathed")	Star near Ecliptic ("dock-pathed")	Star Well North of Ecliptic ("doubly visible")
VMR	VMR	VMR
VMS	VER	VES
VER	VMS	VER
VES	VES	VMS

first visible setting has already taken place, so the star's setting is visible. But the last visible rising has not yet occurred, so the star's rising is also visible. Therefore, during this period, the star's whole path from horizon to horizon is visible. The Geminos *parapēgma* treats the Dog Star (Sirius) as night-pathed. (See appendix 2.)

Stars sufficiently near the ecliptic are called "dock-pathed" (*kolobodiexodos*) by Ptolemy, because of their behavior between VER and the VMS.[152] The last visible rising has already occurred, so the star's rising is no longer visible. The first visible setting has not yet occurred, so the star's setting is not yet visible. Thus, the star crosses the sky in the night, but neither the rising nor the setting can be seen. The ends of its path across the sky are docked, or cut off. The Pleiades provide a good example of the dock-pathed.

Night-pathed and dock-pathed stars all have a period of complete invisibility (between the VES and the VMR), when they are too near the Sun and therefore cross the sky in the daytime. The classic example is the Pleiades, of which Hesiod says, "For forty days and nights they hide themselves."[153]

Finally, Ptolemy calls stars that are sufficiently far north of the ecliptic "doubly visible," or "seen on both sides" (*amphiphanēs*), because of their behavior between the VMR and the VES.[154] (Geminos uses the same term at xiv 11.) The star appears in the western sky after dusk. Then it is seen setting; but it rises in the east before the night is over. Ptolemy also calls such a star "visible all year long" (*eniautophanēs*),[155] because it has no period of invisibility. The star is visible at *some* time of night all year long. Arcturus is the classic example.

The standard terms for the phases (true morning rising, visible evening setting, etc.) were introduced by Autolykos and were universally followed. Autolykos discussed the properties of the stars that we have

[152] Ptolemy, *Phaseis 5. Opera*, vol. 2, p. 8.
[153] Hesiod, *Works and Days* 385.
[154] Ptolemy, *Phaseis 6. Opera*, vol. 2, p. 9.
[155] Ptolemy, *Phaseis 6. Opera*, vol. 2, p. 9.

called dock-pathed, night-pathed, and doubly visible, but did not assign these names to them. Since every star that has risings and settings may be assigned to one of these three classes, it is convenient to have names for the groups. This rigorous systemization and naming appears in Ptolemy's *Phaseis*, and may well have been Ptolemy's own contribution. These terms were, of course, used by earlier writers, but less systematically. As we have seen, Geminos speaks of doubly visible stars at xiv 11. He also uses *nuktidiexodos* at xiv 12, but in a sense completely different from Ptolemy's—which seems consistent with the lack of standard terms for the three star classes before Ptolemy's time.

Some Details of Technical Vocabulary

"RISING" AND "SETTING"

The usual noun for the diurnal rising of a star, or of the Sun or Moon, is *anatolē*, and the associated verb is *anatellein*, "to rise." These are the terms Autolykos uses to describe a diurnal rising in *On the Moving Sphere*. The verb is a compound of a simple verb, *tellein*, "to arise," and a preposition, *ana*, "up." But there is another common compound, formed with the preposition *epi*. In compound, this preposition can indicate accompaniment or opposition. The *epitolē* is therefore the rising that the star makes "with" (or, perhaps, "against") the Sun. The verb *epitellein* was used already by Hesiod to indicate the first visible rising of the Pleiades and also the last visible rising of Arcturus (i.e., not mere diurnal risings).

Oddly, there is no similar distinction between the terms used for the two kinds of settings. One verb, *dunein*, serves for both the diurnal setting and the settings in connection with the Sun. The noun is *dusis*, a "setting."

Autolykos follows Hesiod's usage: *epitolē* is reserved for a rising that takes place while the Sun is either rising or setting. By the time of Autolykos, however, the true risings have been distinguished from the visible ones. So a true morning rising is *heōia alēthinē epitolē* (the English is a literal translation of the Greek) and a visible evening setting is *hesperia phainomenē dusis*. Autolykos's system of terminology puts all eight phases on an equal footing: the name of each of the eight possible phases is formed by selecting one word from each of the three pairs true-visible, morning-evening, rising-setting.

Autolykos's system proved to be definitive, but there was occasional sloppiness in usage. Geminos (xiii 3–4) criticizes people who, out of ignorance, use *anotolē* and *epitolē* interchangeably. Such slips are not unknown even among astronomical writers. For example, Autolykos often uses *anatellein* for a heliacal rising, and even Geminos, who makes such a fuss over the distinction, sometimes appears to lapse in the same way, as Aujac has pointed out.[156] But most of these apparent lapses are not really lapses at all. When Autolykos or Geminos uses *anatellein* in a place that seems to require *epitellein*, we usually find that the meaning is made clear by a modifier. "Rises in the morning" makes perfect sense, regardless of the form of the verb that is used.

Once again we find that Ptolemy is the most consistent and logical of all writers on the subject. If a temporal modifier is added, Ptolemy regularly uses *anatellein*, as the modifier completes the sense: "Arcturus rises in the morning" (*anatellei*). But if no modifier is added, Ptolemy always writes *epitellei*, or else the reader would have no way of knowing that a rising with respect to the Sun is meant, and not a mere diurnal rising: "the bright star in the Eagle rises" (*epitellei*). Moreover, Ptolemy's choice between these two styles is not haphazard; rather, it conveys a subtle distinction. The Eagle emerges from a period of complete invisibility, so Ptolemy uses *epitellei*. Arcturus (a doubly visible star) does not have a period of complete invisibility, so it only "rises (*anatellei*) in the morning." In this book, we translate both *anatellei* and *epitellei* simply as "rises."

Sometimes, Greek writers will say that a star "rises" (*epitellei*) or "sets" (*dunei*) without bothering to say whether it is a morning or an evening event. In such cases, the general rule is that a visible morning phase is intended. Exceptions to this rule are not unknown, but usually represent mistakes. The morning rising is the first rising to be visible. The morning setting is the first setting to be visible. First things (the morning phases) naturally are of greater importance, psychologically, than last things (the evening phases).

A variety of usage is apparent in the Geminos *parapēgma*. For example, at days 15, 16, and 26 of Sagittarius, we read: "the Eagle rises" (*epitellei*), "the Eagle rises at the same time as the Sun," "the Eagle rises in the morning." These expressions all indicate the visible morning rising of the Eagle (Aquila), according to the various authorities cited in the *parapēgma*.

"HIDING" AND "APPEARING"

Besides the general terms for risings and settings, there are several more specialized ones that apply only to the visible phases. When a star makes its VES, it is sometimes said that the star "hides itself" (*kruptetai*). But this term indicates more than a simple evening setting: it implies that the

---

[156] Aujac 1975, 68n.

star goes into a period of complete invisibility. (The Sun moves so near the star that the star becomes lost in its rays for a period of time.) So, in the Geminos *parapēgma* we have notices of the hiding of the Pleiades (day 10 of Aries), the Hyades (day 23 of Aries), and the Dog (day 2 of Taurus). Strictly, then, "hiding" should be used only of night-pathed and dock-pathed stars. (Doubly visible stars can never hide themselves.) In the *Phaseis*, Ptolemy is scrupulous in observing this distinction, but the same cannot be said of all ancient writers. For example, in the *Phaseis*, Ptolemy writes for the *klima* of 15 hours:

Gemma "sets in the evening." (Day 10 of Choiak)
Fomalhaut "hides itself." (Day 4 of Tybi)

These are both visible evening settings, but the distinction is deliberate. In the *klima* of 15 hours, Fomalhaut enters into a period of invisibility, but Gemma (a doubly visible star) does not.

Similarly, when a star emerges from its period of invisibility (at its VMR), it may be said that it "appears" (*phainetai*), "shows itself," or "becomes visible" (*ekphanēs ginetai*). For example, in the Geminos *parapēgma*, the Dog is said to become visible in Egypt (day 23 of Cancer) and the Harbinger of the Vintage is said to appear (day 10 of Virgo). These terms can strictly be used only of night-pathed and docked-pathed stars, though not all ancient writers are careful in this regard.

MODERN TERMINOLOGY

In modern writing on star phases, one encounters the terms:

heliacal rising = VMR
acronychal rising = VER
heliacal setting = VES
cosmical setting = VMS

Except for "acronychal rising," none of these terms is used by the ancient Greek astronomers. And even "acronychal" poses a problem. *Akronychos = akron* (tip, extremity) + *nyktos* (of the night). This adjective is used by Greek writers on star phases and, indeed, belongs to the vocabulary of everyday speech. Usually, it means "in the evening." But Theōn of Smyrna points out that the morning is also an extremity of the night, and therefore, logically enough, he applies the same word to both evening risings and morning settings.[157] Thus, it is far better, and clearer, to stick to Autolykos's technical vocabulary (visible evening rising, etc.), which is followed by Geminos and Ptolemy, and which can hardly be improved upon.

---

[157] Theōn of Smyrna, *Mathematical Knowledge* iii 14.

12. Astronomical Applications of Arithmetic Progressions

An *arithmetic progression* is a sequence of terms $a, b, c, d, \ldots$, in which each pair of successive terms differs by the same quantity δ. Thus $b = a + \delta$; $c = b + \delta$; $d = c + \delta$; and so on. Geminos shows that he is familiar with the use of arithmetic progressions in the solution of two closely related problems of spherical astronomy: determining the rising times of the zodiac signs and finding the length of the day. In this section, we give a brief overview of arithmetic progressions in ancient astronomy. Then we shall turn to Geminos's own use of arithmetic progressions.

By the "rising time" of a zodiac sign, we mean the *amount* of time that the sign takes to cross the eastern horizon. A list of the rising times for each sign (or for each 10-degree segment of the zodiac, or for each single degree, depending on the precision desired) is called a *table of ascensions*. One can use a table of ascensions to solve many different kinds of problems in practical astronomy. A prime example is determining the length of the day at any time of year. As Geminos remarks (vii 12), during any daylight period, six signs of the zodiac rise and six set. The six signs that rise in the course of a day are the six beginning with the Sun's position and counted eastward. So if the Sun is at summer solstice (beginning of Cancer), the day lasts for the time it takes for the six signs starting with Cancer to rise. One may therefore calculate the length of the day from a table of ascensions by adding up the rising times of the appropriate six signs. A table of ascensions is also useful for astrologers, for it permits one to determine the horoscopic point (the point of the ecliptic that is on the eastern horizon), given the date and time.[158]

The rising times of zodiac signs were never directly measured in ancient astronomy. Rather, these rising times were always the result of theoretical calculation. Before the development of trigonometry, Greek astronomers were forced to rely upon approximate, *arithmetic* methods, which were based upon a few properties of arithmetic progressions. The arithmetic progression also gives a simple way of interpolating between the maximum and minimum values of a quantity, such as day length, that varies in a nonlinear way. As we shall see, the arithmetic methods often produce very satisfactory results.

Sophisticated application of arithmetic progressions is a hallmark of Babylonian mathematics. Arithmetic solutions of problems involving day lengths are found on Babylonian tablets of the Seleucid period,[159]

---

[158] This point is emphasized by Manilius, *Astronomica* iii 295–97.

[159] For a discussion of the Babylonian material on day lengths, see Neugebauer 1975, 366–71.

but the mathematical methods themselves (outside of the astronomical context) are much older. The Babylonian arithmetic methods entered Greek astronomy by the early second century B.C. The oldest surviving Greek treatise concerned with numerical values for the rising times of the signs is the *Anaphorikos* of Hypsiklēs.[160] Hypsiklēs states and proves several propositions involving the sums of arithmetic progressions, then applies them to the problem of determining the rising times of the individual zodiac signs.

The arithmetic methods became obsolete almost immediately after their introduction into Greek astronomy, for the development of trigonometry soon permitted exact solutions of the traditional problems of rising times and day lengths. Ptolemy provides a table of ascensions in *Almagest* ii 8, gives directions for using it to solve six different kinds of problems, and shows how to calculate the rising times *trigonometrically* for any desired latitude. Although Ptolemy's is the oldest extant trigonometric treatment of the problem, it is likely that the trigonometric solution goes back to Hipparchos. Nevertheless, the arithmetic methods lived on a long while, especially among astrological writers,[161] most of whom never mastered trigonometry, the "new math" of the second century B.C. Geminos's remark (vi 38) that the lengths of the days have constant second differences of course implies an arithmetic progression, and there is nothing in the *Introduction to the Phenomena* to suggest that Geminos was a user of trigonometry. The arithmetic methods continued to exist side-by-side with trigonometric methods well into the second century A.D. Most Greek writers who use arithmetic methods do so without apology, or any indication that the methods are second-best approximations to more exact trigonometric methods. The methods based on arithmetic progressions therefore constituted an independent approach to practical astronomical computation.

Rising Times and Day Lengths in System A

Two different patterns of day lengths (and underlying patterns of rising times) have been called System A and System B by Neugebauer. In System A, the rising times of the signs form a strict arithmetic progression.[162] Most of the Babylonian material on day lengths belongs to this system, and Hypsiklēs follows it as well. The ecliptic makes its shallowest angle

[160] De Falco, Krause, and Neugebauer 1966. For a discussion of Hypsiklēs' *Anaphorikos*, see Neugebauer 1975, 715–18.

[161] Neugebauer 1975, 719–20.

[162] In System B, the rising times of the signs *almost* form an arithmetic progression, but the changes in rising times are twice as big as normal between Gemini and Cancer, and between Sagittarius and Capricorn. There is no preserved Babylonian procedure text for the day lengths in System B, but isolated day lengths associated with System B do appear

with the horizon when the spring equinoctial point is rising, and its greatest angle with the horizon when the fall equinoctial point is rising (Geminos, vii 15–16, 19–20). Thus, it is natural to assume that Aries takes the least time to rise, and that Libra takes the most (vii 23–25). Let T be the amount of time required for Aries to rise, and $T + D$ be the amount of time required for Taurus to rise. Then, according to System A, the time required for Cancer to rise is $T + 2D$; for Leo, $T + 3D$; and so on, as in the second column of Table I.4.

From the rising times of the signs in the second column of table I.4, one may obtain the length of the day when the Sun is at the beginning of any sign (third column). Let the Sun be at winter solstice (beginning of Capricorn). In the course of the day on winter solstice, the half of the zodiac that rises comprises the six signs from Capricorn through Gemini. If we add up the rising times for these six signs, we obtain $6T + 6D$, as indicated in the third column of table I.4.

The fourth column of table I.4 gives the changes in the lengths of the days from one zodiac sign to the next. These changes are computed by taking the differences between successive entries in column 3. These changes form arithmetic progressions. Moreover, the changes in day length stand to one another as successive odd integers: D, $3D$, $5D$, etc. This means, in modern terms, that System A is a parabolic approximation: the day lengths are fitted by a parabola that passes through a maximum at the summer solstice and through 12^h at the equinoxes. (Superscript^h indicates hours; ^{m} indicates minutes.)

Finally, the fifth column gives the second differences in the lengths of the days. The second differences are constant (in keeping with a parabolic approximation), but change sign at the equinoxes.

To apply System A to a particular *klima*, we need only specify the length M of the summer solstitial day. Setting the expression for the length of the summer solstitial day equal to M, we have

$$M = 6T + 24D.$$

---

occasionally. Day lengths associated with System B also turn up in Greek and Roman writers. An especially relevant example is provided by Kleomēdēs (*Meteōra* i 4.18–29), who gives day lengths for the *klima* of 15^h that are in perfect accord with System B. Kleomēdēs does not indicate any awareness of the derivation of day lengths from the rising times of the signs, and may simply have used the pattern of changes in day lengths without knowledge of its theoretical basis. The use of arithmetic patterns without understanding of their origins or their connections with other patterns is common, especially among astrological writers. Manilius gives a list of rising times that follows System A (*Astronomica* iii 275–94) and a list of day lengths that follows System B (*Astronomica* iii 443–82). These are, of course, inconsistent. Aujac (1975, 38n1) and Manitius (1898, 261–62) both cite Kleomēdēs as providing illustrative detail for Geminos's remarks at vi 29 and 38. (Similarly, see Bowen and Todd 2004, 51n3.) But, in fact, Kleomēdēs and Geminos follow different systems. Kleomēdēs follows System B, while, as we shall see, Geminos uses a modification of System A.

TABLE I.4
Rising Times and Day Lengths in System A

Sign	Rising Time of Sign	Length of Day at Beginning of Sign	Diff.	2nd Diff.
Aries	T	$6T + 15D$		
			$5D$	
Taurus	$T + D$	$6T + 20D$		$-2D$
			$3D$	
Gemini	$T + 2D$	$6T + 23D$		$-2D$
			D	
Cancer	$T + 3D$	$6T + 24D$		$-2D$
			$-D$	
Leo	$T + 4D$	$6T + 23D$		$-2D$
			$-3D$	
Virgo	$T + 5D$	$6T + 20D$		$-2D$
			$-5D$	
Libra	$T + 5D$	$6T + 15D$		0
			$-5D$	
Scorpio	$T + 4D$	$6T + 10D$		$2D$
			$-3D$	
Sagittarius	$T + 3D$	$6T + 7D$		$2D$
			$-D$	
Capricorn	$T + 2D$	$6T + 6D$		$2D$
			D	
Aquarius	$T + D$	$6T + 7D$		$2D$
			$3D$	
Pisces	T	$6T + 10D$		$2D$
			$5D$	

Also, the equinoctial day must be 12 hours, so

$$12^{h} = 6T + 15D.$$

The length m of the winter solstitial day is

$$m = 6T + 6D.$$

Subtracting the third equation from the first gives

$$D = (M - m)/18.$$

Substitution of this expression for D into the second equation gives

$$T = 2^{h} - (5/36)(M - m).$$

TABLE I.5
Rising Times for the *Klima* of 15^h

Sign	Rising Time (System A)	Rising Time (Trigonometric)
Aries	1^{h}10^m	1^{h}10^m
Taurus	1 30	1 25
Gemini	1 50	1 55
Cancer	2 10	2 24
Leo	2 30	2 34
Virgo	2 50	2 33
Libra	2 50	2 33
Scorpio	2 30	2 34
Sagittarius	2 10	2 24
Capricorn	1 50	1 55
Aquarius	1 30	1 25
Pisces	1 10	1 10

Hypsiklēs applies his method to the *klima* of Alexandria ($M = 14^h$). For our example, we will use $M = 15^h$. Thus we obtain

$$D = 1/3 \text{ hour} \quad \text{(for System A, } klima \text{ of } 15^h)$$
$$T = 7/6 \text{ hour.}$$

If these parameters are inserted in the expressions in the second column of table I.4, we obtain numerical values for the rising times shown in the second column of table I.5.

Thus, in System A, for the *klima* of 15 hours, the rising times of the signs increase by constant differences of 20^m. For comparison, the third column of table I.5 gives the rising times for the same *klima* that result from exact, trigonometric computation: these values are taken from Ptolemy's table of ascensions in *Almagest* ii 8.[163]

To obtain day lengths, we insert the numerical values for T and D into the expressions in the third column of table I.4. The results are given in the second column of the table I.6. So, in System A, for the *klima* of 15^h, the day lengths at the beginnings of successive signs have constant

[163] The numerical results in the third column of table I.5 reflect Ptolemy's value for the obliquity of the ecliptic. Both Hypsiklēs and Ptolemy express their rising times in terms of *degrees of time*: 360° correspond to one rotation of the celestial sphere, i.e., 24 hours of sidereal time. We have converted Ptolemy's figures to hours by dividing the angles by 15. Strictly, therefore, the rising times in table I.5 are expressed in terms of sidereal time, rather than mean solar time. To convert to mean solar time, one should subtract 1^m for every 6^h. This small adjustment is without significance for the present discussion.

TABLE I.6
Day Lengths for the *Klima* of 15ʰ

Sign	Day Length at Beginning of Sign (System A)	Day Length at Beginning of Sign (Trigonometric)
Cancer	15ʰ00ᵐ	15ʰ00ᵐ
Leo	14 40	14 31
Virgo	13 40	13 22
Libra	12 00	12 00
Scorpio	10 20	10 38
Sagittarius	9 20	9 29
Capricorn	9 00	9 00

second differences of 40ᵐ. For comparison, the third column of Table I.6 gives the day lengths obtained from trigonometric computation.[164]

Babylonian tablets[165] of the Seleucid period (roughly the last three centuries B.C.) preserve a text giving the procedure for computing day lengths, at 30° intervals along the zodiac, in System A. The longest day is taken as 14ʰ24ᵐ. This corresponds to the ratio $M/m = 3/2$, which is common in Babylonian astronomy. The day lengths show constant second differences of 32ᵐ from sign to sign. The preserved tablets give the day lengths directly, and do not derive them from the rising times of the signs. But there is little doubt that these day lengths were understood by the Babylonian scribes as the rising times of semicircles of the ecliptic. As already mentioned, Hypsiklēs applies System A to the *klima* of Alexandria ($M = 14ʰ$) and explicitly states the connection between the day lengths and the rising times.

It is clear that Geminos, too, used and understood the arithmetic methods, from his remark (vi 38) that the day lengths have constant second differences. Further evidence appears in Geminos's assertion (vii 36) that, for any sign, the sum of the rising time and the setting time is equal to 4 equinoctial hours. In reality, this is approximately, but not rigorously, true; but this condition is *exactly* satisfied in the arithmetic systems (both A and B). This is easily seen in the second column of table I.4. Aries sets while the diametrically opposite sign, Libra, rises. So the sum of the rising and setting times for Aries is $T + (T + 5D) = 2T + 5D$. The

---

[164] That is, by using Ptolemy's table of ascensions in *Almagest* ii 8. The day lengths according to Ptolemy can also be obtained from his rising times in table I.5, by adding the rising times for the appropriate six successive signs.

[165] Neugebauer [1955], vol. 1, 187 (ACT 200, section 2) and i 214 (ACT 200b, section 2). These are discussed in Neugebauer 1975, 369–71.

same total applies to every sign. If the expressions for T and D, given above, are substituted, we see that the sum of the rising and setting times for any sign, and for any *klima*, is exactly 4 equinoctial hours.

Day Lengths Day by Day

So far, we have discussed arithmetic methods for finding the length of the day when the Sun is at the beginning of a zodiac sign. What about other times of year? One could proceed in several different fashions. The crudest possible treatment of the length of the day would suppose a constant daily difference (i.e., constant *first* differences). The length of the day would then increase by equal increments for each day between the winter solstice and the summer solstice. This crude method was applied in Egypt as early as the twelfth century B.C.[166] That it was still in use among Greek writers in the second century B.C. is demonstrated by the *Celestial Teaching* of Leptinēs.[167] Leptinēs assumes that the length of the day increases by 1/45 hour for each of 180 days between winter and summer solstice. The Sun spends 2 days at the summer solstice and 3 days at the winter solstice without any change in the length of the day, thus filling out the 365 days of the year. As we have seen, the same scheme was used in the papyrus *parapēgma* of about 300 B.C. shown in fig. I.19.

But suppose we wish to obtain day lengths within the Babylonian scheme, in which the rising times of the signs are modeled by an arithmetic progression. As we have seen, this leads to day lengths at the beginning of successive signs that show constant second differences. The switch from a theory of constant differences in the length of the day to a theory of constant second differences represents a marked advance, not only in mathematical sophistication, but also in the accuracy with which the theory models the phenomena. So, it might indeed be beneficial to find a way of treating every successive day within the scheme of Hypsiklēs and the Babylonians. How should we proceed in this case?

Most simply, one could interpolate linearly between the day lengths for the beginnings of two successive zodiac signs. This rule is given in one of the Babylonian procedure texts.[168] Linear interpolation in the day lengths amounts to assuming that the individual degrees of a given sign all take equal times to rise. This implies a step-function for the rising times of individual degrees.

A more consistent approach would extend the premise about rising times from whole signs to individual degrees. This method is explicitly used by Hypsiklēs for System A. Thus, Hypsiklēs assumes that the rising

---

[166] Neugebauer 1975, 706.
[167] Blass 1887; Tannery 1893, 283–94, on p. 284.
[168] Neugebauer [1955], vol. 1, 187 (ACT 200, section 2).

times of the individual degrees form an arithmetic progression. Suppose that the first degree of Aries (from Aries 0° to Aries 1°) rises in time τ. The next degree (Aries 1° to 2°) rises in $\tau + \delta$, the next in $\tau + 2\delta$, and so on, to the last degree of Virgo, which takes $\tau + 179\delta$ to rise. In the remaining six signs, there is a descending arithmetic progression. The first degree of Libra takes $\tau + 179\delta$ to rise, and the last degree of Pisces takes τ.[169]

The day length can be obtained for any desired day by summing the rising times for the appropriate 180°. Hypsiklēs does not go so far as to calculate individual day lengths, but this represents only a simple extension of his method. Let Λ denote the Sun's longitude, *measured eastward from the winter solstice*. Then it is straightforward, though tedious, to show that, according to Hypsiklēs' premises,

$$\text{Length of day} = m + \Lambda^2\,\delta \qquad (\Lambda \text{ a whole number, } 0 \leq \Lambda \leq 90)$$

which displays the quadratic growth in the length of the day as we move from winter solstice ($\Lambda = 0$) to spring equinox ($\Lambda = 90$). Putting $\Lambda = 0$ and $\Lambda = 1$ successively, we see that the change in the length of the day from one day to the next at the winter solstice is δ. But putting $\Lambda = 89$ and $\Lambda = 90$ successively, we find that the change in the length of the day from one day to the next at the spring equinox is 179 δ. That is, in the degree-by-degree version of System A, the daily change in the length of the day at equinox is 179 times larger than the daily change in the length of the day at the solstice.

Geminos's Day Lengths, Day by Day

Now we are prepared to interpret Geminos's remarks about day lengths. Geminos follows Hypsiklēs in thinking about day lengths on a degree-

---

[169] As we did above when working with whole signs, we can obtain conditions on τ and δ by requiring that the length of the day be correctly represented at the equinox and the two solstices. The sums over the appropriate semicircles of the ecliptic give:

$$12^h = 180\,\tau + 16,110\,\delta$$
$$M = 180\,\tau + 24,210\,\delta$$
$$m = 180\,\tau + 8,010\,\delta.$$

These equations determine the parameters δ and τ:

$$\delta = (M - m)/16,200.$$
$$\tau = 1^h/15 - (179/32,400)(M - m).$$

With these two parameters, the rising time may be found for any single degree of the ecliptic. The rising times of whole signs may be found by summation. These will agree exactly with the rising times given in the calculation above for System A, in which we worked with whole signs.

TABLE I.7
Day Lengths according to Geminos's Rule

Day n	Length of Day	Diff.	2nd Diff.
0 (winter) solstice)	m		
		x	
1	$m + x$		x
		$2x$	
2	$m + x + 2x$		x
		$3x$	
3	$m + x + 2x + 3x$		
. . .	. . .	. . .	. . .
88	$m + x \cdots + 88x$		
		$89x$	
89	$m + x \cdots + 88x + 89x$		x
		$90x$	
90 (equinox)	$m + x \cdots + 88x + 89x + 90x$		

by-degree or day-by-day basis. But rather than deriving his day lengths from first principles (rising times), Geminos simply invokes the "fact" that the changes in the length of the day form an arithmetic progression.

In vi 38, Geminos says that the daily increase in the length of the day around the equinox is about 90 times the daily increase around the solstices. That is, the difference between the equinoctial day and the following day is 90 times larger than the difference between the solstitial day and its following day. Moreover, he says that the second differences are constant. His way of thinking seems to require the scheme for the day lengths illustrated in table I.7. Note that table I.7 satisfies both of Geminos' assertions: (1) the daily change in the length of the day around the equinox is 90 times the daily change around the solstice[170] and (2) the daily changes form an arithmetic progression, i.e., have constant differences of x.[171]

[170] In contrast, a strict application of System A would lead, as we have seen, to 179 rather than 90.

[171] The general expression for the length of day n in Geminos's scheme is thus

$$\text{Length of day} = m + \tfrac{1}{2}\, n(n + 1)x.$$

This should be compared with the corresponding expression from System A:

$$\text{Length of day} = m + \Lambda^2 \delta.$$

Geminos's rule will reproduce the 12^h day at equinox ($n = \Lambda = 90$) if

$$x = (180/91)\, \delta,$$

in which case the two systems are virtually identical.

TABLE I.8
Day Lengths in Geminos's Scheme for the *Klima* of 15ʰ

Day	*n*	Day Length in Geminos's Scheme
Summer solstice	0	15ʰ00.0ᵐ
	30	14 39.6
	60	13 39.6
Equinox	90	12 00.0

This scheme may be used to determine numerical values for the lengths of all the days of the year, once *m* is specified. The procedure would be very similar to that already illustrated for System A. So, rather than going through it again, we merely present some results by way of example. It turns out that the parameter $x = (M - m)/8190$. For the *klima* of 15ʰ this becomes $x = 3^h/4095$. Table I.8 gives some numerical results for this *klima*. These figures may be compared with those in table I.6. Geminos's scheme is nearly, though not exactly, equivalent to the degree-by-degree version of System A.

To sum up, Geminos appropriated a conclusion from System A—that the changes in day length form an arithmetic progression from solstice to equinox—and used the arithmetic progression as a way of interpolating between the equinoctial and solstitial day lengths. But he did not bother to go back to first principles, i.e., to derive his day lengths from the rising times of individual degrees of the ecliptic. Rather, he simply postulated that the daily change in day length should be 90 times larger at the equinox than at the solstice.

13. Lunar and Lunisolar Cycles

In chapter viii Geminos takes up lunisolar cycles, notably the 8- and 19-year cycles. Scholars have disagreed in their assessments of how much of Geminos's treatment represents faithful historical reporting and how much is merely his own pedagogy. Nevertheless, Geminos's account remains our most detailed discussion of this topic in Greek. In chapter xviii, Geminos returns to lunar cycles in his discussion of the *exeligmos*, a cycle useful in the prediction of eclipses.

The Octaetēris

The *octaetēris* is an 8-year lunisolar cycle. In principle, the *octaetēris* contains a whole number of tropical years and a whole number of syn-

TABLE I.9
First New Moons of the Years 1961–90

Year	Date of First New Moon	Year	Date of First New Moon
1961	January 16	1976	January 1
62	January 6	77	January 19
63	January 25	78	January 9
64	January 14	79	January 28
65	January 2	80	January 17
66	January 21	81	January 6
67	January 10	82	January 25
68	January 29	83	January 14
69	January 18	84	January 3
70	January 7	85	January 21
71	January 26	86	January 10
72	January 16	87	January 29
73	January 4	88	January 19
74	January 23	89	January 7
75	January 12	90	January 26

odic months. (The tropical year is the time from one summer solstice to the next, i.e., the time for the Sun to make one trip around the zodiac. The synodic month is the time from one new Moon to the next.) Thus, after 8 years, both the Sun and the Moon return, very nearly, to their original positions in the zodiac. The discussion can be made more concrete by the use of table I.9, which gives the date of the first new Moon for each year from 1961 to 1990.

The tropical year does not contain a whole number of synodic months. Thus, the date of the new Moon of January does not repeat from one year to the next. Rather, as Geminos says (viii 28), 12 synodic months contain only 354 days, while the solar year contains 365¼. Thus, there is an 11¼-day difference. So, between 1963 and 1964, the date of the January new Moon suffers an 11-day shift. Table I.9 shows that the shift from one year to the next may be 10, 11, or 12 days. Because 12 synodic months are 11¼ days *too short* by comparison with the solar year, the new Moon of January occurs progressively *earlier* in the month (until it reaches the beginning of the month and we start over).

STRUCTURE OF THE *OCTAETĒRIS*

In 8 years, the 11¼-day annual difference amounts to $8 \times 11¼ = 90$ days, or very nearly 3 synodic months (viii 29). Thus, we might expect 8 years to contain nearly a whole number of synodic months. There would be 8×12 plus the 3 embolismic (or "leap") months, for a total of 99 months (viii 27). The existence of an 8-year cycle is confirmed by examination of table I.9. If we pick any 2 years that are 8 years apart, we find that the January new Moon falls on nearly the same date in each. Here are the essential features of Geminos's *octaetēris*:

THE OCTAETĒRIS ACCORDING TO GEMINOS (VIII 27-31)

8 years = 99 months = 2,922 days, comprising
5 years of 12 months (354 days each) and
3 years of 13 months (384 days each).
There are 51 full and 48 hollow months:
the ordinary months are alternately full and hollow,
but all the embolismic months are full.
Adopted length of year: 365 ¼ days.
Resulting length of month: 29 + 1/2 + 1/66 days.

Here and in what follows, "adopted length" identifies a parameter selected for the construction of the cycle. "Resulting length" is a value that falls out as a result of computation from the premises on which the cycle is based. The resulting length of the month in the *octaetēris* is not mentioned by Geminos, but he is aware that this month is too short, and he calculates the necessary correction at viii 37–39.

DEFECTS OF THE *OCTAETĒRIS*

Table I.9 shows that after 8 years the date of the January new Moon does not repeat exactly, but shifts forward by 1 or 2 days. Thus, the first new Moon of 1961 fell on January 16, and the first new Moon of 1969 fell on January 18. Geminos discusses the defects of the *octaetēris* at viii 37–38. To begin, he introduces a more accurate value for the length of the synodic month. According to Geminos, the synodic month is not 29½ days, but 29 + 1/2 + 1/33 days (≈29.53 days, which is a good value). So 99 synodic months are therefore (29 + 1/2 + 1/33) × 99 = 2,923½ days. But 8 solar years come to only (365¼) × 8 = 2,922 days. Thus, 99 synodic months are longer than 8 solar years by 1½ days. Because the 99 synodic months are *longer* by 1½ days in comparison with the 8 solar years, the date of the January new Moon comes this much *later* at the beginning of the next 8-year cycle, as may be confirmed by examination of table I.9. As Geminos remarks, the *octaetēris* treats the solar year accurately, but involves an error in reckoning the months.

The 16-Year Period

Geminos therefore proposes (viii 39) a modification of the *octaetēris*: 1½ days must be added every 8 years, or 3 days every 16 years, in order to get a month of the right length. This results in a 16-year period, which consists of 2 *octaetērides* plus 3 extra days:

GEMINOS'S FIRST MODIFICATION OF THE OCTAETĒRIS:
16-YEAR PERIOD
16 years = 198 months = 5,847 days
= 2 × 99 months = 2 × 2,922 + 3 days,
comprising 2 *octaetērides* plus 3 days.
Adopted length of month: 29 + 1/2 + 1/33 days.
Resulting length of year: 365 + 1/4 + 3/16 days.

The 16-year cycle is constructed to get the month exactly right (i.e., in agreement with the adopted value of 29 + 1/2 + 1/33). But in the process the year has been made too long. Geminos does not mention the resulting length of the year, but immediately shows how to correct it.

The 160-Year Period

Geminos proposes (viii 40–41) a second modification of the *octaetēris*, which results in a 160-year cycle. In the *octaetēris*, the year was represented correctly. Then 3 days were added to each 16-year period to get the months right. Thus the 16-year period makes the year too long by 3 days in 16 years. Therefore, in 160 years, 30 days too many will have been added, and it will be time to drop an entire month of 30 days, to restore harmony with respect to the Sun:

GEMINOS'S SECOND MODIFICATION OF THE OCTAETĒRIS:
160-YEAR PERIOD
160 years = 1,979 months = 58,440 days
= 10 × 198 − 1 months = 10 × 5,847 − 30 days,
comprising ten 16-year periods minus 1 full month.
Adopted length of year: 365 + 1/4 days.
Resulting length of month: 29 + 1/2 + 119/3,958 days.

The 160-year period is again based on a year of 365¼ days. The month (which was correctly represented in the 16-year period) has been thrown off, but only by a little. Geminos does not mention the resulting length of the month.

Table I.10 shows the cycles of 8, 16, and 160 years as progressive (and progressively more complicated) approximations to Geminos's adopted lengths for the year and the month. At each stage, one of these two is represented correctly and the other is off a bit. But by the third iteration (the 160-year period), both the year and the month are very close to the adopted values. At this stage (viii 42), Geminos throws up his hands and

Table I.10
Year and Month Lengths in the *Octaetēris* and Related Cycles

	Length of Year (days)	Length of Month (days)
Geminos's adopted value	365.25	29 + 1/2 + 1/33 = 29.53030
Octaetēris	365.25	29 + 1/2 + 1/66 = 29.51515
16-year period	365 + 1/4 + 3/16 = 365.6875	29 + 1/2 + 1/33
160-year period	365.25	29 + 1/2 + 119/3958 = 29.53007

declares the *octaetēris* a total failure, then passes on to a discussion of the 19-year cycle.

Geminos's account of the development of the 8-, 16-, and 160-year cycles should not be taken as history. Rather, his discussion is simply good pedagogy. He sets out the basic parameters of the *octaetēris*, discusses the defects of this cycle, then shows how one might improve it by successive refinements. These successive refinements do not depend on successively better observations. Rather, they are successively better arithmetical approximations to previously adopted lengths for the synodic month and the tropical year.

Origin of the Octaetēris

Geminos does not attribute the invention of the *octaetēris* to any particular writer. Diogenēs Laertios says that Eudoxos wrote on the *octaetēris*.[172] Censorinus[173] (third century A.D.), says that the *octaetēris* was commonly attributed to Eudoxos, but that some writers ascribed it to Kleostratos (late sixth century B.C.?). Censorinus goes on to say that others have presented their own versions of the *octaetēris*, based upon different patterns for the embolismic months, and mentions Harpalus, Nauteles, and Menestratus.[174] Censorinus concludes his list of attributions by saying that the *octaetēris* of Eudoxos is frequently attributed to Dositheus (late third century B.C.). And Geminos (viii 24) knew of a treatise on the *octaetēris* by Eratosthenēs. All we can be sure of is that

---

[172] Diogenēs Laertios, *Lives and Opinions* viii 87. For the fragments of Eudoxos's work on this subject, see Lasserre 1966, 75ff.

[173] Censorinus, *On the Day of Birth* 18.5.

[174] Nauteles and Menestratus are otherwise unknown. For a summary of the speculations on the identity and dates of Harpalus, see Samuel 1972, 39.

the *octaetēris* was introduced earlier than 432 B.C., the date associated with the beginning of Metōn's 19-year cycle. On these grounds, the only viable candidate of those mentioned by Censorinus is Kleostratos. Pliny[175] claims that after Anaximandros discovered the obliquity of the zodiac, it was Kleostratos who divided it into signs. Of course, this makes little sense, for we know that the equal-sign zodiac was borrowed by the Greeks from the Babylonians. From all this, it is clear that by Censorinus's time, and even by the time of Geminos, no one had any sure knowledge of the early history of the *octaetēris*.[176] The only safe course is to leave the question of its origin unanswered.[177]

The 19-Year Period

The existence of a 19-year period is easily discerned in table I.9. In any 2 years separated by 19 years, the date of the January new Moon repeats almost exactly:

NINETEEN-YEAR PERIOD ACCORDING TO GEMINOS (VIII 50–54)

19 years = 235 months = 6,940 days, comprising
12 years of 12 months and
7 years of 13 months.
There are 125 full and 110 hollow months.
Resulting length of year: 365 5/19 days.
Resulting length of month: 29 + 1/2 + 3/94 days.

Geminos does not mention the resulting length of the month, but mentions the length of the year at viii 51.

Geminos asserts (viii 55) that the arrangement of full and hollow months should be as uniform as possible. There are 6,940 days in the 19-year period and 110 hollow months. If all the months were temporarily considered as full, it would therefore be necessary to remove a day after every run of 63 days (6,940/110 ≈ 63). That is, every 64th day number would be removed. According to Geminos, the 30th day of the month is not always the one scheduled for removal. Rather, the hollow month is produced by removing whichever day falls after the running 63-day count. Such a procedure would have enormously complicated the construction of a calendar. Neugebauer[178] therefore doubts that this rule was ever followed. Geminos, however, is unambiguous on this point,

---

[175] Pliny, *Natural History* ii 6.31.

[176] A summary of the speculations of modern writers on the origin of the *octaetēris* is given in Samuel 1972, 39–40. See also Heath 1913, 287–92.

[177] Neugebauer 1975, 620–21.

[178] Neugebauer 1975, 617.

and two efforts at a reconstruction of the Metonic cycle have taken him at his word.[179]

HISTORY OF THE 19-YEAR PERIOD

Geminos (viii 50) attributes the 19-year period to "the astronomers around Euktēmōn, Philippos, and Kallippos." Philippos and Kallippos both lived about a century too late to have been involved in the original construction of the 19-year period. Perhaps Geminos means only to say that they wrote on this subject. The name Kallippos is, of course, associated with the 76-year period, made up of four 19-year periods. The construction of the 19-year cycle is attributed by most modern writers to Metōn, whom Geminos does not mention, or to Metōn and Euktēmōn.[180] Modern writers often refer to this as the "Metonic cycle." The ancient testimony is reasonably convincing. Diodōros[181] says that Metōn introduced the 19-year period at Athens, beginning at the summer solstice of the year 432 B.C. Ptolemy[182] says that this summer solstice was observed by "those around Metōn and Euktēmōn." Censorinus[183] attributes the 19-year cycle to Metōn of Athens.

Knowledge of the fundamental relation (19 years = 235 synodic months) is attested in Babylonia from about 490 B.C.[184] In contrast to the Greeks, who never adopted a fixed rule for the embolismic months in their civil calendars, the Babylonians actually used the 19-year period as the regulating principle for their own calendar, certainly by 380 B.C. (in the reign of Artaxerxes II), and perhaps as early as 497 (Darius I). We do not know whether the Greek astronomers borrowed the 19-year cycle from the Babylonians or arrived at it independently. Borrowing may be considered likely in view of other demonstrated debts of Greek astronomy to Babylonian practice. On the other hand, the fundamental relation (235 months = 19 years) is very simple, and we know from the existence of the older *octaetēris* that the Greek astronomers of the fifth century were interested in lunisolar cycles. The *octaetēris*, at least, appears to be indigenous to the Greeks, for there is no evidence for an 8-year cycle in the preserved Babylonian material.[185]

---

[179] Fotheringham 1924; van der Waerden 1960.

[180] There is a large modern literature on the 19-year, or Metonic, cycle. For an introduction, see the following: Neugebauer 1975, 622–24; Samuel 1972, 42–49; Toomer, article "Meton" in Gillispie 1970–80; Bowen and Goldstein 1989.

[181] Diodōros of Sicily, *Historical Library* xii 36.2.

[182] Ptolemy, *Almagest* iii 1; Heiberg, i p. 205.

[183] Censorinus, *On the Day of Birth* 18.8.

[184] Neugebauer 1975, 354–57.

[185] Neugebauer 1975, 354.

TABLE I.11
Year and Month Lengths in the 19- and 76-Year Periods

	Length of Year (days)	Length of Month (days)
Geminos's adopted value	365.25 days	29 + 1/2 + 1/33 = 29.53030
19-year period	365 + 5/19 = 365.2632	29 + 1/2 + 3/94 = 29.53191
76-year period	365.25	29 + 1/2 + 29/940 = 29.53085

The 76-Year Period

Modern writers often refer to this period as the Kallippic Cycle. In the 19-year cycle, the solar year turns out to be 365 5/19 days, which is longer than the adopted length of the year (365¼ days) by 1/76 day. Therefore, after 76 years it will be necessary to omit 1 day in order to restore harmony with the Sun. This can be done by forming a period from four successive 19-year cycles, leaving out one day from one of the 19-year cycles:

THE 76-YEAR PERIOD ACCORDING TO GEMINOS (VIII 59–60)

76 years = 940 months = 27,759 days,
  = 4 × 235 months = 4 × 6,940 − 1 days,
comprising four 19-year periods, less 1 day.
Adopted length of year: 365 + 1/4 days.
Resulting length of month: 29 + 1/2 + 29/940 days.

Table I.11 shows the cycles of 19 and 76 years as progressive approximations to Geminos's adopted lengths for the year and the month.

Historical Purpose of the 19- and 76-Year Cycles

In Babylonia, the 19-year cycle was used to regulate the civil calendar, perhaps as early as the fifth century B.C. In contrast, there is no evidence that the Athenian calendar was ever regulated by any fixed lunisolar cycle. This raises the question of what role these cycles played in Greek astronomy. An accurate lunisolar cycle had applications in three domains: the regulation of a parapēgma, the design of artificial astronomical calendars, and the evaluation of parameters important in the lunar theory. Of these three applications, the regulation of a parapēgma probably presented the earliest stimulus for the investigation of lunisolar cycles.

The early Greek calendars were lunisolar, and the first day of each month was, in principle, the new Moon. For the Greeks, this meant the first appearance of the crescent Moon in the west after sunset. A particular calendar year could contain either 12 or 13 months. Thus, a given star phase in the annual cycle (e.g., the morning rising of the Pleiades) did not occur on a fixed calendar date. Moreover, each major city used a different calendar, with different month names and different starting points for the year. Worse yet, the civil authorities had substantial freedom to intercalate days and months at will. Thus the calendar was often significantly out of step with the Moon. Athenian writers therefore sometimes distinguished between "the new Moon according to the goddess" (Selēnē, the Moon) and "the new Moon according to the archon" (the head magistrate of the city).[186] We might call these the actual and the calendrical new Moons.

Tying a particular calendar year to a *parapēgma* was therefore a delicate task.[187] This is how a lunisolar cycle might be of use. Toomer[188] has argued that when Metōn introduced his 19-year cycle in Athens, around 432 B.C., he published it together with a *parapēgma* that covered 19 years. The lunisolar cycle served to connect two completely different classes of phenomena on which Greek time-reckoning was based: the cycle of heliacal risings and settings (used for telling the time of year) and the cycle of Moon phases (used for telling the time of month). In Metōn's cycle, suppose that, in a certain year, the peg goes into the first hole of the *parapēgma* on the day of the first new Moon after the summer solstice (the traditional beginning of the year at Athens). Nineteen years later, the peg will again go into the same hole on the first new Moon after summer solstice.[189]

From the use of a lunisolar cycle in the regulation of a *parapēgma* to the construction of an independent astronomical calendar is but a small step. This step may well have been taken by Metōn.[190] About a century

---

[186] Pritchett and Neugebauer 1947 is the most thorough treatment of the Athenian calendar.

[187] A similar difficulty was experienced in Babylonian applications of luni solar cycles. Thus, in the *parapēgma* that forms a part of MUL.APIN (seventh century B.C.), the dates are expressed in terms of Babylonian month names; but these must be interpreted as fictitious months of an ideal or artificial year—not the months of any particular civil calendar year. See Hunger and Pingree 1989, 40–47, 139–40.

[188] G. J. Toomer, article "Meton" in Gillispie 1970–80.

[189] It seems unlikely that Metōn would have set up an inscription covering 19 years. But one should consider the example of the bronze calendar of Coligny (second century A.D.), which carried a 5-year cycle showing the correspondences between the Roman calendar and the Gallic or Celtic calendar, complete with holes for each of 1,835 days. See McClusky 1998, 54–57, or, for more detail, Duval and Pinault 1986.

[190] In *Almagest* iv 11 (Toomer 1984, 211–13), Ptolemy cites three lunar eclipses from the fourth century B.C., details of which he got from Hipparchos. These three eclipses, though observed at Babylon, are all dated in terms of Athenian month names and the years of Athenian archons. Toomer (1984, 12n18) suggests that these dates should be understood as

later, it certainly was taken by Kallippos. In Ptolemy's *Almagest*, a number of the older observations cited are dated in terms of the Kallippic cycle. For example, Ptolemy says that Timocharis, at Alexandria, observed an occultation of the Pleiades by the Moon

> in the 47th year of the First Kallippic 76-year period, on the 8th of Anthestēriōn, which is Athyr 29 in the Egyptian calendar, toward the end of the 3rd hour [of the night]. . . .[191]

In Kallippos's calendar, the years were expressed in terms of their position in the running count from 1 to 76 in the 76-year cycle. The month names were borrowed from the Athenian calendar. But the calendar being used was an ideal, standardized calendar, designed by astronomers, and never used by any city for civil purposes. Ptolemy's dual dating of these observations (i.e., his conversion of the Kallippic date to an equivalent in the Egyptian calendar) provides most of the concrete information we have about the functioning of Kallippos's calendar. Scholarly attempts to reconstruct Kallippos's astronomical calendar have been based on the general overview provided by Geminos (chapter viii) and the handful of equivalent dates provided by Ptolemy.[192]

Ptolemy gives full Kallippic dating (year of Kallippic cycle plus Athenian month name and day) for only four observations of the third century B.C. In addition, he gives some seventeen observation reports, gleaned from the writings of Hipparchos, in which the year is given in terms of its position in the Kallippic cycle, but no Athenian month name is given. The last of these is for an observation in 128 B.C. For this reason, it seemed that the Kallippic calendar was abandoned by the end of the second century B.C. However, several new Kallippic dates on papyrus, recently published, show that the Kallippic calendar was still used (always in an astronomical context) as late as the middle of the first century A.D.[193]

The final application of lunisolar cycles was in the evaluation of parameters of mean motion for the lunar theory. Lunisolar cycles played

---

given in terms of the Metonic calendar. Hipparchos, or some other Greek astronomer, was presumably responsible for converting the original dates, expressed in terms of the Babylonian calendar, to equivalent dates in the Metonic calendar, which was an idealized version of the Athenian calendar.

[191] Ptolemy, *Almagest* vii 3 (Toomer 1984, 334). For a convenient list of all the dated observations in the *Almagest*, see Pedersen 1974, 408–22.

[192] Samuel 1972, 42–49, provides a good overview. Fotheringham 1924 and van der Waerden 1960 make ambitious attempts to reconstruct the Kallippic calendar. One's judgment of their success depends greatly on one's faith in Geminos. Toomer (1984, 13) regards Geminos's account as "fiction." We would call it "pedagogical," but agree with Toomer that one should not take it literally.

[193] Jones 2000.

just such a role in Babylonian astronomy. In Greek astronomy, there probably was no lunar theory with predictive capacity until the time of Hipparchos, and this application is therefore rather late. Hipparchos began his own lunar theory with the *exeligmos*, then passed on to improved parameters. Both the *exeligmos* and the improved parameters were derived from Babylonian sources.[194]

The Exeligmos (Geminos, Chapter XVIII)

The *exeligmos* is a period of 669 synodic months that plays a role in eclipse prediction. Two lunar eclipses separated by one *exeligmos* will have very similar circumstances. In particular, the two eclipses will occur at the same node of the Moon's orbit, will have about the same magnitude and duration, and will even occur at roughly the same time of day. Thus, eclipses may be grouped into cycles that repeat, at least to a fair approximation, after an *exeligmos*. The Greek word *exeligmos* means "revolution." Toomer[195] translates it as "turn of the wheel." Hildericus, in his Latin translation of Geminos (1590), used *evolutio*, a "rolling." The term thus indicates a cycle, a restoration of things to their original state.

For the discussion that follows, it is necessary to distinguish several kinds of month.

SIDEREAL MONTH

The average time required for the Moon to go from one fixed star, all the way around the zodiac, and back to the same star is called the *sidereal month*. This is the period that must be used in dynamical calculations involving Newton's law of motion. Although the sidereal month may seem the most natural to many modern readers, it was rarely used by the Greek astronomers. It is not used, for example, in Ptolemy's *Almagest*.[196] Geminos (i 30) says that the Moon moves around the zodiac in 27⅓ days. However, Leptinēs gives a round figure of 27 days.[197]

SYNODIC MONTH

The time from new Moon to the next new Moon is called the *synodic month*. This is always what Geminos means when he refers to the

---

[194] Our source for Hipparchos's treatment of the Moon's mean motions is Ptolemy, *Almagest* iv 2. Ptolemy, however, did not appreciate the full extent of Hipparchos's dependence on the Babylonian tradition for his numerical parameters. See Toomer 1984, 176n10, and Neugebauer 1975, 309–12.

[195] Toomer 1984, 175n8.

[196] Pedersen 1974, 160.

[197] Leptinēs, *Celestial Teaching*, in Tannery 1893, 286.

"month" or the "monthly period." If the Moon starts out in conjunction with the Sun, it must travel more than 360° with respect to the fixed stars in order to catch up with the Sun again, for in the course of a month, the Sun moves nearly 30° forward on the ecliptic. So, as Geminos says (viii 2), in the course of a synodic month the Moon moves through approximately 13 zodiac signs. Thus, the synodic month is roughly 2 days longer than the sidereal month.

TROPICAL MONTH

The time required for the Moon to go from the summer solstitial point, all the way around the zodiac, and back to the same solstitial point is called the *tropical month*. If the equinoxes and solstices were fixed with respect to the stars, the tropical and sidereal months would be the same. But the equinoxes and solstices precess westward at the rate of 1° in 72 years (1° in 100 years according to Hipparchos and Ptolemy). Therefore, if the Moon starts at the summer solstitial point, it has to travel a tiny bit less than 360° with respect to the fixed stars in order to make it back to the solstitial point. Thus, the tropical month is a little shorter than the sidereal month. Because modern astronomers, like their Greek predecessors, use the spring equinoctial point as the zero of longitude, all practical calculations of lunar positions must make use of the tropical month. For example, Ptolemy's tables for calculating the Moon's longitude from deferent-and-epicycle theory are based on the tropical month. But Geminos never mentions precession, and thus makes no distinction between the sidereal and tropical months.

ANOMALISTIC MONTH

The motion of the Moon on its path around the zodiac is, as Geminos says, nonuniform. In the course of one 24-hour day, the Moon may move as much as 15° or as little as 11°, in rough, round numbers. The time between least motion and the next least motion is the *anomalistic month*. Geminos explains this clearly at xviii 2. When he mentions a "return" of the Moon, he always means the anomalistic month. From a modern point of view the anomalistic month may be explained as follows. The Moon moves most slowly when it is at apogee on its Keplerian ellipse. If the elliptical orbit were fixed in space, the anomalistic month would be the same as the sidereal month. But the gravitational influence of the Sun perturbs the orbit and causes the ellipse to rotate slowly in its own plane. The line of apsides (the line drawn through the apogee, the Earth, and the perigee), makes one complete rotation in about 9 years. It turns eastward, i.e., in the same direction as the Moon moves on its orbit. Consequently, when the Moon is at apogee, it has to travel slightly more than 360° with respect to the fixed stars in order to reach apogee again. Thus, the anomalistic month is a little longer than the sidereal month.

DRACONITIC MONTH

Eclipses can occur only when the Moon is in the plane of the ecliptic, or nearly so. Since the Moon's orbit is inclined to the ecliptic by about 5°, the orbit crosses through the ecliptic plane at two points called the *nodes* of the orbit. This is why there are not eclipses of the Moon at every full Moon. At most full Moons, the Moon is either above or below the plane of the ecliptic and so does not fall into the Earth's shadow. A lunar eclipse can occur only if the Moon happens to be full when it is near one of the two nodes. The time required for the Moon to go from one node of its orbit, all the way around the Earth, and back to the same node is called a *draconitic* (or *nodical*) *month*. If Moon's orbit were fixed in space, the draconitic month would be the same as the sidereal month. But a perturbation by the Sun causes the plane of the orbit to precess slowly about the pole of the ecliptic, at the rate of one revolution in about 18 years. The nodes move westward, opposite to the Moon's motion in its orbit. Consequently, the draconitic month is a bit shorter than a sidereal month. Though Geminos does not allude to the draconitic month, it is an important part of the full expression of the *exeligmos* eclipse cycle, both in Ptolemy and in Babylonian astronomy.

Some numerical values (for the epoch A.D. 1900) may help to illustrate the relationships among the five months:[198]

draconitic month	27.212 220 days
tropical	27.321 582
sidereal	27.321 661
anomalistic	27.554 551
synodic	29.530 589

Geminos on the Exeligmos

Geminos takes up the *exeligmos* in chapter xviii. The essential statement, from which all the rest of that chapter follows, is at xviii 3. Geminos says that the *exeligmos* contains a whole number of synodic months and a whole number of anomalistic months. He also gives the number of days in the cycle and the motion of the Moon in longitude during this period:

$$669 \text{ synodic months} = 717 \text{ anomalistic months}$$
$$= 19{,}756 \text{ days}$$
$$= 723 \text{ tropical revolutions plus } 32°.$$

[198] Smart 1977, 420.

The synodic month and the anomalistic month represent two periodic phenomena with slightly different periods. In such a case, one can always find a longer time interval that contains very nearly a whole number of both periods. It is a remarkable circumstance that the *exeligmos* also contains very nearly a whole number of *draconitic months*:

669 synodic ≈ 717 anomalistic ≈ 726 draconitic months.

It is curious that Geminos does not mention the equality with 726 draconitic months, since this is of central importance for the use of the *exeligmos* in eclipse prediction. Ptolemy, in his own discussion of the *exeligmos*, gives all the numerical information that Geminos (xviii 3) does, but also includes the equality to 726 draconitic months.[199]

Astronomical Significance of the Exeligmos

Suppose the Moon is in eclipse. Since *exeligmos* contains a whole number of synodic months, after one *exeligmos* the Moon will again be full, or nearly so. The *exeligmos* also contains a whole number of draconitic months, so, after one *exeligmos*, the Moon will again be near the same node of its orbit. Thus, we have a very good chance of finding an eclipse. One possible difficulty is the Moon's variable speed in its orbit. Since the anomalistic month is of a different length than either of the other two, it is possible that the Moon will be somewhat ahead of or behind the place it needs to be to produce an eclipse. Because of the variation in speed, the Moon, in some parts of its orbit, can be more than 6° ahead of or behind the place it would occupy if it moved uniformly.[200] But this difficulty is largely removed by the fact that the *exeligmos* contains a whole number of anomalistic months. Thus the conditions of a lunar eclipse should repeat almost exactly after one *exeligmos*. Moreover, since an *exeligmos* also contains a whole number of days, the eclipse will even occur at about the same time of day as before. But because the Moon moves 32° in mean longitude, over and above complete cycles, during an *exeligmos*, the second eclipse will occur approximately one zodiac sign farther east than did the first one.

The *exeligmos* is three times as long as a period that today is usually called the *saros*. Although the saros cycle is ancient, the term "saros" is not: Edmond Halley seems to have been the first to apply this term to an

---

[199] Ptolemy, *Almagest* iv 2.

[200] In modern celestial mechanics, the leading term of the Moon's equation of center is (in radian measure) $2e \sin(2\pi t/T_A)$, where t is the time elapsed since the mean Moon passed perigee, T_A is the anomalistic month, and e is the eccentricity of the Moon's ellipse. Since e is about 0.054, the maximum value of this expression is about 6.2°, and obtains when the mean Moon is ±90° from perigee.

eclipse cycle, in 1691.[201] In the *Almagest*, Ptolemy refers to the saros as the "periodic time" (*periodikos chronos*) and gives it the following properties:[202]

$$223 \text{ synodic} = 239 \text{ anomalistic} = 242 \text{ draconitic months}$$
$$= 6,585\tfrac{1}{3} \text{ days}$$
$$= 241 \text{ revolutions in longitude plus } 10\tfrac{2}{3}°.$$

Origin of the Saros and Exeligmos

Ptolemy ascribes the saros to unnamed astronomers "more ancient" than Hipparchos. Then, as Ptolemy says, in order to obtain a period with a whole number of days, they (the ancients), multiplied the saros by 3, which yields the *exeligmos*. Geminos is nearer the mark. For although he does not explicitly attribute the *exeligmos* to the Chaldeans (Babylonian astronomers), he does (xviii 9) mention them in connection with the value of the mean daily motion deduced from the *exeligmos*. Indeed, the saros is well attested in Babylonian astronomy.[203]

Geminos's Application of the Exeligmos to the Lunar Theory

The bulk of Geminos's chapter xviii concerns the application of the *exeligmos* to the Babylonian lunar theory.[204] First (xviii 6–9), Geminos deduces the mean daily motion m in longitude:

$$m = (723 \times 360° + 32°)/19,756^d$$
$$= 13;10,35°/d, \text{ to the nearest second of angle.}$$

Next (xviii 10), Geminos computes the length T_A of the anomalistic month:

$$T_A = 19,756^d/717 = 27;33,20 \text{ days.}$$

More accurate arithmetic gives a substantially different value in the second sexagesimal place: 27;33,13 days. The reason for Geminos's unusual rounding is immediately (xviii 11) made clear: he wants a round number for $\tfrac{1}{4}T_A$:

$$\tfrac{1}{4}T_A = 27;33,20^d/4 = 6;53,20 \text{ days.}$$

---

[201] Neugebauer 1975, 497n2.

[202] Ptolemy, *Almagest* iv 2.

[203] Neugebauer 1975, 497–505.

[204] On the Babylonian lunar theory, see Neugebauer 1975, 474–540, and Neugebauer [1955], 41–85.

If the accurate value $T_A = 27;33,13^d$ were used, the quotient $\frac{1}{4}T_A$ would not truncate at the second sexagesimal place. Moreover, Geminos is following the Babylonian lunar theory of system B, in which the value $T_A = 27;33,20$ days is standard. In a time equal to $\frac{1}{4}T_A$, the Moon is assumed to go from least motion to mean motion, or from mean motion to greatest motion.

Geminos also assumes that the Moon changes its speed on the zodiac in a linear way. Let s be the least angle through which the Moon moves in the course of a day. The next day, the Moon is assumed to move $s + x$; the next day, $s + 2x$; then $s + 3x$, and so on. Thus the daily movements form an arithmetic progression with constant differences x. The remaining portion of Chapter xviii is devoted to finding x, the daily change in the Moon's daily motion. Indeed, x is the number that "one must find," subject to the conditions that Geminos states in xviii 16–17.

Let us define a few quantities:

s, the Moon's smallest daily motion
m, the Moon's mean daily motion
g, the Moon's greatest daily motion
T_A, the anomalistic month
x, the change in the Moon's daily motion from one day to the next.

Now we may state the conditions of the problem posed by Geminos:

$$m = 13;10,35° \hspace{3cm} \text{(xviii 9)}$$
$$TA = 27;33,20^d \hspace{3cm} \text{(xviii 10)}$$

$$11° < s < 12° \hspace{3cm} \text{(xviii 5)}$$
$$15° < g < 16°.$$

Because the daily change in motion is constant, s, m, and g form an arithmetic progression. Thus,

$$s + g = 2m. \hspace{3cm} \text{(xviii 13)}$$

Now, the time from mean motion to greatest motion is $\frac{1}{4}T_A$, a quarter of the anomalistic month. If we multiply this time by x and add the product to the mean motion, we will have the greatest motion:

$$g = m + \frac{1}{4}T_A \, x. \hspace{3cm} \text{(xviii 16–17)}$$

The least motion may be expressed in similar terms:

$$s = m - \frac{1}{4}T_A \, x.$$

The problem, then, is to determine x so that both s and g lie in the ranges supposed for them, subject to the condition that $s + g = 2m$. The latter condition will be automatically satisfied if we operate with the above formulas, which assume an arithmetic progression.

If we insert the numerical values for m and T_A into the expressions for g and s, and then apply the inequalities, we have

for g: $15° < 13;10,25° + 6;53,20^d\ x < 16°$
and, for s: $11° < 13;10,25° - 6;53,20^d\ x < 12°$.

The expressions for g and for s each result in an upper and a lower bound for x:

From the expression for g: $0;15,53 < x < 0;24,36$.
From the expression for s: $0;10,15 < x < 0;18,57$.

(The units of x are degrees per day.) Combining all four conditions, we see that only two matter:

$$0;15,53° < x < 0;18,57°.$$

If we restrict ourselves to solutions with whole numbers of minutes of angle, there are three possibilities:

$$x = 0;16 \text{ or } 0;17 \text{ or } 0;18.$$

Geminos chooses 0;18 (in accord with system B of the Babylonian lunar theory), but does not say why the other two choices should be excluded. It is clear, then, that his discussion of this problem is motivated by pedagogy and has nothing to do with the actual historical development. Geminos means only to explain how the parameters of the lunar theory may, in principle, be derived from the data of the *exeligmos*.

Now that x is known, it is easy to calculate the smallest and greatest daily motions:

$$s = m - \tfrac{1}{4}T_A\ x = 11;6,35°$$
$$g = m + \tfrac{1}{4}T_A\ x = 15;14,35°. \qquad\qquad \text{(xviii 18)}$$

It should be noted here that the original unusual rounding of $T_A\ /4$ guarantees that $x(T_A\ /4)$ contains a whole number of minutes of arc. Thus, as Geminos says (xviii 18), the daily motion changes by $2;4°$ between least and mean motion. Geminos concludes by summarizing all the parameters for the Moon's motion (xviii 19): least motion, mean motion, greatest motion, and daily augmentation. Of course, these values by themselves would not permit one to calculate lunar *positions*. One would also need a set of epoch values—the position and daily motion of the Moon on a particular starting date. Geminos's scheme therefore represents only the basic pattern of the Moon's motion. But, with epoch values in hand, one could easily spin out a day-by-day list of theoretical lunar positions.

Historical Roots of Geminos's Lunar Scheme

The scheme of letting the Moon's daily motion increase by equal increments of 18' from one day to the next, up to a maximum value, after which it falls by equal increments of 18', is an example of what Neugebauer called a "linear zigzag function." These functions are among the most characteristic features of Babylonian planetary theory. Indeed, Geminos's zigzag function for the Moon's daily motion is well attested in system B of the Babylonian lunar theory. Cuneiform tablets from the second and first centuries B.C. give the Moon's positions at 1-day intervals, based on a linear zigzag function for the daily motion with exactly the properties mentioned by Geminos ($s = 11;6,35°$, $g = 15;14,35°$, and $x = 18'$).[205] Another tablet, ACT 190, does not give lunar *positions*, but only lists the values of the Moon's daily motion, day by day, for 248 days.[206] This tablet is not associated with any particular year. Rather, it is a "template," which could have served as a calculating aid for a scribe who wished to generate day-by-day positions. The parameters s, g, and x are easily deduced from this tablet. The minimum and maximum values (s and g) of the daily motion do not actually occur, because the cycle happens to start a bit off minimum. But the maximum and minimum values are easily established by extrapolation and interpolation. The motivation for listing the Moon's daily displacements for 248 straight days is that the pattern then repeats (248 days = exactly 9 anomalistic months, using the value $T_A = 27;33,20^d$). With a starting position of the Moon taken from some other source, a scribe could use ACT 190 to spin out a set of day-by-day positions for any run of 9 anomalistic months.

Geminos's discussion in chapter xviii is the earliest extant treatment in a Greek source of the motion of a heavenly body in terms of linear

---

[205] ACT 191, 192, 193, and 194, in Neugebauer [1955], are from the 190s and 180s B.C. A similar text, ACT 194a, is for 69/68 B.C.

[206] ACT 190, in Neugebauer [1955], vol. 1, p. 179–80.

zigzag functions. His treatment was probably based on his acquaintance, not with Babylonian sources directly, but rather with Greek calculations employing Babylonian methods, which must already have been in circulation. A considerable body of evidence exists on papyrus for skilled use by Greeks of Babylonian lunar and planetary theories.[207] A few such papyri are of the late first century B.C., but the bulk of them are from the first century A.D. and later. The distribution of dates partly reflects the accidents of preservation. Much of the material came from the ancient garbage dumps at Oxyrhynchus in Egypt. Papyri older than the first century A.D. lay below the damp level and usually did not survive. Geminos's testimony allows us to push Greek familiarity with the details of Babylonian zigzag functions somewhat earlier into the first century B.C.

A Greek papyrus of the second century A.D., probably from Tebtunis, contains a "template" table for the Moon's progress around the zodiac that embodies the same pattern that Geminos describes.[208] This papyrus gives the total motion of the Moon, at 1-day intervals, from a starting point exactly at the minimum value of the zigzag function. Although only a small portion of the papyrus is preserved, it probably tabulated the individual daily movements as well, and probably for a run of 248 days. This papyrus represents a sort of connecting link between the Babylonian template ACT 190 and the scheme described by Geminos. The underlying parameters of all three texts are the same.[209] In common with ACT 190, the papyrus (probably) gave data for one complete cycle of 9 anomalistic months. But in common with Geminos, the papyrus table started off exactly at the minimum of the zigzag function. Though this papyrus is substantially later than Geminos's text, it provides a plausible example of the sort of documentary evidence on which Geminos might have based his account. We see that Geminos was a faithful reporter of astronomical details that were current in his day.[210] He was, as well, a natural teacher, who wanted to justify and explain the useful zigzag pattern for the Moon's daily motion, though he could not really know exactly how the Babylonians had invented it.

---

[207] See Jones 1999b. For a detailed discussion of "templates" and the other categories of astronomical tables on papyrus, see Jones 1999c.

[208] P. Fay. ined. Gc36, in Jones 2001a.

[209] At least this is probably the case. In the papyrus, positions are rounded off to whole minutes; but Jones's reconstruction of the table, based on $s = 11;6,35°$, is convincing.

[210] After Geminos's day, the Greeks in Egypt refined the Babylonian lunar theory by introducing improved parameters for the zigzag function. The improved system is well documented by papyri from the first through the fourth centuries A.D. See Jones 1983, Jones 1997, and Jones 1999b, vol. 1, p. 321–42.

14. ON THE TEXT AND TRANSLATION

Editions and Translations of Geminos

Edo Hildericus [Edo Hilderich van Varel]. *Gemini probatissimi philosophi ac mathematici elementa astronomiae Graecè, & Latinè interprete Edone Hilderico.* Altdorf, 1590. Leiden, 1603. Editio princeps of the Greek text, accompanied by a Latin translation.

Dionysius Petavius [Denis Petau]. *Uranologion sive systema variorum authorum, qui de sphaera ac sideribus eorumque motibus Graecè commentati sunt. Cura & studio Dionysii Petavii Aurelianensis è Societate Iesu.* Paris, 1630. Edition of the Greek text, with revision of the Latin translation of Hildericus. Geminos is found on pp. 1–70.

[Nicolas B.] Halma. *Table chronologique des règnes, prolongée jusqu'à la prise de Constantinople par les Turcs; apparitions des fixes de C. Ptolémée, Théon, et Introduction de Géminus aux phénomènes célestes.* Paris, 1819. Geminos is found in Part II, with separate pagination. Greek text with French translation. Halma worked from a Greek manuscript in the Bibliothèque du Roi (now Bibliothèque Nationale, Paris, Ms. Grec 2385), which ended in the middle of the chapter on months (our chapter viii). For the remainder of the text, he relied upon the edition of Petau.

[Jacques Paul] Migne. *Patrologiae cursus completus, seu bibliotheca universalis.* Series Graeca, Tomus 19. Paris, 1857. Columns 747–868. Reprint of the Greek text and Latin translation in Petau's *Uranologion*, with corrections.

Carolus [Karl] Manitius, ed. and trans. *Gemini Elementa astronomiae.* Leipzig: Teubner, 1898. Critical edition of the Greek text accompanied by a German translation.

Germaine Aujac, ed. and trans. *Géminos, Introduction aux phénomènes.* Association Guillaume Budé. Paris: Les Belles Lettres, 1975. Critical edition of the Greek text accompanied by a French translation.

The *parapēgma* at the end of Geminos's *Introduction to the Phenomena* was edited along with others in:

Curtius Wachsmuth. *Ioannis Laurentii Lydi liber de ostentis et Calendaria Graeca omnia.* Leipzig: Teubner, 1863. The Greek text of the Geminos *parapēgma* is on pp. 180–195 (second edition, 1897).

Text Chosen for This Translation

The Greek text adopted for this translation is that of Aujac 1975. But we have also constantly had before us the edition of Manitius 1898. These two texts, the only ones that have been produced in accordance with modern standards of scholarship, differ to the extent of perhaps two or three words or phrases per page. But only a small fraction of these disagreements are substantial enough to bear on Geminos's meaning. In general, Aujac shows a tendency to stay closer to the manuscript readings, even when these are problematical. She also had the advantage of using the important manuscript A (see below), which was unavailable to Manitius. Manitius, in accordance with the practice of his day, corrects the manuscripts much more freely in matters of spelling, grammar, usage of particles, etc. We have not judged it worthwhile to note every place where the texts of Manitius and Aujac differ. But wherever the disagreement would affect the meaning of the translation, we explain our choice in the **Textual Notes (appendix 1)**. On a few occasions, we have departed from both texts, again with an explanation in the Textual Notes.

The Manuscripts

Full descriptions of the manuscripts of Geminos's text can be found in the editions of Manitius and Aujac. Here, we wish only to explain for the general reader the basis for our knowledge of the text. We have personally examined only one of the manuscripts and have studied microfilm copies of half a dozen others. The descriptions below therefore rely heavily upon Manitius and Aujac. In referring to the manuscripts, we use Aujac's sigla.

Geminos wrote the *Introduction to the Phenomena* probably in the first century B.C. The three oldest surviving Greek manuscripts of the *Introduction* date from the decades immediately after 1300. Any of these three is probably separated from the archetype by several intervening copies. In these three manuscripts Geminos's text is evidently complete and more or less carefully written. The three oldest and best manuscripts are:

A: *Constantinopolitanus Palatii veteris Graecus* 40. (Called C by Manitius.)
Topkapi Palace Museum, Istanbul.

B: *Vaticanus Graecus* 381. (Called V² by Manitius.)
Biblioteca Apostolica Vaticana, Vatican City.

C: *Vaticanus Graecus* 318. (Called V¹ by Manitius.)

To these should be added:

V: *Vindobonensis Phil. Graecus* 89. (Called V by Manitius.)
Österreichische Nationalbibliothek, Vienna.
This manuscript derives, perhaps directly, from A. It was copied around 1500.

Manitius based his text principally upon V, B, and C. He knew of the existence of A, but was unable to consult it. Aujac based her text upon A, B, and C.

Aujac also made use of ten other, later Greek manuscripts. In most of these, Geminos's text is incomplete—the copyist has contented himself with making excerpts, or has copied only the first half or so of the text. These later, incomplete versions are less significant for the reconstruction of Geminos's text, but occasionally contribute insights.

A number of Greek manuscripts include excerpts from Geminos's *Introduction to the Phenomena* under a spurious title, *The Sphere of Proklos*. The material excerpted comes from our chapters iii, iv, v, and xv and concerns the axis and the poles, the circles of the sphere, the terrestrial zones, and the constellations. The oldest of these manuscripts dates from the fifteenth century. These manuscripts have little or no significance for the reconstruction of Geminos's text.[211] However, after the development of printing *The Sphere of Proklos* had a rich life of its own and went through many editions. In all, Aujac describes 27 Greek manuscripts that contain all or part of Geminos's text.

That there were one or more medieval Arabic translations of Geminos's *Introduction to the Phenomena* is certain, but very little is known about them. Geminos's name does not appear in the standard Arabic bio-bibliographies, including that of the great bibliophile, Ibn al-Nadim.[212] The modern historian and bibliographer Sezgin discusses him under the name Aghaniyus (although this identification is controversial).[213] Aujac suggests that the Arabic translation of the *Introduction to the Phenomena* was done as part of a project of creating a corpus of works known to the Arabic writers as "The Middle Books," which served as a preliminary to the study of Ptolemy's *Almagest*.[214] Although this is possible, the books known to have been included in this corpus are much more mathematical in character than Geminos's work.[215] Moreover, since the work is not mentioned by any Arabic author known to us—and certainly is not part of the collection of Middle Books compiled by Naṣīr al-Dīn al-Ṭūsī in the thirteenth century—Aujac's conjecture must remain exactly that.

---

[211] For the manuscript tradition of *The Sphere of Proklos*, see Todd 1993 and Todd 2003b.

[212] Dodge 1970.

[213] Sezgin 1974, 157–58 (s.v. "Aghaniyus"). Sezgin, p. 157, says that, despite Tannery's earlier view, opinion is coming around to accept the identification of the unknown Aghaniyus with Geminos. But he does not mention Geminos in vol. VI, which is devoted to astronomy.

[214] Aujac 1975, civ–cv. Her supposition that the work was translated into Arabic in the ninth or tenth centuries is a likely one, though the eighth century is possible.

[215] Representative of the works in the corpus are Euclid's *Phenomena*, Aristarchos's *On the Size and Distance of the Sun and Moon*, and Hypsiklēs' *On Rising Times (Anaphorikos)*.

Around A.D. 1170, at Toledo, Gerard of Cremona made a Latin translation from the Arabic version. But in the Arabic manuscript that Gerard worked from, Geminos's book was mistakenly attributed to Ptolemy! The mistaken title was included in the list of Gerard's translations that was compiled by his students after his death and appended to Gerard's translation of Galen's *Tegni*. It is listed there as *Liber introductorius Ptolemei ad artem spericam* (Ptolemy's Introduction to the Art of Spherics).[216] Gerard's Latin translation is extant in several copies (not all complete), of the late thirteenth and early fourteenth centuries, and the Latin manuscripts carry such titles as *Ptolemy's Introduction to the Almagest*, or *Ptolemy's Introduction to the Art of Spherics*.[217] So Geminos's work circulated not only under Proklos's name, but also under Ptolemy's. As Manitius pointed out, that the Latin version was translated from the Arabic language is evident from many things, but mainly from the names of the stars that occur in the third chapter, as well as a highly literal translation style indicative of an Arabic source. In the manuscripts of Gerard's Latin translation, the chapter order is not the same as in the Greek manuscripts of the *Introduction to the Phenomena*.[218]

Both Manitius and Aujac found the medieval Latin to be of occasional help in constructing a text of the *Introduction to the Phenomena*, most notably in filling the long lacuna at i 9–10. Although the Latin is separated from the original Greek by an intermediate language, the Arabic text on which the Latin was based must have been translated from a Greek manuscript older than any we now possess. Aujac gives details of six Latin manuscripts,[219] but she relied principally upon

Dresdensis Db 87.
Sächsische Landesbibliothek, Dresden.
This manuscript was written in Germany early in the fifteenth century.

A Hebrew translation was made, also from the Arabic, by Moses ibn Tibbon, in 1246, probably at Naples. Again, the dependence on an Arabic original is clear from the constellation names and other details; moreover, the Hebrew version includes a specific mention that it was made from the Arabic. In one of the Hebrew manuscripts, Geminos's text carries the title *Book of the Science of the Stars: Abridgement of*

---

[216] Grant 1974, 36.

[217] Todd 2003b provides an excellent study of the fortunes of Geminos's text in the Middle Ages and the Renaissance, with full details on the manuscripts of the medieval translations.

[218] Aujac 1975, cv.

[219] Aujac 1975, cvi–cviii. Todd 2003b, 17, describes an additional Latin manuscript at St. Petersburg.

Ptolemy. In this case, the chapter order is the same as in the Greek manuscripts.[220]

Division of the Text into Chapters, Sections, and Statements

Manitius's division of the Greek text into chapters, followed by Aujac, has become established. This translation adopts the division used by these two editors, in spite of the fact that it is in some respects arbitrary. Hildericus, in the *editio princeps*, stayed close to the manuscript evidence for chapter divisions and chapter titles. His division was followed by all other editors before Manitius. Manitius bifurcated three of Hildericus's chapters; as a result, he had to supply his own titles for his chapters ii, vii, and xiv. So, while Hildericus recognized sixteen chapters, we count eighteen, plus the *parapēgma*. The correspondence between our chapters and those of Hildericus is set out in the list below.

Hildericus's chapters	Our chapters	Principal topics
i	i and ii	the zodiac, zodiacal aspects
ii	iii	constellations
iii	iv	axis and poles
iv	v	circles of the sphere
v	vi and vii	day and night, risings of the twelve signs
vi	viii	months
vii	ix	phases of the Moon
viii	x	eclipse of the Sun
ix	xi	eclipse of the Moon
x	xii	eastward motion of planets
xi	xiii and xiv	risings and settings of stars, paths of the stars
xii	xv	terrestrial zones
xiii	xvi	geographical places
xiv	xvii	weather signs from the stars
xv	xviii	the *exeligmos*
xvi	not numbered	*parapēgma*

Each of the three breaks introduced by Manitius does, indeed, correspond to a shift in subject matter, but these shifts are not always abrupt or dramatic. For example, chapter ii, devoted to zodiacal aspects, is a natural continuation of Geminos's discussion of the zodiac in chapter i.

[220] On the Hebrew version, which is extant, see Aujac 1975, cviii–cix, and Todd 2003b, 9n8.

The section headings (printed in small capitals) within each chapter do not occur in the Greek manuscripts. They have been added by the translators as an aid to the reader.

Manitius was the first editor to assign reference numbers to individual statements throughout the text. His numbering was followed by Aujac, and we have followed it, too.

Diagrams

Two diagrams are referred to in the text of chapter i, but the manuscript tradition for the diagrams is very poor. In many manuscripts, the diagrams are absent, no space even having been left for them. However, the three oldest Greek manuscripts all show evidence for the diagrams that accompany chapter i. A contains two diagrams in chapter i, but these are identical copies of one another, and seem to correspond to Geminos's description of his second diagram. In B, two rectangular spaces were left for the diagrams in the text of chapter i, but the diagrams were never drawn. In C, the second diagram to chapter i is preserved, but corners of three pages have been torn away, entailing (probably) the loss of the first of the diagrams to chapter i.

The first diagram for chapter i as printed here therefore lacks good manuscript authority, but the diagram is described quite clearly in words at i 34–35. Some features of the first diagram, as printed here, have been borrowed from the second diagrams in A and C: the Sun symbol occurs in both diagrams of A as well as in the second (only surviving) diagram of C. The lengths of the seasons are written in the second diagram of C, but with the lengths of fall and winter mistakenly reversed. In both A and C, the copyists failed to draw the Sun's circle with its apogee in the right position. For example, in the diagram of C, the apogee is in the direction of the summer solstice (on the boundary between Gemini and Cancer), rather than in the sign of Gemini, as the text states.

The diagrams that accompany chapter ii are of questionable authenticity, for the text makes no mention of them. Neither A nor B has any diagrams in chapter ii. C does offer some evidence in favor of the diagrams of chapter ii. C has one diagram, which illustrates the trine aspect, carefully drawn in a space left for it amid the text. A torn-away corner may originally have borne a diagram illustrating one or more of the other aspects.

Manitius prints five diagrams in his text for chapter ii—one each for the four aspects (opposition, trine, quartile, syzygy) and one illustrating the mistaken notion of syzygy for which Geminos criticizes "the ancients." There appears to be no authority for the last. Aujac prints no diagrams in the text of chapter ii. Aujac (1975, xcvi), however, does speculate that Geminos's text originally contained six diagrams: two in chapter i illustrating the eccentricity of the Sun's circle, three in chapter ii illustrating the aspects of opposition, trine, and quartile, and one additional diagram in ii showing the parallel circles on the sphere. The evidence for the last is poor. C does, indeed, contain such a diagram, but it is squeezed into the margin as an afterthought and is not referred to in the text. It is worthy of note that C contains two obviously spurious diagrams, unrelated to the text, drawn in the margins. Thus we should not count too much on the testimony of C except where its diagrams appear in spaces intentionally left for them.

M has a very nice set of diagrams for chapters i and ii. This manuscript, of the late fourteenth century, was carefully written, but contains only the first two chapters of the text. It includes the usual two diagrams to chapter i, illustrating the eccentricity of the Sun's circle. In chapter ii, it includes five, illustrating the aspects opposition, trine, quartile, syzygy, and sextile (i.e., the hexagon). All of these, even the last, which is surely spurious, are carefully drawn in spaces that were deliberately left for them. Thus we cannot discount the possibility of a copyist embroidering the text with diagrams. The more frequent failing was, however, to leave them out. The diagrams printed for chapter ii in this book have been taken from M. Though we cannot be sure that they really reflect Geminos's intentions, they do at least provide good examples of the diagrams in medieval manuscripts of elementary treatises on astronomy, so we have thought it worthwhile to include them.

Finally, C has two not very satisfactory diagrams, illustrating eclipses, drawn in the margins near the text for chapter xi. Because we believe these are conventional illustrations added by a copyist, we have reproduced them in the commentary to chapters x and xi, rather than in the translation of the text itself.

Our Approach to Translation

We have tried to keep our translation as true to the Greek as is possible, while striving for clear and readable English. Modern English cannot always duplicate the sentence structure of ancient Greek, because the Greek writer had available a full set of cases for declinable nouns, as well as syntactic structures that do not exist in English. Nevertheless, we feel that our translation captures the rhythm and flow of Geminos's prose, which moves back and forth between logical argument and casual informality.

Sometimes two different Greek words (e.g., *boreios* and *arktos*) occur in the text when there is really only one natural, modern English equivalent (in this case "north"). In such cases, we have not hesitated to translate two different Greek words by one English word, rather than stretch to preserve a fine distinction by using some English archaism. Nor have we shied away from giving simple English equivalents of Greek phrases that could be translated more literally by longer and less idiomatic English expressions. For example, we write simply, "fixed stars" and "the Milky Way," rather than "non-wandering stars" and "the circle of milk." Also, when a Greek word has an English derivative that is in common use today but that has a quite different meaning, we have used instead an English word that reflects the meaning of the Greek. For example, Geminos consistently uses *astrologia* where we would use not "astrology" but "astronomy," and *geographia* where we would use not "geography" but "world map." The Glossary of Technical Terms (appendix 3) provides Geminos's original Greek expressions.

In a few cases we have decided not to translate, but simply to transliterate Greek words (indicated by italics) when suitable English equivalents do not exist. Examples are *sphairopoiïa*, *klima*, and *oikoumenē*. In all such cases, we explain the term with a comment at the first occurrence of the term in the text, and sometimes with a longer discussion in the Introduction.

Spelling of Greek Names

In keeping with what seems to be a trend in classical studies, we have made Greek names look Greek by following a common convention for transliteration of the Greek alphabet. The long vowels eta (ē) and omega (ō) are marked with macrons to distinguish them from epsilon (e) and omicron (o). Usage is in flux, however, and many scholars still adhere to the practice inherited from our Renaissance predecessors, whereby Greek names were Latinized. In the older practice, the Greek masculine, nominative ending *os* became Latin *us*. Greek diphthongs were transformed into their Latin phonetic equivalents. Thus Greek *ei* became Latin *i*. So we write the name of the Stoic philosopher of the first century B.C. as Poseidōnios, while some scholars spell it Posidonius. Usually, this will cause no confusion. In a few cases (notably when Greek kappa is represented by *k* rather than *c*), alphabetical ordering (as in an index or encyclopedia) can be affected. (Thus, a reader seeking more information on Kratēs, the ancient literary critic, may need to look under Crates in an encyclopedia.) However, we make exceptions to this practice for people whose names are so well known in their Anglicized spelling that transliteration would appear pedantic (e.g., Homer, Aristotle, Plato, Ptolemy). Another exception involves familiar ethnic groups or locations. Thus we choose the common English versions (Egyptians, Rome) over the unfamiliar Greek versions (*Aiguptioi, Hrōmē*).

Numbers

When a numeral appears in the text it is because the equivalent Greek numeral appears in Aujac's Greek text. (It should be noted that the manuscripts are not completely uniform in their manner of representing numbers, and that Aujac has regularized the notation.) In printing sexagesimal (base-60) numerals we follow Neugebauer's convention: a semicolon separates the integer from the fractional part, and commas are used to separate the sexagesimal places. Thus

$$29;31,50,8,20 = 29 + 31/60 + 50/60^2 + 8/60^3 + 20/60^4.$$

Star and Constellation Names

In this book the twelve zodiac constellations are called by their Latin names, as in modern English usage. For other constellations and stars, we have adopted the following conventions. If the name in modern English use descends directly from the Greek, the modern name is used in its standard form (e.g., Procyon, Canopus). In all other cases a literal translation of the Greek name is used (e.g., Bird for Geminos's *Ornis*, which corresponds to the modern constellation Cygnus). See the tables at the end of chapter iii.

Editorial Symbols and Abbreviations

Aujac	Germaine Aujac, *Géminos, Introduction aux phénomènes*. Paris: Les Belles Lettres, 1975.
Manitius	Carolus Manitius, *Gemini Elementa astronomiae*. Leipzig: Teubner, 1898.
Wachsmuth	*Curtius Wachsmuth. Ioannis Laurentii Lydi liber de ostentis et Calendaria Graeca omnia*. Leipzig: Teubner, 1863.
A	Constantinopolitanus Palatii veteris graecus 40 (c. 1300).
B	Vaticanus graecus 381 (c. 1300).
C	Vaticanus graecus 318 (c. 1330).
M	Marcianus graecus 323 (c. 1380). Biblioteca Nazionale Marciana, Venice.
V	Vindobonensis phil. graecus 89 (c. 1500).
Latin	Latin version translated from the Arabic in the twelfth century = Dresdensis Db 87 (fifteenth century).
LSJ	H. G. Liddell and R. Scott, revised by H. S. Jones, *A Greek-English Lexicon*, 9th ed. with New Supplement (Oxford: Clarendon Press, 1996).

< >   Pointed brackets indicate conjectural restorations of the Greek text, where, for example, one or more words have dropped out of the text.

[ ]   Square brackets enclose words added by the translators for clarity but without any corresponding words in the Greek text.

Translation and Commentary

I. On the Circle of the Signs

I 1 The circle of the signs[a] is divided into 12 parts, and each of the sections is designated both by the common term "twelfth-part" and by a particular name taken from the stars that it contains and by which each sign is formed. 2 The twelve signs are: Aries, Taurus, Gemini, Cancer, Leo, Virgo, Libra, Scorpio, Sagittarius, Capricorn, Aquarius, Pisces.

SIGNS AND CONSTELLATIONS

I 3 The word "sign" is used in two ways.[1] According to one way it is a twelfth-part of the zodiac circle, that is, a certain interval of space demarcated by stars or points. According to the other [way], it is an image formed from the stars, based on resemblance and the position of the stars. 4 Thus, the twelfth-parts are equal in size: for the circle of the signs has been divided into 12 equal parts by means of the *dioptra*.[2] But the constellation-signs are not equal in size, nor are they formed from equal numbers of stars, nor do they all exactly fill the spaces assigned to the twelfth-parts. 5 Rather, some fall short, such as Cancer, for it occupies only a small part of its assigned space. Others, such as Leo,[b] spill over and occupy a certain part of the preceding and following signs.[3]

Notes signaled by superscript letters are given in Appendix 1, Textual Notes.

[1] The word "sign" is used in two ways. Each sign of the zodiac is 30° long by definition. The sign of Aries is the first 30° of the zodiac, beginning with the vernal equinox. The sign of Taurus is the next 30°, and so on. The *signs* were convenient idealizations of the more ancient zodiacal *constellations*. In the ancient Greeks' day, each zodiacal constellation lay roughly in the sign named for it, but in our day this is no longer true. Because of precession, all the stars advance eastward through the zodiac at the rate of 1° in 72 years, so in the nearly 2,100 years that have elapsed since Geminos's time, the stars have advanced by almost 30°. Thus, the stars of Aries are now in the sign of Taurus. Although Hipparchos discovered precession in the second century B.C., Geminos nowhere alludes to this phenomenon. Ptolemy (*Almagest* vii 1–3) suggests that Hipparchos had left several questions about precession unresolved or inadequately proven, and it appears that Hipparchos's lost work on precession never circulated very widely. Indeed, there are no mentions of precession outside of Ptolemy until late Antiquity, when Ptolemy's influence was clearly felt.

[2] *dioptra*. The *dioptra* was a sighting instrument. See sec. 8 of our Introduction.

[3] preceding and following signs. Preceding = "to the west." Following = "to the east." This terminology probably derives from the daily rotation of the heavens from east to west.

Moreover, certain of the 12 signs do not wholly lie on the zodiac circle. Rather, some lie north of it, such as Leo; others lie south of it, such as Scorpio.

DIVISION INTO DEGREES

I 6 Further, each of the twelfth-parts is divided into 30 parts; and each piece is called a degree. Thus the whole circle of signs contains 12 signs, or 360 degrees. 7 The Sun passes through the zodiac circle in a year: for it is in a year's time that the Sun travels around the zodiac circle and, [having started] from a certain point, returns to the same point. This time is 365¼ days: for in just so many days does the Sun pass by the 360 degrees, so that the Sun moves very nearly a degree in one day. 8 However, a degree is one thing, and a day is another. For a degree is a certain distance, being 1/30 of a sign, while a day is a time period, being very nearly 1/30 of the monthly period. Moreover, the degree is 1/360 of the zodiac circle, while the day is very nearly 1/365¼ part of the annual period. All the signs are thirty degrees long, but not all are thirty days.[4]

EQUINOXES, SOLSTICES, AND SEASONS

I 9 The year's time is divided into 4 parts: spring, summer, fall, and winter.[5] Spring equinox occurs around the height of the flowering time, [when the Sun is] in the first degree of Aries. Summer solstice occurs

---

Aries, lying west of Taurus, precedes it in the march toward the western horizon. Alternatively, some scholars derive this terminology from the direction of the Sun's annual motion. The Sun, traveling eastward, passes through Aries before Taurus. Thus, Aries may be said to precede Taurus in the zodiac. The former explanation is to be preferred, in view of the fact that Euclid (c. 300 B.C.) uses these terms in his *Phenomena*, a work that makes no reference to the annual motion of the Sun.

[4] not all are thirty days. Because of its variable apparent speed, the Sun takes more than 30 days to traverse some signs and less than 30 to traverse others. In making this remark, Geminos may have in mind the example of *parapēgmata* divided into zodiac signs, which often began each sign with a notice of the number of days required for the Sun to traverse that sign. See the Geminos *parapēgma*, below, as well as fig. I.18 on p. 60.

[5] spring, summer, fall, and winter. Geminos divides the seasons by means of the equinoxes and solstices, as we do today. Although this division was usual among astronomers, it was by no means universal in everyday life. The heliacal risings and settings of the fixed stars, especially Arcturus and the Pleiades, were often used as signs of the seasons: "Winter lasts from the setting of the Pleiades to the spring equinox, spring from the equinox to the rising of the Pleiades, summer from the Pleiades to the rising of Arcturus, autumn from Arcturus to the setting of the Pleiades" (Hippokratēs, *Regimen III* 68).

around the intensification of the heat, in the first degree of Cancer. <Autumnal equinox[6,c] is during the season of fruits, when the Sun enters the first degree of Libra. Winter solstice is at the time when the state of the cold reaches its ultimate, when the Sun enters the first degree of Capricorn.

The two solstices and the two equinoxes occur, in the way of thinking of the Greek astronomers, in the first degrees of these signs; but in the way of thinking of the Chaldeans,[7] they occur in the eighth degrees of these signs.[8] The days on which the two solstices and the two equinoxes occur are the same days in all places, because the equinox occurs in all places at one time, and similarly the solstice. And again, the points on the circle of signs at which the two solstices and the two equinoxes occur are exactly the same points for all astronomers. There is no difference between the Greeks and the Chaldeans except in the division of the signs, since the first points of the signs are not subject to the same convention for them: among the Chaldeans, they precede by 8 degrees. Thus, the summer solstitial point, according to the practice of the Greeks, is in the first part of Cancer; but according to the practice of the Chaldeans, in the eighth degree. The case goes similarly for the remaining points.

I 10 The vernal equinox occurs when the Sun, in the course of its climb from south to north, is on the equatorial circle: at that time the day becomes equal to the night. For day and night are not always constantly equal, but on certain days the day is longer than the night, and on certain others the night is longer than the day. Night and day are not equal except on two days in the year, which are the days of vernal

---

[6] <Autumnal equinox. Here begins a long lacuna in the Greek text, which Aujac has filled from the medieval Latin translation of Geminos. The text supplied from the Latin is bounded by the marks < >. See textual note c.

[7] Chaldeans. This is one of several references that Geminos makes to Chaldean, or Babylonian, astronomical practice. On the term "Chaldean," see sec. 5 of our Introduction, pp. 13–14.

[8] eighth degrees of these signs. Geminos places the equinoctial and solstitial points at the beginnings of their signs, as we do today. Thus, the spring equinox is, as Geminos says, "at the first degree" of Aries. By this he means the *beginning* of the first degree, or what we would call the zeroth degree. This was also the convention of Greek mathematical astronomy in the tradition of Hipparchos and Ptolemy. But a Babylonian convention placed the equinoctial and solstitial points 8° within their signs. As Geminos points out, the equinoctial and solstitial points are the same for everyone. Night and day are equal when the Sun reaches the equinoctial point. The Sun's place in the zodiac where this occurs was called by the Greeks the beginning of Aries, and by the Chaldeans 8° within Aries. Thus the Babylonian signs begin 8° in advance of the corresponding Greek signs. The 8° norm persists in many Roman writers, including Vitruvius (*On Architecture* ix 3), Pliny (*Natural History* xviii 221), and Columella (*De re rustica* xi 2.31). Besides the 8° norm with which Geminos was familiar, the Babylonians sometimes used a 10° norm. See Neugebauer 1975, 593–600.

equinox and autumnal equinox. Summer solstice occurs when the Sun arrives closest to the zenith in our region and is elevated to its greatest elevation above our horizon,[d] when > it describes its most northerly circle and produces the longest day of all those in the year, and the shortest night. The longest day is equal to the longest night, and the shortest day is equal to the shortest night. For the *klima*[9] at Rhodes the longest day is 14½ equinoctial hours.[10]   **11** Autumnal equinox occurs when the Sun, passing from north to south, is again located on the equator circle and makes the day equal to the night.   **12** Winter solstice occurs when the Sun is farthest from our region. Then it describes the circle that is most southerly and lowest relative to the horizon, and makes the longest night of all those in the year, and the shortest day. For the *klima* at Rhodes, the longest night is 14½ equinoctial hours.

The Inequality of the Seasons

I 13 The times between the solstices and the equinoxes[11] are divided in the following way. From spring equinox to summer solstice there are

---

[9] *klima*. A *klima* is a band on the Earth lying near a single parallel of latitude. Often, *klimata* were designated in terms of the length of the longest day: Rhodes is in the *klima* of 14½ hours, because at Rhodes on the summer solstice the day lasts 14½ hours. The root meaning of the word is "slope" or "inclination": the *klima* is determined by the inclination of the axis of the cosmos to the horizon.

[10] equinoctial hours. The *equinoctial hour* is the hour we use today, 1/24 of the diurnal period. The equinoctial hour was often used by Greek astronomers, for it provided an invariable unit of time. But in everyday life the Greeks used the *seasonal hour*, which is 1/12 of the time between sunrise and sunset. Thus the seasonal day hour is long in the summer and short in the winter. All surviving Greek sundials are calibrated in seasonal hours. See our Introduction, sec. 8, pp. 34–35.

[11] The times between the solstices and the equinoxes. Geminos's values for the lengths of the seasons are due to Hipparchos. Hipparchos's treatise on the solar theory has not survived, but it is discussed by Ptolemy (*Almagest* iii 4). Ptolemy mentions only Hipparchos's lengths for spring and summer. However, in an eccentric-circle theory such as Hipparchos used, it must be the case that spring + fall = summer + winter. And, of course, all four seasons must add to 365¼ days. Thus the lengths for fall and winter are determined, and are as Geminos has them. Geminos, who cites sixteen different authorities in the course of his work, does not cite Hipparchos at this juncture. Indeed (if we exclude the *parapēgma*, which may not be by Geminos), Hipparchos is mentioned by name only twice in the *Introduction to the Phenomena*, and then only in connection with the names of constellations (iii 8 and 13). Geminos probably did not have a firsthand acquaintance with Hipparchos's technical treatises. Theōn of Smyrna (*Mathematical Knowledge Useful for Reading Plato* iii 26) gives the same season lengths, also without mentioning Hipparchos. Kleomēdēs (*Meteōra* i 4.44–48) also fails to cite any authority for his season lengths. He gives Hipparchos's values for spring and summer, but modifies fall and winter to 88 and 90¼ days, respectively.

94½ days, for in just so many days the Sun passes through Aries, Taurus, and Gemini; then, arriving at the 1st degree of Cancer, it produces summer solstice. **14** From summer solstice to autumnal equinox there are 92½ days, for in just so many days the Sun passes through Cancer, Leo, and Virgo; then, arriving at the 1st degree of the Claws,[12] it produces autumnal equinox. **15** From autumnal equinox to winter solstice there are 88⅛ days, for in just so many days the Sun passes through the Claws, Scorpio, and Sagittarius; and when the Sun arrives at the 1st degree of Capricorn it produces winter solstice. **16** From winter solstice to spring equinox there are 90⅛ days, for in just so many days the Sun passes through the remaining three signs, Capricorn, Aquarius, and Pisces. **17** All the days of these four seasons, when added up, make 365¼, which was just the number of days in the year.

FUNDAMENTAL HYPOTHESIS OF ASTRONOMY

I **18** A question arises in this connection: the fourth parts of the zodiac circle being equal, how is it that the Sun, moving always at constant speed, runs through equal arcs in unequal times? **19** For the hypothesis that underlies the whole of astronomy[13] is that the Sun, the Moon, and the 5 planets move circularly and at constant speed in the direction opposite to that of the cosmos.[14] The Pythagoreans, who first ap-

---

[12] the Claws. The early Greeks did not refer to the sign between Virgo and Scorpio as a balance (*zugos*, our Libra), but rather as the Claws (*chēlai*) of the Scorpion, and counted it as a distinct part of a double constellation. Aratos (*Phenomena*, c. 270 B.C.) and Eratosthenēs (*Catasterisms*, c. 230 B.C.) always refer to these stars as the Claws. Hipparchos, in the *Commentary on the Phenomena of Eudoxos and Aratos* (c. 160 B.C.), uses Claws throughout, mentioning a Balance only once (iii 1.5). Geminos is one of the earliest Greek writers to make frequent use of the Balance. In one passage (vii 25), where he uses "Balance" and "Claws" in the same sentence, Geminos attributes the latter to "the ancients." Ptolemy (c. A.D. 147) uses Claws throughout the text of the *Almagest*; however, the Balance appears in the headings of the tables in his Book ii. Ptolemy's only use of the Balance in the text proper (*Almagest* ix 7) is in connection with a "Chaldean" observation of Mercury. As the Babylonians indeed designated these stars a Balance (Gössmann 1950, 72), it is clear that the Balance was introduced into Greek astronomy from Babylonian sources.

[13] the hypothesis that underlies the whole of astronomy. Geminos attributes to the Pythagoreans the principle that uniform circular motion is proper to celestial things. Simplikios (sixth century A.D.) ascribed this principle to Plato. (See the discussion in sec. 10 of our Introduction.)

[14] direction opposite to that of the cosmos. The whole cosmos rotates from east to west once each day about the axis through the celestial poles. The Sun, Moon, and planets have an additional motion that carries them from west to east along the zodiac, opposite to the motion of the cosmos.

proached such investigations, hypothesized that the movements of the Sun, Moon, and the 5 wandering stars are circular and uniform. **20** For they did not accept, in things divine and eternal, such disorder as moving sometimes more quickly, sometimes more slowly, and sometimes standing still. These [standings] are called "stations"[15] for the 5 wandering stars. One would not accept such anomaly of movement in the goings of an orderly and well-mannered man. **21** The business of life is often the cause of slowness or of swiftness for men. But in the case of the incorruptible nature of the stars, it is not possible to adduce any cause of swiftness or of slowness. For this reason, they put forward the question: how would the phenomena be accounted for[16] by means of uniform and circular motions? **22** We shall explain elsewhere[17] the cause [of the variable motion] in the case of the remaining stars. But now, for the case of the Sun, we shall indicate the cause by which it runs through equal arcs in unequal times while moving at constant speed.

POSITIONS AND PERIODS OF THE CELESTIAL BODIES

I 23 Highest of all is the sphere called the sphere of the fixed stars, including the representations of all the zodiacal constellations. One must not suppose that all the stars lie on one surface, but rather that some are higher, some lower[18]; but because vision extends to equal distances, the

---

[15] stations. Like the Sun and Moon, the planets generally move eastward along the zodiac, each with its own period. Mars requires a little less than 2 years to make one circuit of the zodiac; Jupiter, 12 years; Saturn, 30 years. But each of the planets occasionally reverses direction and travels *westward* along the zodiac for a few weeks or a few months, depending on the planet, before reverting to its usual eastward motion. The motion to the west is called "retrograde motion." While going into or coming out of retrograde motion, the planet remains nearly in one place with respect to the fixed stars for days or weeks at a time. This standing still is called a "station." (From the modern point of view, the apparent retrogradations and stations of the outer planets are due to the motion of the Earth.)

[16] how would the phenomena be accounted for? When Geminos asks how "the phenomena would be accounted for," this expression carries the same sense as the more famous prescription to "save the phenomena," but Geminos's version of the saying may well be the older. See the discussion in sec. 10 of the Introduction, p. 50.

[17] we shall explain elsewhere. Geminos's promise to explain the motion of the planets is not kept. The explanation would have entailed an introduction to the theory of epicycles. That this can be done in a text at the elementary level is demonstrated by Theōn of Smyrna. Perhaps Geminos is referring to a work he had planned for the future.

[18] some are higher, some lower. That is, some are more, some less, distant from the Earth. Some scholars read Geminos as saying that the *fixed stars* are not all at the same distance from us (Aujac 1975, 6n2; Russo 2004, 88). But this is not the case: Geminos only means to introduce his account of the planets (also called *stars*), which do lie closer to us than the sphere of fixed stars.

difference of height is unperceived.[19] **24** Beneath the sphere of the fixed stars lies Phainōn, called the star of Kronos[20]; this one passes

[19] the difference of height is unperceived. Geminos is attempting to explain why our eyes cannot tell which planets are higher and which are lower, even though, according to astronomical theory, they do not all lie at the same distance from the Earth. But we must confess that we do not understand his argument. On the Greek theories of vision, see Smith 1999, 23–47.

[20] Phainōn, called the star of Kronos. Each planet had a particular name in Greek and was also said to be the star of a particular god, as in table 1.1. The early Greeks knew Venus, which is often spectacular, but probably not the other planets. For Venus they had two different names—*Hesperos* ("of the evening," Homer, *Iliad* xxii 317; Sappho 95) when it appeared as an evening star, and *Heōsphoros* ("bringer of morning," *Iliad* xiii 226) when it appeared as a morning star. The discovery that these are one and the same star was attributed by various Greek authorities to Pythagoras or Parmenidēs. The name *Phōsphoros* ("light bringer") for Venus appears a little later. A complete set of five planet names among the Greeks is not attested before the fifth century B.C., when the divine names came into use all at once. The divine associations must have been new in the time of Plato, for their appropriateness is discussed in *Epinomis* 986A–988E. (The authenticity of the *Epinomis* has been challenged, and in Antiquity this work was sometimes attributed to Plato's student, Philip of Opus, c. 350 B.C.) In any case, Aristotle uses the divine names (*Metaphysics* xii 8) in his discussion of the homocentric spheres of Eudoxos and Kallippos. The divine associations of the Greeks were modeled on those of the Babylonians. Thus the Babylonians associated the planet Jupiter with Marduk, the dominant god of their pantheon. Moreover, Venus was associated with Ishtar, a goddess of fertility, and Mars with Nergal, the god of war and pestilence.

In the Hellenistic period, multiple versions of the divine associations arose. For example, in Pseudo-Aristotle, *On the Cosmos* (392a24–29), we learn that some associate the planet Venus with Hera rather than with Aphrodite, and the planet Mercury with Apollo rather than with Hermes. On the coronation horoscope of Antiochos of Commagene (62 B.C.), Mars is associated with Herakles and Mercury with Apollo. (See Evans 1999, 286–87.) In the *Celestial Teaching* of Leptinēs (Tannery 1893, 287), the planet Saturn is associated with Helios, perhaps because Saturn stands in for the Sun at night. Cumont argued that the instability in the divine associations came about because Greek contact with the gods of Egypt and the East resulted in a syncretism that obscured the correspondences between the Babylonian gods and the gods of the old Greek pantheon. In order to a have stable set of planetary names, Greek astronomical writers began to use a set of secular or "scientific" names based on the visual properties of the planets. (These secular names, too, appear to have Babylonian roots.) The oldest datable use is in the description of observations of "Stilbōn" (Mercury) in 262 and 265 B.C., cited by Ptolemy (*Almagest* ix 7 and ix 10) in conjunction with an astronomer named Dionysios. The double list of names we find

TABLE 1.1
Divine and Secular Names for the Planets

Secular Name	Translation	Divine Ruler	English Name
Phainōn	shiner	Kronos	Saturn
Phaëthōn	bright one	Zeus	Jupiter
Pyroëis	fiery one	Ares	Mars
Phōsphoros	light bringer	Aphrodite	Venus
Stilbōn	gleamer	Hermes	Mercury

through the zodiac circle in very nearly 30 years,[21] or through one sign in 2 years and 6 months. **25** Beneath Phainōn, and lower than it, moves Phaëthōn, called the star of Zeus; it passes through the zodiac circle in 12 years, or through one sign in a year. **26** Beneath it lies Pyroëis, the star of Ares; it goes through the zodiac circle in two years and six months, or through a sign in two and a half months. **27** The next space is held by the Sun, which passes through the zodiac circle in a year, or through a sign in very nearly one month. **28** Below it lies Phōsphoros, the star of Aphrodite; this one moves very nearly at the same speed as the Sun. **29** Below it lies <Stilbōn>,[e] the star of Hermes, which also moves at the same speed as the Sun. **30** And lower than all moves the Moon, which passes through the zodiac circle in 27⅓ days, or through a sign in 2¼ days, very nearly.

INTRODUCTION TO THE SOLAR THEORY

I **31** If the Sun really were moving among the zodiacal constellations, the times between the solstices and the equinoxes would certainly be

---

in Geminos is found (sometimes with variants) in other elementary writers, such as Leptinēs (Tannery 1893, 286–87), Thēon of Smyrna (*Mathematical Knowledge* iii 15), and Kleomēdēs (*Meteōra* i 220–42). By Ptolemy's time the secular names had become obsolete, but occasionally recurred in archaizing writers. The only extensive study is Cumont 1935.

[21] [Saturn] passes through the zodiac circle in very nearly 30 years. Geminos gives each planet's tropical period (the time for a circuit of the zodiac). Most of the periods are good round values, adequate for students, but not precise enough for astronomical practice. The puzzle is Geminos's value of 2 ½ years for the tropical period of Mars. The actual tropical period is 1.88 years. Geminos is usually careful with numbers, as in his treatment of lunisolar cycles (chapters viii and xviii). We have no explanation to offer, other than an early corruption of the text.

The order of the planets mentioned by Geminos is one of several proposed in Antiquity. The Moon is seen to occult not only the stars, but also the planets (Aristotle, *On the Heavens* 292a4). Thus all ancient schemes agree in placing the Moon nearest to us, and in placing the fixed stars farthest away. But since the distances of the planets could not be measured, their order could only be decided on cosmological or physical principles. A principle adopted at an early date is that the distances are correlated with the speeds, i.e., that the planets farthest away are those with the longest tropical periods (Aristotle, *On the Heavens* 291a32–b9). Thus, Saturn is nearest the fixed stars, with Jupiter next below, and then Mars. The difficulty was what to do with the Sun, Venus, and Mercury, as all three have a tropical period of 1 year. The order mentioned by Geminos became standard somewhat before his time and remained so down to the time of Copernicus. For Ptolemy's justification of this order, see *Almagest* ix 1. For other orders of the planets, see Neugebauer 1975, 690–93, and Dreyer 1906, 167–70.

A natural inference to draw from the principle that distance and speed are correlated would be that the Sun, Mercury, and Venus are all at the same mean distance. The deferent circles of Mercury and Venus would then coincide with the Sun's circle, and their epicycles equal to one another; for while traveling the equal arcs at a constant speed, it would be bound to complete them in equal times. **32** Similarly, if the Sun traveled along lower than the zodiac circle but [still] moved around the same center—that is, the center of the zodiac circle—then, in this case too, the times between the solstices and the equinoxes would be equal; for all circles described about the same center are divided in the same way by diameters.[22] **33** And thus, since the zodiac circle is cut into 4 equal parts by the diameters joining the solstitial and equinoctial points, the solar circle, too, would necessarily be divided into 4 equal parts by the same diameters; therefore the Sun, moving at constant speed upon its own sphere, would render the times of the quadrants equal.

---

would be centered on or near the Sun. This partially heliocentric system was proposed in Antiquity (Theōn of Smyrna, *Mathematical Knowledge* iii 33). The name of Herakleidēs Pontikos (fourth century B.C.) has sometimes, incorrectly, been associated with this system. See Neugebauer 1975, 694–96, and Eastwood 1992.

[22] divided in the same way by diameters. Refer to fig. 1.1. If the Sun's circle were smaller than the zodiac, but still centered on the Earth, a diameter of the zodiac circle would also bisect the Sun's circle. Thus, the lines through the equinoctial and solstitial points, which divide the zodiac into four equal arcs, would divide the Sun's circle also into four equal parts. And so, if the Sun moved uniformly on its circle, the four seasons would be of equal length.

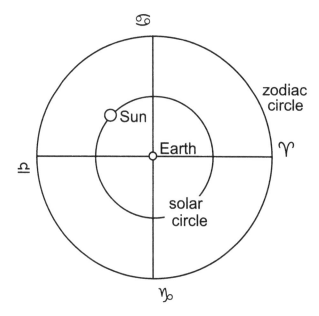

Fig. 1.1. If the Sun's circle were centered on the Earth, the Sun's uniform motion around its circle would produce four equal seasons.

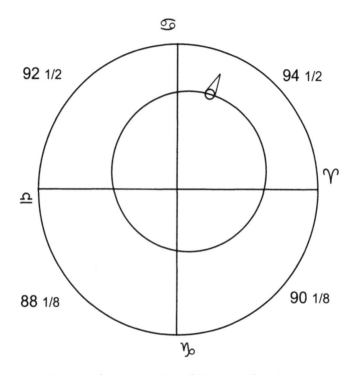

Diagram 1. Conjectural reconstruction of Geminos's first diagram, showing the Sun's circle eccentric to the zodiac. The Earth is at the intersection of the two lines. Symbols for the equinoxes and solstices have been added as a convenience. These symbols first occur in the Middle Ages, and would not have been a part of Geminos's original diagram. The solar symbol, however, does occur in ancient Greek astronomical papyri.

I 34 Now, the Sun does indeed travel along lower, but it moves upon an eccentric circle, as has been sketched.[23] For the centers of the solar circle and the zodiac are not the same; rather, the sphere of the Sun lies to one side. Because of such a position, the course of the Sun is divided into 4 unequal parts. 35 The longest arc[24] is the one that falls beneath the quadrant of the zodiac circle reaching from the 1st degree of Aries to

---

[23] as has been sketched. Drawings that occur in the Greek manuscripts of Geminos's *Introduction to the Phenomena* will be called *diagrams*. Illustrations designed by the translators to clarify points made in the commentary will be called *figures*. The diagram printed here (diagram 1) lacks good manuscript authority, but it is described clearly in words at i 34–35. See sec. 14 of the Introduction for more detail.

[24] the longest arc. The placement of the longest arc of the Sun's circle beneath the spring quadrant of the zodiac is consistent with the season lengths given at i 13–16. In Antiquity, spring was the longest season.

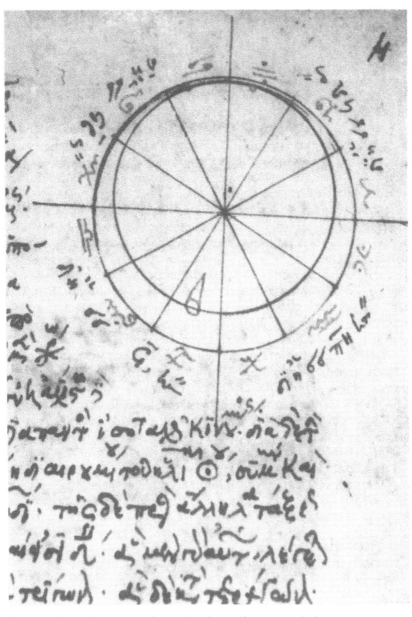

Diagram 2. An illustration of the solar theory, from an early fourteenth-century manuscript of Geminos's *Introduction to the Phenomena*. The diagram shows the zodiac (outer circle) divided into signs and labeled with their symbols, and the eccentric solar circle (inner circle). The cone-shaped symbol is a common sign for the Sun. The spring equinoctial point is at right. Around the outside, the copyist has labeled the four quadrants with the lengths of the four seasons. They are, starting at the upper right quadrant and proceeding counterclockwise, $94\frac{1}{2}$, $92\frac{1}{2}$, $90\frac{1}{8}$, and $88\frac{1}{8}$ days. The copyist has mistakenly reversed the lengths of fall and winter: the $90\frac{1}{8}$ and $88\frac{1}{8}$ *should be* $88\frac{1}{8}$ and $90\frac{1}{8}$, respectively. Finally, the copyist has mistakenly put the Sun's apogee at the summer solstice, rather than in the sign of Gemini, as Geminos's text calls for. The symbol for the Sun is attested in Antiquity, but the symbols for the zodiac signs are of medieval, and not ancient, origin. Bibliotheca Apostolica Vaticana, Vaticanus gr. 318, folio 4.

the 30th degree of Gemini; the shortest arc is the one that lies beneath the quadrant reaching from the 1st degree of Libra to the 30th degree of Sagittarius.

I 36 Thus it is reasonable that the Sun, moving at a constant speed on its own circle, passes through the unequal arcs in unequal times and traverses the longest in the longest time, and the shortest in the shortest time. 37 When it is completing the longest arc on its own circle, it is passing through the quadrant of the zodiac from spring equinox to summer solstice; and when it is moving along the shortest arc on its own circle, it is passing through the quadrant of the zodiac from fall equinox to winter solstice. 38 Therefore, since unequal arcs of the solar circle lie under equal arcs of the zodiac circle, it is necessary that the times between the solstices and the equinoxes be unequal, and that the longest be that from spring equinox to summer solstice; and the shortest, that from fall equinox to winter solstice. 39 Thus the Sun moves always at constant speed, but because of the eccentricity of the solar sphere, it passes through the quadrants of the zodiac in unequal times.

I 40 For the same reason, the Sun passes through the equal signs in unequal times. If we draw straight lines from the ends of the twelfth-parts to the center of the zodiac circle, as has been sketched,[25] the circle of signs will be divided into 12 equal parts; but, on account of the eccentricity, the circle of the Sun will be divided into 12 unequal parts, the longest arc falling beneath Gemini, and the shortest falling beneath Sagittarius. 41 For this reason the Sun passes through Gemini in the greatest time,[26] and through Sagittarius in the least time, although moving always at constant speed. Since the solar circle, on account of its eccentricity, is divided into unequal parts, it results that the times of the signs are also unequal.

[25] as has been sketched. See sec. 14 of the Introduction for a discussion of the diagrams in the manuscripts. The diagram printed here (diagram 2) is taken from ms. C.

[26] the Sun passes through Gemini in the greatest time. This is consistent with the solar theory of Hipparchos, who placed the Sun's apogee at 65½° of longitude, that is, 5½° within Gemini (Ptolemy, *Almagest* iii 4).

II. [Aspects of the Zodiacal Signs]

II 1 Four positions and arrangements of the 12 signs with respect to one another are distinguished. They are called in opposition, in trine, in quartile, and in syzygy, [which is called] by some *antiskian*.[1, a]

OPPOSITION

II 2 Signs that lie on the same diameter are said to be in opposition.[2] They are the following: Aries and Libra, Taurus and Scorpio, Gemini and Sagittarius, Cancer and Capricorn, Leo and Aquarius, Virgo and Pisces. 3 For these it happens that, when the other one is rising, the one diametrically opposite sets, and vice versa. This proposition [applies to] the twelfth-parts, and not to the constellations.[3] 4 When Aries is rising, Libra sets; when Taurus is rising, Scorpio sets. The same principle [extends] to the rest of the signs in opposition.

II 5 Signs in opposition are considered by the Chaldeans in connection with sympathies[4] in nativities. For those born diametrically oppo-

[1] in opposition, in trine, in quartile, and in syzygy, [which is called] by some *antiskian*. In the translation, *opposition*, *trine*, and *quartile* are borrowed from modern astrological parlance. Literal translations of the Greek for these three terms are: "on a diameter," "in a triangle," "in a square." Syzygy and *antiskian* are taken directly from the Greek. *In syzygy* means "yoked together" or "in a pair." *Antiskian* means "with shadows thrown opposite." (The mss. mistakenly have *antisyzygian* rather than *antiskia*. See textual note a).

"Signs in syzygy" and "*antiskian* signs" are synonymous terms. Two other synonymous expressions are also used for this relationship. Signs in syzygy may also be called "signs that look upon one another" or "equipotent signs," as by Ptolemy (*Tetrabiblos* i 15). The *antiskia* are given special emphasis by Firmicus Maternus (*Mathesis* ii 29). On zodiacal aspects, see Bouché-Leclercq 1899, 165–79.

[2] in opposition. The diagrams of chapter ii are suspect, since the text makes no mention of them. For a full discussion, see sec. 14 of the Introduction. The diagrams printed here for chapter ii are reproduced from ms. M.

[3] not to the constellations. As Geminos has mentioned (i 4), the zodiacal constellations are not all exactly on the ecliptic; nor are they all exactly 30° long. Thus the geometrical propositions involving astrological aspects apply only to the signs, or twelfth-parts, and not to the homonymous constellations.

[4] sympathies. The word was used in Greek music theory of strings that vibrate together by resonance. In Greek astrology, a sympathy is a deep, abiding friendship, to be distinguished from a passing acquaintanceship. See Ptolemy, *Tetrabiblos* iv 7 and Bouché-Leclercq 1899, 453n5, 454n1. In Stoic philosophy, *sympathy* indicates the affinity and interaction between things that arise from the unity of the cosmos. This sympathy is used by the Stoics to provide a rational basis for divination, including astrology. (On the Stoic notion of sympathy, see Reinhardt 1926; Sambursky 1959, 9, 41–42; Gould 1970, 100–101; Kidd 1988, vol. 2, 418, 423–24.)

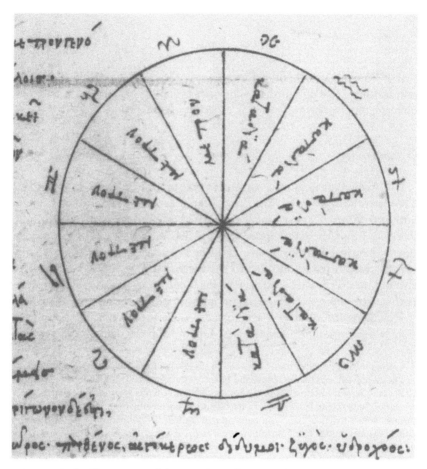

Diagram 3. An illustration of the astrological aspect of opposition, from a late-fourteenth-century manuscript of Geminos's *Introduction to the Phenomena*. In the pie-wedges for each pair of diametrically opposite signs is written *kata diametron*, "in opposition." Bibliotheca Nazionale Marciana (Venice), Marcianus gr. 323, folio 482.

sitely seem to be in sympathy with each other and, as some would say, to lie opposite each other. 6 Also, the positions of stars [i.e., planets] in diametrically opposite signs[5] at the same critical time act together to improve or to worsen nativities, according to the given powers of the stars.

[5] opposite signs. Geminos makes two points about oppositions: first (ii 5), two people "born diametrically oppositely" will have fellow-feeling; second (ii 6), two planets in opposite signs will act together to make a horoscope either better or worse, depending on the natures of the particular planets. Neither of Geminos's rules agrees with the conventions of later Greek astrology. Concerning the first rule: oppositions are most often connected with hostility, as in Ptolemy, *Tetrabiblos*, iv 7. Concerning the second rule: for Ptolemy

TRINE

II 7 In trine are: Aries, Leo, Sagittarius; Taurus, Virgo, Capricorn; Gemini, Libra, Aquarius; Cancer, Scorpio, Pisces—all 4 equilateral triangles. The side of the triangle subtends 4 signs, or 120 degrees.

II 8 The first triangle, the one beginning with Aries, is called northern[6]; for if the north wind begins to blow while the Moon is in one of the three signs [of the first triangle], the same condition lasts for many days. Thus the astronomers,[7] starting from this observation, forecast [these] northerly conditions. 9 For if a northerly condition arises when the Moon is in another sign, the north wind is quite likely to break off. But if the north wind blows in one of the assigned signs in the northern triangle, they foretell that the same condition will last for many days. 10 The next triangle, the one beginning with Taurus, is called southern; for, again, if the

---

(*Tetrabiblos* i 13), trine and sextile are harmonious aspects, while quartile and opposition are disharmonious. The consequences of this classification are far from simple (see, e.g., *Tetrabiblos* iii 4 and 10); however, it is clear that planets in opposition do not simply reinforce one another's effect.

Geminos's remarks about the significance of opposition appear to reflect, as he says, an older, Chaldean tradition. The oppositions and conjunctions of the Sun and Moon were, in Babylonian practice, propitious times for certain medical practices and ritual acts (Reiner 1995, 134–36). In the celestial divination of the neo-Assyrian period (8th and 7th centuries B.C.), before the rise of personal astrology, the oppositions of the Moon were ominous (Koch-Westenholz 1995, 103; Brown 2000, 88). Conjunctions of the planets with one another were ominous (Koch-Westenholz 1995, 140; Brown 2000, 91, 96), but little note seems to have been taken of oppositions of the planets to one another. Personal astrology developed late in the fifth century B.C.—the oldest extant Babylonian horoscope is for 410 B.C. But only one of the thirty-two extant horoscopes actually mentions an opposition, and this only of Saturn to the Sun (Rochberg 1998, 51–55). On the divergence of views about oppositions, see Bouché-Leclercq 1899, 166–69.

[6] The first triangle . . . is called northern. The association of winds with the four triangles is attested in Babylonian celestial divination. See Rochberg-Halton 1984, 121–26; Koch-Westenholz 1995, 167–68. Ptolemy makes a similar association and supports it elaborately (*Tetrabiblos* i 18). For example, the first triangle (Aries, Leo, Sagittarius) is assigned to the Sun and Jupiter, which govern it by day and by night, respectively. "This triangle is preeminently northern because of Jupiter's share in its government, since Jupiter is fecund and windy, similarly to the winds from the north" (Robbins, trans.). Ptolemy, like Geminos, makes the second triangle southern, but, Ptolemy reverses the assignments of the third and fourth triangles, making the third (Gemini, Libra, Aquarius) eastern and the fourth (Cancer, Scorpio, Pisces) western.

[7] astronomers. Geminos's term for an astronomer is *astrologos*. It would be incorrect to translate this as "astrologer." An *astrologos* is one who applies an orderly science to the stars. This is clear from Geminos's other uses of the same word, at viii 6, 50, 59; and xvii 23 and 49. Similarly, we have translated *astrologia* as "astronomy" at i 19; v 13, 14, 15, 17; and xvii 25. In Geminos's sense, *astrologia* includes the theory of the celestial sphere, lunisolar cycles, and planetary motion, as well as weather prediction from the heliacal risings of the stars. On the terms used by Greek astronomers to characterize their science ("astronomy," "astrology," "mathematics") see Tannery 1893, 1–55.

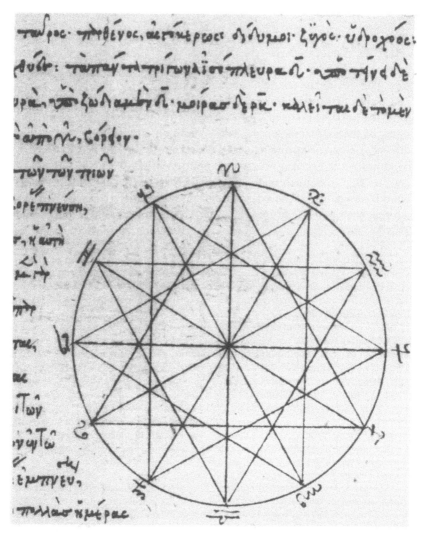

Diagram 4. Trine aspect. Marcianus gr. 323, folio 482.

south wind blows while the Moon is in one of the three signs, the same condition lasts for many days.   11 The next triangle, the one beginning with Gemini, is called western for a similar reason. The final triangle, the one beginning with Cancer, is called eastern for the same reason.

II 12 The triangles, too, are used in connection with sympathies in nativities. For those born in trine seem to be in sympathy with one another; and the situations of the stars [i.e., planets] in the same triangles seem to act together to improve or to worsen the nativities.

II 13 Sympathies arise in three ways[8]: in opposition, trine, and quartile: in any other separation there is no sympathy. 14 And yet it would be logical that there be sympathies from the signs lying closest together; for the outpouring or emanation proceeding from the peculiar power of each star should most of all color and mix with the neighboring signs. 15 Now, just as triangles and squares are inscribed in the circle, so too [are the] hexagon, octagon, and dodecagon. However, there is no sympathy corresponding to the inscription of these, but only for the aforementioned cases, since a certain natural sympathy exists [only] in such intervals.

QUARTILE

II 16 In quartile are Aries, Cancer, Libra, Capricorn; Taurus, Leo, Scorpio, Aquarius; Gemini, Virgo, Sagittarius, Pisces—which is all 3 [possible] squares. The side of the square subtends 3 signs or 90 degrees. 17 Called the first square is the one starting from Aries, in which the [beginnings of the] seasons occur[b]—spring, summer, fall, and winter. Called the second square is the one starting from Taurus, in which the seasons have a midpoint—of spring, of summer, of fall, of winter.[c] Called the third square is the one starting from Gemini, in which the seasons end their times.

II 18 The squares, too, are used, as has been said, for sympathies in the nativities. Moreover, the arrangement of the squares is used by some for another purpose. 19 For they supposed that, when one of the signs of the selfsame square is setting, the next sign culminates[9] <in> the hemisphere above the Earth, <the next rises, and the last culminates in the hemisphere beneath the Earth,>[d] as when Capricorn is setting, Aries culminates, Cancer rises, and Libra culminates beneath the Earth. The same logic applies to the remaining squares.[10]

[8] Sympathies arise in three ways: Geminos recognizes only three aspects as significant for sympathies in horoscopic astrology: opposition, trine, and quartile. Manilius (*Astronomica* ii 270–432) recognizes four: opposition, trine, quartile, and sextile. He considers two others, but rejects them as insignificant (neighboring signs and signs that are five signs apart). Ptolemy (*Tetrabiblos* i 13) recognizes the same four aspects as Manilius does. In explaining why these are the only significant aspects, Ptolemy appeals to a theory of harmonies based upon musical intervals. He treated this subject in more detail in his *Harmonics*. See Neugebauer 1975, 931–34 and Swerdlow 2004.

[9] culminates. This is the standard term in modern astronomy. A star "culminates" when it crosses the celestial meridian, i.e., reaches the highest point of its passage across the visible hemisphere. Geminos's verb is *mesouranein*, "to mid-heaven."

[10] the remaining squares. Thus, for any square of signs, if one of the signs is setting, the next sign of the four will be culminating, the next will be rising, and the last will be crossing the meridian beneath the ground. As Geminos goes on to explain, this proposition is only very roughly correct. However, "some people" applied it as if it were strictly accurate.

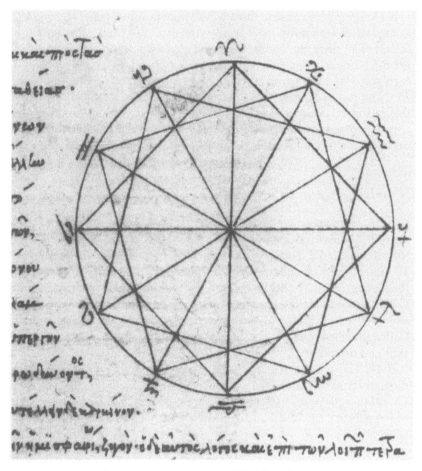

Diagram 5. Quartile aspect. Marcianus gr. 323, folio 483.

II 20 Such a proposition is on the whole in harmony with the phenomena when applied to the one square comprising the equinoxes and
solstices; but, in strict logic, it is in disaccord [with the phenomena].   21
For when the first degree of Capricorn is setting, the first degree of Aries
culminates, the first degree of Cancer rises, and the first degree of Libra
culminates below the Earth[11,e]; for the circle through the middles of the
signs is divided into 4 equal parts by the colure circles,[12] so that in this

---

[11] the first degree of Libra culminates below the Earth. This whole statement (ii 21) is
corrupt in all the manuscripts, for the astronomy is wrong. See textual note e, p. 258.

[12] the colure circles. There are two colure circles. One (the *solstitial colure*, in modern
astronomical language) passes through the celestial poles and through the solstitial points
of the ecliptic. The other (the *equinoctial colure*) passes through the poles and through the

situation the distances from the culminating [point] to the rising and setting [points] of the zodiac are equal: for each of them is 3 signs.

II 22 But in the remaining positions of this square, and of the remaining [squares], it does not happen that the zodiac circle is divided into 4 equal parts [by the horizon and the meridian]. Because of this, the distances from the culminating [point] to the rising and setting [points] are not always equal, the distances being taken on the zodiac circle. 23 In the case of a parallel circle, of course, the distances from the culminating [point] to the rising and setting [points] are always equal; and thus for the Sun, which moves each day on parallel circles, the course from rising to culmination is equal to the one from culmination to setting. 24 But if the distances are taken on the zodiac circle, the distance from the culminating to the rising [point] is unequal to the distance from the culminating to the setting because of the obliquity of the zodiac circle. There are times when, of the six signs that are always above the Earth, three and a half are cut off from the culminating to the rising [point], and two and a half [from the culminating] to the setting.[13]   25 Indeed, because

---

equinoctial points. The two colures divide the ecliptic into four equal 90° segments, which correspond to the four seasons of the year. These circles are said to be *kolouroi* ("dock-tailed") because their tails are cut off by the antarctic circle. That is, a portion of each colure lies permanently below the horizon and is never seen. The colures have little theoretical importance. They are, however, important structural members of an armillary sphere.

[13] two and a half . . . to the setting. The horizon and the meridian generally divide the ecliptic into two equal long segments and two equal short segments, the disparity between the long and the short segments depending upon both the geographical latitude and the orientation of the ecliptic. At temperate latitudes, the disparity is greatest approximately (though not exactly) when the solstitial points are on the horizon. If, therefore, we assume that the winter solstitial point is setting, we find that Geminos's example (the meridian divides the portion of the ecliptic that is above the horizon into parts that are 2½ and 3½ signs long) applies at latitude 28°.

The state of mathematics in Geminos's day would have permitted calculation of the arc between the setting and the culminating point of the ecliptic for a given latitude. But in an elementary work such as the *Introduction to the Phenomena*, it was sufficient to state the case in words. A teacher could demonstrate the proposition to students with an armillary sphere or a celestial globe, and Geminos probably used just such an instrument in composing his example.

The problem of calculating the arc of the ecliptic between the setting and the culminating point may also be solved with a *table of ascensions*. Composing such a table in turn requires the calculation of the time required for each sign of the zodiac to rise above the horizon. The oldest complete mathematical treatment of this problem is that of Hypsiklēs of Alexandria (second century B.C.), who showed how to find the rising times of the signs by means of approximate arithmetical (rather than exact trigonometrical) methods. (See sec. 12 of the Introduction.) The trigonometric version of the problem probably was solved later in the same century, and perhaps by Hipparchos, although the oldest extant trigonometric treatment is Ptolemy's in *Almagest* ii 7–8. Geminos knows only the arithmetical methods, which he uses for related problems elsewhere in the text (e.g., vi 38), and we have no evidence to suggest that he was a user of trigonometry.

of the differences of the *klimata*, it [the zodiac] is [sometimes] divided by the meridian circle into parts that are more unequal; and it can happen that, of the 180 degrees that are always above the horizon, 120 degrees are cut off from the culminating to the rising [point], and 60 [from the culminating] to the setting,[14] or vice versa. **26** As there is such variation in the division of the zodiac circle, the error is entirely plain. For when Aquarius is setting, Taurus does not culminate, but rather will be an entire sign from the culminating [point],[15] and sometimes even more; and Scorpio does not culminate beneath the Earth, but rather will be an entire sign from the meridian, and sometimes even more. And thus, in general, the theory of the squares [mentioned in ii 18–19] is quite mistaken.

SYZYGY

II 27 Said to be in syzygy are signs that rise from the same place and set in the same place; these are [signs] contained[f] by the same [two] parallel circles. **28** Now the ancients reckoned the syzygies in the following fashion. They postulated Cancer to have no syzygy with any other sign, but to rise farthest north and to set farthest north, and acquiesced, plausibly enough, in some such [reasoning] as this: **29** Since the summer solstices occur in Cancer, and the Sun is farthest north at the summer solstices, they therefore supposed Cancer to rise farthest north, and to set in the same way. **30** The same logic [applies] also to Capricorn; for they supposed this one to rise farthest south and to have a syzygy with no other sign. **31** For since the winter solstices occur in Capricorn, and the Sun is farthest south at the winter solstices, they therefore supposed that Capricorn rises farthest south, and that no other sign rises from the same place and sets <at the same place> as Capricorn. **32** They set out the remaining syzygies thus[16]: Leo with Gemini, Virgo

---

[14] 60° [from the culminating] to the setting. When the winter solstitial point is on the western horizon, 60° of the ecliptic will be cut off between the setting and the culminating point at a latitude of approximately 46°.

[15] an entire sign from the culminating [point]. Let the beginning of Aquarius be on the western horizon. The next sign in the square is Taurus, three signs farther east. According to the mistaken doctrine of the squares, which Geminos refutes, the beginning of Taurus should be on the meridian. However, if the latitude is chosen properly, it can happen that the beginning of Taurus will be an entire sign from the meridian (and so the beginning of Aries will be culminating). This occurs at latitude 55°.

[16] They set out the remaining syzygies thus. Geminos appears to be referring to an older division of the zodiac. According to this scheme, the equinoctial and solstitial points were placed at the *midpoints* of their signs (fig. 2.1). In this older way of doing things, Gemini and Leo are in syzygy, as are Taurus and Virgo, but Cancer and Capricorn are not in syzygy with any sign. Hipparchos (*Commentary* i 1.15) says that such a division of the zodiac was adopted by Eudoxos: "Eudoxos made the division thus, so that these points [the

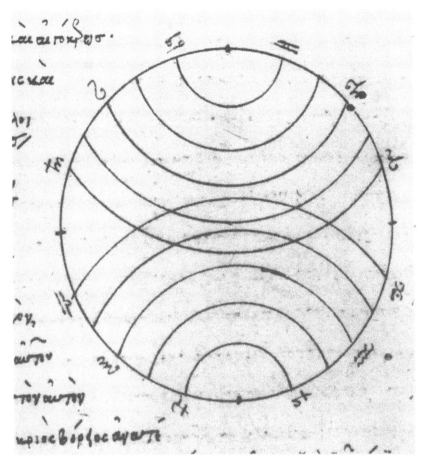

Diagram 6. Signs in syzygy. Marcianus gr. 323, folio 484v.

solstitial and equinoctial points] are the middles of Cancer and Capricorn in the one case, and of Aries and the Claws in the other." However, Jean Martin (1998, vol. 1, 124–31) argues that Hipparchos was mistaken in attributing this convention to Eudoxos. See Neugebauer 1975, 598–600 for other instances of this convention in Babylonian and Greek sources. In Geminos's day (as in our own), convention placed the equinoctial and solstitial points at the *beginnings* of the signs (fig. 2.2).

It is curious that Geminos (ii 33–43, vi 44–49) denounces the older convention so vociferously. He has already explained the conventional character of the zodiac signs and has given an example of a convention that differs from that of the Greeks—the "Chaldean" convention, according to which the equinoctial and solstitial points are 8° within their respective signs (i 9). It appears likely that Geminos was unaware of Eudoxos's use of the 15° norm. This was Manitius's view (1898, 255n6); but see also Neugebauer 1975, 583. In any case, Geminos was striving to correct a common error made by laymen of his own day, who were probably not followers of Eudoxos's outdated convention, but were merely careless in their use of the term "syzygy."

with Taurus, Libra with Aries, Scorpio with Pisces, Sagittarius with Aquarius.

II 33 But it happens that such an account is completely erroneous. For solstices do not occur in the whole of Cancer; rather, there is one certain point, perceivable through reason,[17, g] at which the Sun makes its turning; for the solstices take place in a moment's time. 34 The whole twelfth-part of Cancer is situated in the same way as Gemini, and each of them is equally far from the summer solstitial point. 35 For this reason, the lengths of the days are equal in Gemini and Cancer, and on the sundials[18] the curves described by the gnomons [when the Sun is] in

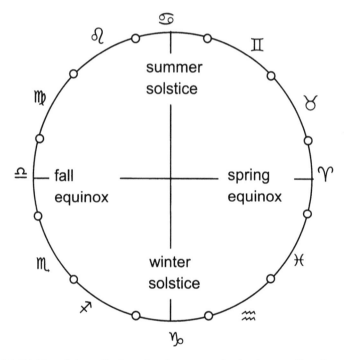

Fig. 2.1. Division of the zodiac into signs in a system that has been attributed to Eudoxos.

---

[17] one certain point, perceivable through reason. One cannot see the theoretical summer solstitial point in the sky, but one can, as we would say, "imagine it."

[18] on the sundials. Greek sundials were often marked, not only with hour lines, but also with solstitial and equinoctial curves (Gibbs 1976, 86). See the figures in sec. 8 of the Introduction. On summer solstice, the tip of the gnomon's shadow traces out the curve corresponding to the summer tropic. If the Sun were in the middle of Gemini or in the middle of Cancer, it would in either case be half a sign from the summer solstitial point. The curve

Cancer and in Gemini are equally distant from the summer tropic [curve]; 36 For the two twelfth-parts are situated equally with respect to the summer [solstitial] point. Hence they are also contained by the same [parallel] circles, because of which Gemini and Cancer rise from the same place and, similarly, set in the same place.

II 37 The same reasoning [applies] also to Capricorn. For it is not this [entire sign] that is southernmost, but rather one certain point, perceivable through reason, which is common to the end of Sagittarius and the beginning of Capricorn; on which account it [Capricorn] is situated equally to and has the same distance from the winter solstitial point as Sagittarius. 38 And thus the lengths of the days and of the nights are the same in Sagittarius and Capricorn, and the tip of the gnomon on the

---

traced out by the shadow tip would then lie a little inside the arc for summer solstice. In his explanation of the syzygies, Geminos draws on properties of sundials that would have been familiar to his readers.

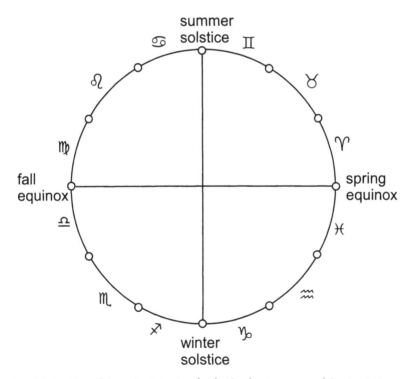

Fig. 2.2. Division of the zodiac into signs by the Greek astronomers of Geminos's time. This is the same as the modern convention.

sundials describes the same curves.   **39** The two twelfth-parts of Sagittarius and Capricorn are also contained by the same parallel circles; and, because of this, Sagittarius and Capricorn rise from the same place and set in the same place. Thus Sagittarius and Capricorn are in syzygy.

II 40 In the same way, it follows that the remaining syzygies are erroneous. Clearest [of all] is the error concerning the syzygy of Aries. For they declare Aries to be in syzygy with Libra, on the grounds that these signs rise from the same place and set in the same place.   **41** But Aries rises and sets in the north, for it lies north of the equatorial circle; while Libra rises and sets in the south, for it lies south of the equatorial circle. **42** How then can Aries be in syzygy with Libra? For they rise from different places and set in the same way. Thus it cannot be that these signs are contained by the same parallel circles.   **43** In the same way, the remaining syzygies also do not match up. For they [the ancients] have not perceived that they have taken circumstances relating to the first degrees for signs in syzygy and have extended them to the whole signs: it would have been better by far to keep to circumstances that apply to the whole twelfth-parts in the description and the propositions.

II 44 There are, then, in truth, 6 syzygies: Gemini with Cancer, Taurus with Leo, Aries with Virgo, Pisces with Libra, Aquarius with Scorpio, Capricorn with Sagittarius. And so these rise from the same place and set in the same place; they are contained by the same parallel circles; and they are situated equally with respect to the solstitial points.   **45** And thus in these the lengths of the days and of the nights are equal, and the tips of the gnomons on the sundials describe the same curves.

III. On the Constellations

III 1 The constellations[1] are divided into three groups. Some lie on the zodiac circle, some are said to be northern, and some are called southern.

THE ZODIACAL CONSTELLATIONS

III 2 The ones lying on the zodiac circle are the 12 signs, whose names we have already mentioned.

And in the 12 signs are certain stars deemed worthy of proper names because they are indicators[2] for them. 3 Thus the stars lying on the back of Taurus and 6 in number are called the Pleiades[3]; while the stars lying on the head of Taurus, 5 in number, are called the Hyades. 4 The leading star of the feet of Gemini is called Propus.[4] Those in Cancer resembling a nebulous mass are called the Manger; and the two stars lying near it are called the Asses. 5 The bright star lying on the heart of Leo, named for the place it occupies, is called Heart of the Lion; but it is called by some Basiliskos,[5] because those who are born around this place seem to have a kingly geniture. 6 The bright star lying on the tip of the

---

[1] the Constellations. See tables 3.1–3.6 at the end of this chapter for the correspondences between Geminos's star and constellation names and modern nomenclature. For the Babylonian constellations, see: Gössmann 1950; Hunger and Pingree 1999, 271–77; Van der Waerden 1974. Allen 1899 is still useful as an entrée to Greek and Roman sources (but not reliable for Arabic star names—for these, see Kunitzsch and Smart 1986). For Eratosthenēs' *Catasterisms*, see Charvet 1998. Also useful are Le Boeuffle 1989; Bakhouche 1996; Boll and Gundel 1924–37.

[2] indicators. *Episēmasia* could also be translated "sign," except that this word has already been used for *zōdion*, a sign of the zodiac. *Episēmasia* is also applied to stars whose heliacal risings and settings serve as signs of the seasons and of the weather, as in the *parapēgma* that follows Geminos's *Introduction to the Phenomena*. See Lehoux 2000, 130–45.

[3] Pleiades. Aratos (*Phenomena* 262–63) names all seven, though he says only six are visible.

[4] Propus. It is so called because it is the "forward foot"—the leading (or western) foot of the leading Twin.

[5] Basiliskos = "little king." Aratos does not mention the star, and neither Eratosthenēs nor Hipparchos uses this name, which must have been fairly new among the Greeks in Geminos's day. The Babylonians called Regulus LUGAL, "king" (Gössmann 1950, 89; Hunger and Pingree 1999, 273; van der Waerden 1974, 73). Ptolemy usually calls it "the star on the heart of Leo," but in the star catalog of the *Almagest* he adds that it is "called Basiliskos."

left hand of Virgo is called the Wheat Ear[6]; and the small star lying beside the right wing of Virgo is named Harbinger of the Vintage. The 4 stars lying on the tip of the right hand of Aquarius are called the Pitcher. 7 The stars lying in a row starting from the tail parts of Pisces are called the Cords.[7] There are 9 stars in the southern cord and 5 stars in the northern cord; the bright star lying at the end of the cord is called the Knot.

THE NORTHERN CONSTELLATIONS

III 8 The northern [constellations] are those that lie north of the circle of the signs. They are the following: the Great Bear, the Little [Bear], Draco between the Bears, the Bear Keeper, the Crown, the Kneeling Man, Ophiuchus, the Serpent, the Lyre, the Bird, the Arrow, the Eagle, the Dolphin, the Forepart of a Horse according to Hipparchos, the Horse, Cepheus, Cassiopeia, Andromeda, Perseus, the Charioteer, the Triangle, and the constellation later established by Kallimachos, the Lock of Berenikē.

III 9 And, again, in these [constellations], certain stars have proper names because they are important indicators for them. The notable star lying between the legs of the Bear Keeper is named Arcturus. 10 The bright star lying near the Lyre, named in the same way as the whole constellation, is called Lyra.[8] The middle one of the three stars in the Eagle,

---

[6] Wheat Ear. Virgo holds a wheat ear in her left hand. For the Babylonians, this constellation was usually the "furrow" (Hunger and Pingree 1989, 138; Hunger and Pingree 1999; Koch-Westenholz 1995, 207). But on one famous Babylonian astrological tablet, the constellation was pictured as a woman holding a sheaf of wheat. See van der Waerden 1974, 81, Plate 11c.

[7] the Cords. This may also be translated as "fishing lines." The two fish of Pisces are tied together by cords that come from their tails. The two cords form a V-shape and are joined in a knot at α Psc. In Table 3.4 we list the identifications according to Ptolemy's description in the *Almagest* star catalog (Toomer 1984, 379–80). However, Ptolemy puts ten stars in the southern cord and four in the northern, which differs from Geminos's count. Geminos is not clear about whether the Knot (α Psc) is included in his count of five for the northern cord. The Babylonians also had the cords (or lines or ribbons), but instead of two fishes, they had a fish and a swallow (Hunger and Pingree 1999, 148; Hunger and Pingree 1989, 138).

[8] Lyra. Aratos, while mentioning the constellation Lyra in several places, never singles out its brightest star. Eratosthenēs mentions the "white and bright star" but gives it no name. Hipparchos (*Commentary on the Phenomena of Eudoxos and Aratos* i 6.15) refers to the star as "the bright one in the Lyre." Ptolemy (*Almagest* vii 5), like Geminos, calls the star Lyra. The association of the constellation name with the star may therefore be late. The constellation is small and contains only one bright star—circumstances that undoubtedly facilitated this transformation.

named in the same way as the whole[a] constellation, is called Aëtos [Eagle].[9] 11 The stars lying on the tip of the left hand of Perseus are called the Gorgon's Head[10]; the small, closely packed stars at the tip of the right hand of Perseus are set in the Sickle.[11] 12 The bright star lying on the left shoulder of the Charioteer is called the Goat; and the two small stars lying at the tip of his hand[b] are called the Kids.

THE SOUTHERN CONSTELLATIONS

III 13 The southern [constellations] are those that lie south of the circle of the signs. They are the following: Orion and Procyon, the Dog, the Hare, Argo, Hydra, Crater, the Raven, Centaurus, the Wild Animal that Centaurus is holding, and the Thyrsus-lance that Centaurus is holding according to Hipparchos,[c] the Censer, the Southern Fish, the Sea Monster, the Water coming from Aquarius, the River coming from Orion, the Southern Crown, called by some the Canopy, [and] the Caduceus according to Hipparchos.[d]

III 14 And again, certain stars in these [constellations] deserve to have their own names. The bright star in Procyon is called Procyon.[12] The bright star in the mouth of the Dog, which seems to cause the intensification of the heat, is called the Dog,[13] in the same way as the whole constellation. 15 The bright star lying on the tip of the Rudder of Argo is named Canopus.[14] In Rhodes it is hard to see, or is seen

[9] Aëtos [Eagle]. Ptolemy (*Almagest* vii 5), like Geminos, gives the star the same name as the constellation. Aratos and Hipparchos do not mention this star. Eratosthenēs mentions it but gives it no name.

[10] the Gorgon's Head. This is the head of Medusa, carried in Perseus's left hand. The story is told by Eratosthenēs (Charvet 1998, 107–108).

[11] the Sickle. Perseus used it to cut off Medusa's head.

[12] Procyon. It is called Procyon ("before the Dog") because it makes its heliacal rising in advance of the Dog Star. Aratos (*Phenomena* 450, 595, 690) uses Procyon only for the entire constellation. It is a small constellation, however, and contains only one prominent star. Thus the transfer of the constellation's name to the star parallels the cases of the Lyre and the Eagle. Eratosthenēs and Ptolemy call the star Procyon. The constellation Procyon was also considered a dog: Eratosthenēs makes it the dog of Orion (Charvet, 1998, 189).

[13] the Dog (*Kuōn*). Homer calls it the "dog of Orion" (*Iliad* xxii 29). But the name Sirius (*Seirios*) is also ancient (Hesiod, *Works and Days* 417, 587, 609).

[14] Canopus (or Canobus). The bright star on the steering oar of Argo. Canopus was invisible in Greece and, as Geminos says (iii 15), was barely visible from the latitude of Rhodes. The name Canopus is used by Hipparchos, but not by Aratos or Eratosthenēs. Hipparchos (*Commentary* i 11.6) quotes Eudoxos as placing this star upon the antarctic circle, and makes it clear that Eudoxos referred to it simply as "the star visible in Egypt." Strabo (*Geography* i 1.6) mentions Canobus and the Lock of Berenikē as recently invented star names. Ptolemy's use of the name in the star catalog of the *Almagest* helped secure the tradition.

completely [only] from high places; but in Alexandria it is completely evident, for it appears approximately one fourth part of a sign above the horizon.[15]

GEMINOS'S STARS AND CONSTELLATIONS: A SUPPLEMENT TO CHAPTER III

Many readers will want to have a quick way of matching Geminos's names for the constellations and stars (most of which were standard for his day) with their modern counterparts. To avoid the need for a large number of footnotes to Geminos's chapter iii, we have grouped this information into the six tables below. Tables 3.1, 3.2, and 3.3 deal with constellation names, and are followed by remarks on a few constellations with respect to which the ancient Greek writers expressed important differences. Tables 3.4, 3.5, and 3.6 deal with the stars that Geminos mentions by name.

TABLE 3.1
Geminos's Constellations: The Zodiac

Translation	Geminos's Greek
Aries	Krios
Taurus	Tauros
Gemini	Didumoi
Cancer	Karkinos
Leo	Leōn
Virgo	Parthenos
Libra	Zugos
Scorpio	Skorpios
Sagittarius	Toxotēs
Capricorn	Aigokerōs
Aquarius	Hydrochoös
Pisces	Ichthues

[15] one-fourth part of a sign above the horizon. That is, at Alexandria, when Canopus culminates, it is one forty-eighth of a circle (7½°) above the horizon. In the first century B.C., Poseidōnios used this datum, and the fact that the star was on the horizon when observed from Rhodes, to calculate the circumference of the Earth. Poseidōnios assumed that Rhodes and Alexandria are on the same meridian, and that the distance between them is 5,000 stades. Thus he obtained 5,000 × 48 = 240,000 stades for the circumference of the Earth. See Kleomēdēs, Meteōra i 7.7–48.

TABLE 3.2
Geminos's Northern Constellations

Translation	Geminos's Greek*	Modern Name
Great Bear	*Megalē Arktos*	Ursa Major
Little Bear	*Mikra Arktos*	Ursa Minor
Draco	*Drakōn*	Draco
Bear Keeper	*Arktophulax*	Boötes
Crown	*Stephanos*	Corona Borealis
Kneeling Man	*Engonasin*	Hercules
Ophiuchus	*Ophiouchos*	Ophiuchus
Serpent	*Ophis*	Serpens
Lyre	*Lyra*	Lyra
Bird	*Ornis*	Cygnus
Arrow	*Oïstos*	Sagitta
Eagle	*Aëtos*	Aquila
Dolphin	*Delphis*	Delphinus
Forepart of a Horse	*Protomē Hippou*	Equuleus
Horse	*Hippos*	Pegasus
Cepheus	*Kēpheus*	Cepheus
Cassiopeia	*Kassiepeia*	Cassiopeia
Andromeda	*Andromeda*	Andromeda
Perseus	*Perseus*	Perseus
Charioteer	*Hēniochos*	Auriga
Triangle	*Deltōton*	Triangulum
Lock of Berenikē	*Berenikēs Plokamos*	Coma Berenices

*Geminos's list of northern constellations is almost identical with the list given by Aratos, and probably standardized a century earlier by Eudoxos. The order is different, however, for Aratos mixes the zodiacal constellations among the dnorthern ones. Only the following points should be noted.

Pleiades. Geminos, like most later writers, puts the Pleiades in Taurus. Aratos (*Phenomena* 254) appears to give the Pleiades a separate status and does not mention them in connection with Taurus, thus following Homer and Hesiod, who always mention the Pleiades in their own right. Among the Greeks the Pleiades—so singular in appearance—were probably older than Taurus. Eratosthenēs gives them a separate paragraph, but says plainly that they are on the back of Taurus, where they have been ever since.

Bear Keeper. This is the only name that Geminos uses. But Aratos knew the constellation by two names, for he speaks of the "Bear Keeper, whom men also call Boötes" (*Phenomena* 91). He is a Bear Keeper because he follows close behind the Great Bear. But the Bears were also visualized as wagons or the carriages of plows (*hamaxai*) (Homer, *Iliad* xviii 487; Aratos, *Phenomena* 27). In this case the Bear Keeper is better thought of as *boötēs*, a "plowman" (Aratos, *Phenomena* 92). Homer uses only Boötes (*Odyssey* v 272), so this may be the older name.

Kneeling Man. Geminos follows Aratos (*Phenomena* 66 and 669), who did not identify the constellation with any mythological figure. Eratosthenēs identifies the Kneeling Man with Heraklēs (Charvet 1998, 41). But most Greek astronomical writers (including Hipparchos and Ptolemy) do not.

Forepart of a Horse. A *protomē* could be the "forepart" of anything, the "face" of an animal, or a "bust" in statuary. Ptolemy (*Almagest* vii 5) lists the "constellation of the forepart [or bust] of a horse," which includes four stars, all faint. This is our modern

Equuleus, the colt. Ptolemy and Geminos are virtually alone among ancient writers in mentioning it. Geminos tells us that the name was devised by Hipparchos, which seems possible, since it is not mentioned by Aratos or Eratosthenēs. It appears, then, that Ptolemy followed Hipparchos in recognizing the new constellation. Hipparchos, however, does not mention the *Protomē* in his *Commentary on the Phenomena of Eudoxos and Aratos*, nor does it appear in any of the several versions of the list of constellations attributed to Hipparchos.[16] Thus Geminos's attribution of the new constellation to Hipparchos cannot be confirmed.

Lock of Berenikē. This constellation is due to the Alexandrian astronomer Konōn, and the story goes like this. Berenikē was the cousin and wife of Ptolemaios III Euergetēs, the third of the Macedonian kings of Egypt (ruled 247–22 B.C.). When her husband departed for war in Syria she vowed to make an offering to the gods of a lock of her hair if he should return safely. When her husband did return safely to Alexandria, she cut off a lock and placed it in a temple, from which it mysteriously disappeared. Konōn, the court astronomer, consoled her by designating a new constellation the Lock of Berenikē. The story has come down to us because it inspired Kallimachos of Kyrēnē, who also worked at Alexandria at this time, to compose a poem on the subject, which survives in a fragmentary state.[17] (A Latin version made two centuries later by Catullus is intact.[18]) The new constellation had an unsettled history. Although it appears in the *Catasterisms* of Eratosthenēs (in the paragraph on Leo), it was not included by Ptolemy among his 48 constellations. Ptolemy does, however, mention the "lock" in his description of the unconstellated stars around Leo.

TABLE 3.3
Geminos's Southern Constellations

Translation	Geminos's Greek*	Modern Name
Orion	Ōriōn	Orion
Procyon	Prokuōn	Canis Minor
Dog	Kuōn	Canis Major
Hare	Lagōos	Lepus
Argo	Argō	Carina, Puppis, and Vela
Hydra	Hydros	Hydra
Crater	Kratēr	Crater
Raven	Korax	Corvus
Centaurus	Kentauros	Centaurus
Wild Animal	Thērion	Lupus
Thyrsus-lance	Thursologchos	part of Centaurus
Censer	Thumiatērion	Ara
Southern Fish	Notios Ichthus	Piscis Austrinus
Sea Monster	Kētos	Cetus
Water	Hydōr	part of Aquarius
River	Potamos	Eridanus
Southern Crown	Notios Stephanos	Corona Australiis
Caduceus	Kērukeion	—

*Again, Geminos's list is fairly standard: Aratos mentions all of these constellations except for the Thyrsus-lance and the Caduceus. For these and a few others, the following comments may be helpful.

[16] For a brief description of these manuscripts, see Neugebauer 1975, 285. For the texts, see Maass 1898, 136–39; Boll 1901; Weinstock 1951, 189–90.

[17] Kallimachos, *Aetia* 110.

[18] Catullus, poem 66.

Wild Animal and Thyrsus-Lance. A thyrsus, used in Bacchic rites, is a branch wreathed in ivy and vine-leaves with a pine cone at the top. Aratos (*Phenomena* 440–42; see also 662–63) says that Centaurus stretches his right hand toward the altar and holds in it the Wild Animal, but does not mention the Thyrsus. Eratosthenēs says that Centaurus (identified with Chiron) holds a wild animal, but that "some people say it is a wine-skin (*askos*) . . . , from which he pours an offering to the gods upon the Altar. He holds it in his right hand and the thyrsus in his left." Although Aratos mentions the Wild Animal only in conjunction with the Centaur, he seems to recognize it as a separate constellation. This separate status is clearer in Eratosthenēs. For although Eratosthenēs mentions the Wild Animal at the close of his paragraph on the Centaur, he gives a separate total for the number of stars that it contains: there are twenty-four stars in the Centaur and eight in the Wild Animal. In contrast, the Thyrsus is treated merely as a part of Centaurus.

It seems from Geminos's remark that Hipparchos made the Thyrsus, or Thyrsus-lance, into a separate constellation. If so, he was not followed by Ptolemy, who kept to the older tradition: In Almagest viii 1, the Thyrsus is part of Centaurus, but the Wild Animal has the status of a separate constellation, which is the way things have remained. Unfortunately, Hipparchos's designation of the Thyrsus-lance as a new constellation cannot be confirmed. The list of constellation names attributed to Hipparchos includes "the Wild Animal that Centaurus holds in his right hand," but makes no mention of the Thyrsus. In Hipparchos's Commentary (ii 5.14; iii 5.6) the Thyrsus is plainly considered to be a part of Centaurus.

Water. This is the water that is poured out by Aquarius, the Water-Pourer. Geminos treats the Water as a separate constellation, "the Water from the Water-Pourer." Separate or semi-separate status for the Water is attested in Aratos (*Phenomena* 395–99). Eratosthenēs mentions "the Pouring of the Water" in the course of his discussion of Aquarius, but he gives a separate tally for the number of stars in the Water, which demonstrates its semi-separate status in his arrangement. Hipparchos (*Commentary* ii 6.3) plainly makes it a part of Aquarius. Ptolemy's star catalog (*Almagest* viii 1) fixed the tradition by placing the water among the stars of Aquarius. The Water is an asterism of considerable size: in Ptolemy's catalog it contains twenty of Aquarius's forty-two stars and stretches over 22° in latitude.

River. Although Geminos merely describes this constellation as a "river," the identification with Eridanus already occurs in Aratos (*Phenomena* 360).

Southern Crown. This constellation appears to have been established rather late, and its tradition was long unsettled. Aratos (*Phenomena* 401) refers to it merely as a few stars "turned in a circle" (*dinōtoi kuklōi*) beneath the feet of Sagittarius. Eratosthenēs does not mention it, nor does Hipparchos in his *Commentary*. But it does figure in the list of constellation names attributed to Hipparchos, where it is called the "Crown beneath Sagittarius."[19] Geminos says that some called it a Canopy (*Ouraniskos*). For Ptolemy it is *Stephanos Notios*, the Southern Crown (*Almagest* viii 1).

Caduceus. A *kērukeion* was a herald's wand, of the sort sometimes carried by Hermes, often with two serpents wound around it. It is not known for certain where this constellation might have been located, nor can Geminos's attribution of it to Hipparchos be confirmed, for Hipparchos breathes not a word of it in the *Commentary*. Some scholars read this passage as if *kērukeion* were an alternative name for the Southern Crown (e.g., Tannery 1893, 271; Allen 1899, 172), but considering the absence of any linking particle, the final phrase seems rather to indicate a separate constellation. Boll (1899) made a good case for placing the *kērukeion* in Orion's left hand, citing texts in which Orion carries a sword and a *kērukeion*, rather than the usual staff (or club) and pelt (e.g., Vettius Valens, *Anthologies*, i 2; Bara 1989, 54).

---

[19] Maass 1898, 138; Weinstock 1951, 190.

TABLE 3.4
Geminos's Stars: Zodiacal Constellations

Translation	Geminos's Greek	Modern Identification	Early Mentions*
Pleiades	*Pleiades*	Pleiades	Il. xviii 486. Od. v 272. WD 383, 572. A 262
Hyades	*Hyades*	Hyades, α, θ, γ, δ, ε Tau	Il xviii 486
Propus	*Propous*	η Gem	H iii 2.10, 4.12
Manger	*Phatnē*	The cluster M44 Beehive	A 892
Asses	*Onoi*	γ and δ Cnc	A 898
Heart of Lion, or Basiliskos	*Kardia Leontos, Basiliskos*	Regulus, α Leo	H i 10.11; ii 5.7
Wheat Ear	*Stachus*	Spica, α Vir	A 97
Harbinger of the Vintage	*Protrugētēr*	Vindemiatrix, ε Vir	A 138
Pitcher	*Kalpis*	γ, ζ, η and (?) π Aqr	H ii 6.5; iii 1.9, 3.11, 4.8
Cords	*Linoi*	Northern: ο, π, η, ρ and (?) α Psc Southern: 41, 51, δ, ε, ζ, 80, 89, μ, ν and ξ Psc	A 243 H ii 6.1; iii 3.9
Knot	*Sundesmos*	α Psc	A 245

* Il. = Homer, *Iliad.* Od. = Homer, *Odyssey.* WD = Hesiod, *Works and Days.* A = Aratos, *Phenomena.* H = Hipparchos, *Commentary on the Phenomena of Eudoxos and Aratos.*

TABLE 3.5
Geminos's Stars: Northern Constellations

Translation	Geminos's Greek	Modern Identification	Early Mentions
Arcturus	*Arktouros*	Arcturus, α Boo	WD 566, 610
Lyra	*Lyra*	Vega, α Lyr	See comment 8 to ch. 3
Eagle	*Aetos*	Altair, α Aql	See comment 9 to ch. 3
Gorgon's Head	*Gorgonion*	β, ω, ρ, π Per	H ii 3.27, 6.15; iii 5.19
Sickle	*Harpē*	galactic clusters 869, 884	H ii 5.15, 6.1; iii 1.1, 4.8
Goat	*Aix*	Capella, α Aur	A 157
Kids	*Eriphoi*	ζ and η Aur	A 158

TABLE 3.6
Geminos's Stars: Southern Constellations

Translation	Geminos's Greek	Modern Identification	Early Mentions
Procyon	*Prokuōn*	Procyon, α CMi	See comment 12 to ch. 3
Dog	*Kuōn*	Sirius, α CMa	Il. xxii 29
Canopus	*Kanōbos*	Canopus, α Car	H i 11.7; iii 2.14

IV. On the Axis and the Poles

IV 1 The cosmos being of spherical form, the diameter of the cosmos around which the cosmos turns is called the axis.[1] The extremities of the axis are called poles of the cosmos.[2] **2** Of the poles, one is called northern, the other southern. The northern [pole] is the one that is always visible for our region; the southern [pole] is the one that is always

[1] axis. *axōn*, originally an "axle," as of a cart. Geminos seems to be saying that, since the cosmos is round, it is natural to speak of an axle.

[2] poles of the cosmos. The axis of the cosmos passes through the center of the Earth. If the axis is extended, as in fig. 4.1, it pierces the celestial sphere at the north and south celestial poles (the "poles of the cosmos"), which lie directly above the corresponding poles of the Earth. The horizon plane is tangent to the surface of the Earth. In fig. 4.1, for clarity, we show an Earth of finite size. But the Earth is a mere point with respect to the vast sphere of the cosmos. So, for strict astronomy, one should imagine shrinking the Earth to a point, so that the horizon plane rigorously bisects the sphere of the cosmos. In subsequent figures, we will always show the horizon as passing through the center of the cosmos.

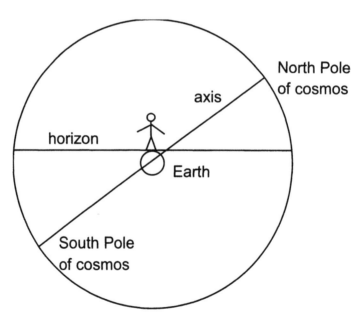

Fig. 4.1. The Earth surrounded by the sphere of the cosmos. The axis passes through the poles of the cosmos and is (for an observer in Greece) inclined to the horizon approximately as shown.

invisible for our horizon. **3** There are, however, some places on the Earth where the pole that is always visible for us happens to be invisible for them, while the one that is invisible for us is visible for them. **4** And again there is a certain place on the Earth where the two poles lie, equally, on the horizon.[3]

[3] on the horizon. Geminos describes three possible orientations of the celestial sphere with respect to the local horizon. In the northern hemisphere ("our region"), the north celestial pole is always above the horizon, as in fig. 4.2. The angle marked θ is the altitude of the pole. The angle φ between the equator and the local zenith is, by definition, the geographical latitude of the place of observation. It is a simple matter to show from the diagram that $\theta = \varphi$. That is, the altitude of the pole is equal to the latitude of the place of observation. So, in Athens (latitude 38° N), the north celestial pole is seen 38° above the horizon. It follows that the south celestial pole, 38° below the southern horizon, is always invisible in Athens.

In the southern hemisphere of the Earth, the south celestial pole is above the southern horizon, as in fig. 4.3. The altitude θ of the south celestial pole is equal to the south latitude of the place of observation.

At the Earth's equator (latitude zero), both celestial poles lie on the horizon, as in fig. 4.4.

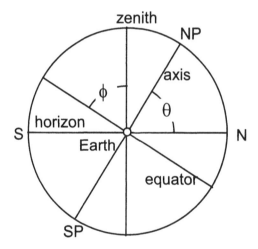

Fig. 4.2. The latitude φ of the observer is equal to the altitude θ of the celestial pole. *N* and *S* mark the north and south points of the horizon. *NP* and *SP* are the north and south celestial poles.

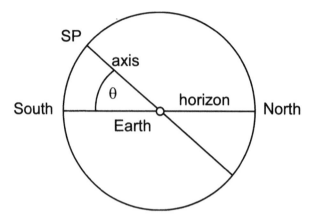

Fig. 4.3. For an observer south of the equator, the north celestial pole is below the horizon. But the altitude θ of the south celestial pole *SP* is equal to the south latitude of the observer.

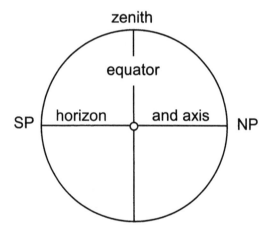

Fig. 4.4. For an observer at the equator, the two celestial poles *SP* and *NP* lie on the horizon.

V. The Circles on the Sphere

V 1 Of the circles on the sphere, some are parallel, some are oblique, and some [pass] through the poles.

The parallel [circles][1] are those that have the same poles as the cosmos. There are 5 parallel circles: arctic [circle], summer tropic, equator, winter tropic, and antarctic [circle].

V 2 The arctic circle[2] is the largest of the always-visible circles, [the circle] touching the horizon at one point and situated wholly above the

---

[1] the parallel circles. See fig. 5.1.

[2] arctic circle. For the Greeks, the arctic circle is a circle on the celestial sphere, with its center at the celestial pole, and its size chosen so that the circle grazes the horizon at the north point. (See fig. 5.1.) The stars within the arctic circle are circumpolar, i.e., they never rise or set but remain above the horizon 24 hours a day. The arctic circle is therefore the dividing line between the stars that have risings and settings and those that do not. The size

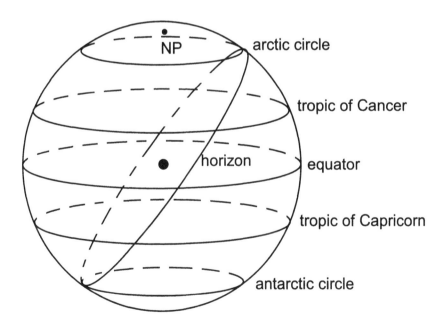

Fig. 5.1. The celestial sphere for an observer at about 30° latitude.

Earth. The stars lying within it neither rise nor set, but are seen through the whole night turning around the pole.   **3** In our *oikoumenē*,[3] this circle is traced out by the forefoot of the Great Bear.[4]

---

of the local arctic circle depends on the latitude of the observer. Figs. 5.2 and 5.3 show the arctic circle for latitudes 40° N and 20° N. Because the size of the arctic circle in the Greek sense varies with the location of the observer, we shall sometimes refer to this as the *local arctic circle*.

The *modern* arctic circle is fixed in size: it is a small circle, centered on the pole, whose radius is approximately 24°.

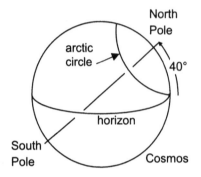

Fig. 5.2. The arctic circle for an observer at 40° latitude.

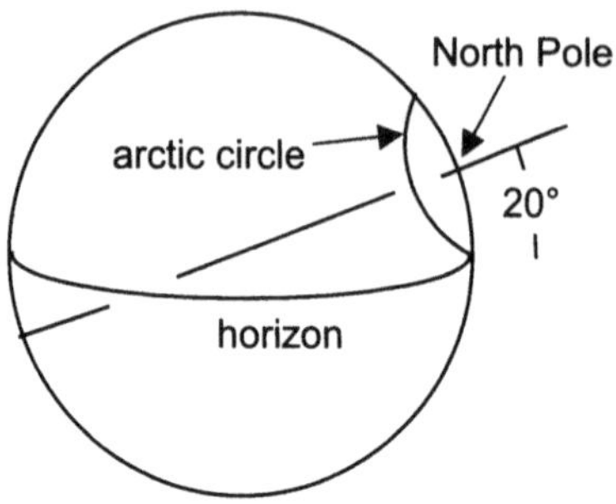

Fig. 5.3. The arctic circle for an observer at 20° latitude.

[3] In our *oikoumenē*. The *oikoumenē* is the "inhabited Earth." This term is used by Greek writers in two senses. It may designate the Greeks' portion of the Earth, as opposed to barbarian lands. But it may also mean the whole known inhabited world, namely Asia, Europe, and Africa (Ptolemy, *Geography* i 7; Berggren and Jones 2000, 64 and 20–22. See also Dilke 1985).

Oikoumenē is a participle meaning "inhabited," but the tacit noun "Earth" may usually be supplied, as pointed out in LSJ. Geminos first uses the term at v 26, 29, 30, 43, and 45, an example being v 29: "the sizes of some of the . . . [five] parallel circles remain the same for the whole *inhabited Earth*." This would include the parts south of the equator if, as some writers allowed as being possible, they should happen to be inhabited.

The first appearance of the longer phrase, *tēi kath' hēmas oikoumenēi* ("the Earth inhabited by us"), is at v 39 and is a fair example of the more restricted meaning of the Greek portion of the Earth, as opposed to barbarian lands, for in this passage Geminos excludes the parts inhabited by, say, Scythians or Egyptians. The same phrase appears in xv 4, where we read that the northern temperate zone is inhabited by those "in the Earth inhabited by us." (We take the following passage on dimensions to refer to the northern temperate zone, not just that part of it "in the Earth inhabited by us.") But the phrase is also found in xvi 2 and 3, where the meaning is clearly the northern inhabited Earth. The passage xvi 5 is the one place in the work where the adjective appears in the neuter (in the phrase "the inhabited part of the Earth"). From the value 2:1 given there for the ratio of the length to the breadth of this part it seems that, in light of the same ratio in xv 4, it means the northern inhabited Earth. Hence, in general, "the inhabited Earth" (with or

V 4 The summer tropic circle is the most northern of the circles described by the Sun during the rotation of the cosmos. When the Sun is on this circle, it produces the summer solstice, on which occurs the longest of all the days of the year, and the shortest night. 5 After the summer solstice, however, the Sun is no longer seen going toward the north, but it turns toward the other parts of the cosmos, which is why [this circle] is called "tropic."[5]

V 6 The equator circle is the largest of the 5 parallel circles. It is bisected by the horizon so that a semicircle is situated above the Earth, and a semicircle below the horizon. When the Sun is on this circle, it produces the equinoxes, both the vernal and the autumnal.

V 7 The winter tropic circle is the southernmost of the circles described by the Sun during the rotation of the cosmos. When the Sun is on this circle it produces the winter solstice, on which occurs the longest of all the nights of the year, and the shortest day. 8 After the winter solstice, however, the Sun is no longer seen going toward the south, but it turns toward the other parts of the cosmos, for which reason this [circle] too is called "tropic."

V 9 The antarctic circle is equal [in size], and parallel to, the arctic circle, being tangent to the horizon at one point and situated wholly beneath the Earth.[6] The stars lying within it are always invisible to us.

V 10 Of the 5 aforementioned circles the equator is the largest, the tropics are next in size, and—for our region—the arctic circles are the smallest.[7] 11 One must think of these circles as without thickness, perceivable [only] by reason, and delineated by the positions of the stars, by observations made with the *dioptra*,[8] and by our own power of thought.

---

without the explicit qualifier "the whole") means "the whole inhabited Earth." However, context (as in xv 4) may restrict that to the northern inhabited Earth. And, by "the Earth inhabited by us," Geminos means the part of the Earth we live in, either "we Greeks" or in the wider sense of "we northerners."

[4] the Great Bear. For an observer in ancient Greece, the forefoot of the Great Bear (Ursa Major) did not quite set, but grazed the horizon in the north. The forefoot of the Bear was therefore situated on the local arctic circle; and, in the course of the night, it traced out this circle in the sky. Our *arctic* is from *arktos*, "bear."

[5] which is why [this circle] is called "tropic." *tropikos* is the adjectival form of *tropē*, a "turn," "turning."

[6] beneath the earth. The antarctic circle—lying tangent to the horizon and below it—is a circle that we can never see. (Refer to fig. 5.1.) Geminos uses beneath "the earth" (as here) or under "the horizon" (v 6) interchangeably.

[7] the arctic circles are the smallest. This is true for ordinary latitudes, e.g., in Greece. However, Geminos will give examples (v 29–34) of situations in which the arctic circles are larger than the tropic circles.

[8] *dioptra*. Several different kinds of sighting instruments were called by this name. For a *dioptra* to sweep out the parallel circles, as Geminos says, it would have to be rotatable about the diurnal axis. For a plausible reconstruction, see sec. 8 of the Introduction, p. 42.

For the only circle visible in the cosmos is the Milky Way; the rest are perceivable by reason.

V 12 Only five parallel circles are inscribed on the sphere, but not because these are the only parallel [circles] in the cosmos. For, as far as the senses are concerned,[9] each day the Sun describes a circle parallel to the equator in the course of the rotation of the cosmos, so that between the tropic circles 182 parallel circles are traced out by the Sun: for there are just this many days between the solstices. 13 Indeed, all the stars are carried on parallel circles in the course of each day. All these circles are inscribed[a] on the sphere because they contribute much to other undertakings in astronomy; 14 for it is not possible to place the stars well on the sphere without all the parallel circles, nor to discover accurately the lengths of the nights and days without the aforementioned circles. However, because they contribute no result to a first introduction to astronomy, they are not inscribed on the sphere.

V 15 But the 5 parallel circles, because they do contribute certain definite results in the first introduction to astronomy, are inscribed on the sphere. 16 For the arctic circle bounds the stars that are always visible; the summer tropic circle contains the <summer> solstice and is the limit of the Sun's motion toward the north; the equator circle contains the equinoxes; the winter tropic circle is the terminus of the Sun's advance toward the south and contains the winter solstice; and the antarctic circle bounds the stars that are not visible. 17 Since, then, they have important and definite results for the introduction to astronomy, they are with good reason inscribed on the sphere.

Division of the Parallels by the Horizon

V 18 Of the 5 aforementioned parallel circles, the arctic circle is situated entirely above the Earth.

V 19 The summer tropic circle is cut by the horizon into two unequal parts: the larger part is situated above the Earth, the smaller part below the Earth. 20 But the summer tropic circle is not cut by the horizon in the same way for every land and city: rather, on account of the variations

---

[9] as far as the senses are concerned. If the Sun were fixed at one place on the celestial sphere, it would trace out a circle as the sphere rotates. But in one day the Sun shifts its position on the zodiac by about 1°, which causes it to move a little north or south on the sphere. Thus, the Sun's actual motion through the sky in the course of a day is not really a circle but one loop of a sort of spherical spiral. For Geminos's present purpose, the Sun's change in position may be ignored, and so the Sun does sensibly trace out a circle in the course of a day.

in the *klimata*,[10] the excess of the [one] part [over the other] differs. **21** For those who live farther north than we do, it happens that the summer <tropic> is cut by the horizon into parts that are more unequal; and the limit is a certain place where the whole summer tropic circle is above the Earth.[11] **22** But for those who live farther south than we do, the summer tropic circle is cut by the horizon into parts more and more equal; and the limit is a certain place, lying to the south of us, where the summer tropic circle is bisected by the horizon.[12]

V 23 <For the horizon in Greece,[b] the summer tropic> is cut <by the horizon> in such a way that, if the whole circle is [considered as] divided into 8 parts, 5 parts are situated above the Earth, and 3 below the Earth. **24** And it was for this *klima* that Aratos seems in fact to have composed his treatise, the *Phenomena*; for, while discussing the summer tropic circle, he says:

> If it is measured out, as well as possible, into eight parts,
> five turn in the open air above the Earth,
> and three beneath; on it is the summer solstice.[13]

From this division it follows that the longest day is 15 equinoctial hours[14] and the night is 9 equinoctial hours.

V 25 For the horizon at Rhodes, the summer tropic circle is cut by the horizon in such a way that, if the whole circle is divided into 48 parts, 29 parts are situated above the horizon, and 19 below the Earth. From this division it follows that the longest day in Rhodes is 14½ equinoctial hours[15] and the night is 9½ equinoctial hours.

V 26 The equator circle, for the whole *oikoumenē*, is bisected by the horizon, so that a semicircle is situated above the Earth, and a semicircle below the Earth. For this reason, the equinoxes are on this circle.

V 27 The winter tropic circle is cut by the horizon in such a way that the smaller part is above the Earth, the larger below the Earth. The inequality of the parts has the same variation in all the *klimata* as was the case with the summer tropic circle, because the opposite parts of the tropic circles

[10] the variations in the *klimata*. In the different *klimata* (or, as we would say, at different latitudes), the axis is inclined differently to the horizon, and so the summer tropic circle is cut differently by the horizon.

[11] the whole summer tropic circle is above the Earth. This is true anywhere north of the modern arctic circle, i.e., for geographical latitudes above approximately 66° N.

[12] the summer tropic circle is bisected by the horizon. This is true at the Earth's equator, latitude 0°.

[13] Aratos, *Phenomena* 497–99.

[14] the longest day is 15 equinoctial hours. If 5/8 of the summer tropic circle is above the horizon, then the length of the summer solstitial day is 24 hours × 5/8 = 15 hours.

[15] the longest day in Rhodes is 14½ equinoctial hours. If 29/48 of the summer tropic is above the horizon in Rhodes, the solstitial day is 24 hours × 29/48 = 14½ hours.

are always equal to one another.[16] For this reason, the longest day is equal to the longest night, and the shortest day is equal to the shortest night.

V 28 The antarctic circle is hidden wholly beneath the horizon.

SIZES, ORDER, AND PROPERTIES OF THE PARALLELS

V 29 The sizes of some of the 5 aforementioned parallel circles remain the same for the whole *oikoumenē*, but the sizes of others change with the *klimata*. 30 For some [*klimata*] the circles are larger, and for some they are smaller. The tropic circles and the equator remain the same size for the whole *oikoumenē*; but the arctic circles change in size and are larger for some [*klimata*] but smaller for others.

V 31 For those who live toward the north, the arctic circles are larger: since the pole appears higher, the arctic circle—that is, the circle tangent to the horizon—must necessarily become always larger and larger. 32 For those who live still farther north, at one place the summer tropic becomes the arctic circle,[17] so that the two circles coincide with one another—the summer tropic circle and the arctic—and occupy a single position. 33 For places still farther north the arctic circles even become larger than the summer tropic circle. 34 And the limit is a certain place, lying toward the north, where the pole is at the zenith,[18] and the arctic circle occupies the position of the horizon and coincides with it during the rotation of the cosmos, and takes the same size as the equator. As a result, the three circles—the arctic [circle], the equator, and the horizon—take the same place and position.

V 35 And, in turn, for those lying to the south of us, the poles are lower and the arctic circles smaller. 36 And the limit is a certain place lying to the south of us—this is [the region] said to be "beneath the equator"[19]—where the poles are on the horizon and the arctic circles are

---

[16] the opposite parts of the tropic circles are always equal to one another. That is, the portion of the summer tropic circle that is above the horizon is equal to the portion of the winter tropic that is below the horizon.

[17] the summer tropic becomes the arctic circle. This is true at the geographical latitude where the summer tropic circle just grazes the horizon in the north, which happens at the terrestrial arctic circle (in the modern sense); i.e., at approximately 66° north latitude. Propositions of this sort are easily demonstrated on a celestial globe or an armillary sphere.

[18] where the pole is at the zenith. The celestial pole is seen at the zenith by observers at the north pole of the Earth. In this case, as Geminos says, the horizon coincides with the celestial equator. The stars do not rise or set, but travel on circles parallel to the horizon. The horizon is therefore also the boundary between the stars that are always visible and those that are never visible, i.e., the horizon serves as arctic circle.

[19] "beneath the equator." At the Earth's equator, the celestial equator circle is vertically overhead. This part of the Earth is therefore said to lie "beneath the [celestial] equator." As

completely suppressed, so that instead of the 5 parallel circles there are 3 parallels, [that is,] the tropics and the equator.

V 37 Therefore, on account of what has been said, one must not suppose that there are 5 parallel circles generally, but rather that the number applies to our *oikoumenē*.[c] For there are certain horizons for which only 3 parallel circles exist. 38 They are the [following] places[20] on the Earth: first, the place where the summer tropic circle is tangent to the horizon and takes the position of the arctic circle; second, the place called "beneath the pole"; and third, the place just spoken about, the one called "beneath the equator."

V 39 Consequently, the order of the 5 parallel circles is not the same everywhere. In our *oikoumenē*, the arctic [circle] is designated first, the summer tropic second, the equator third, the winter tropic fourth, and the antarctic [circle] fifth. 40 But for those who live north of us, the summer tropic circle may be first, the arctic second, the equator third, the antarctic fourth, and the winter tropic fifth: for wherever the arctic circle is larger than the summer tropic, the order just mentioned necessarily arises.

V 41 Similarly, the properties[21] of the 5 parallel circles are not the same for all those dwelling on the Earth. For the circle that is summer tropic where we live is the winter tropic circle for our *antipodes*[22]; and the summer tropic where they live is winter tropic for us. 42 And for those who live beneath the equator,[23] the three circles are, by virtue of

---

Geminos says, in this situation, the two celestial poles lie on the horizon; thus the arctic and antarctic circles collapse to points and only three of the principal parallel circles remain.

[20] places. *oikēseis* is usually to be translated "regions," but is here used in a more specific way, almost as a synonym for *horizōntes* (horizons) or *klimata* (latitudes).

[21] properties. *Dunameis* might be translated more literally as "powers." In this paragraph Geminos discusses which parallel circle acts like a summer tropic circle at a given latitude, and which is most truly like an equator, etc. These properties or powers of the circles are different in significance from mere size and position.

[22] *antipodes*. The *antipodes* are those who stand "with the feet opposite," i.e., those who dwell on the part of the Earth diametrically opposite to the region in question. When Geminos uses this term he does not mean to assert that such people actually exist; rather, he means only that there exists an inhabitable region of the Earth that is diametrically opposite the Greek world. (See xvi 19–20.)

[23] those who live beneath the equator. In this paragraph, Geminos enjoys paradoxical plays on the astronomers' terminology for the parallel circles. His aim is to emphasize how greatly the phenomena at the Earth's equator differ from those at ordinary latitudes. Geminos's points may be paraphrased as follows: (1) At the Earth's equator, the Sun is never more than 24° from the zenith at noon; thus, whether the Sun is on the celestial equator or on either tropic, it is going to be hot: all three parallel circles are in this sense like summer tropics. (2) But in a different sense, it is the celestial equator alone that is most like a summer tropic for the inhabitants at the Earth's equator, since, when the sun is on the celestial equator, it is closest to being overhead at noon (in fact it is straight overhead). In this sense both tropic circles are winter tropics, for the Sun is at its greatest angular distance from the

their properties, summer tropics: for they [i.e., those people] lie under the very path of the Sun. But, by virtue of the relative positions, our equator would be their summer tropic circle, while the two tropics would be winter tropics.   43 For, by [its] nature and universally for the whole *oikoumenē*, the circle that is closest to the region would be said to be the summer tropic.[24] Therefore, for those living beneath the equator, the equator becomes the summer tropic circle: for then the Sun is at the zenith for them.   44 But all the parallels are equator circles for them: where they live it is always equinox, for all the parallel circles are bisected by the horizon.

V 45 The distances of the circles from one another do not remain the same for the whole *oikoumenē*. But in the engraving of the spheres, one makes the division in declination in the following way.[d]   46 The entire meridian circle being divided into 60 parts, the arctic [circle] is inscribed 6 sixtieths from the pole; the summer tropic is drawn 5 sixtieths from the arctic [circle]; the equator 4 sixtieths from each of the tropics; the winter tropic circle 5 sixtieths from the antarctic; and the antarctic [circle] 6 sixtieths from the pole.[25]

V 47 The circles do not have the same separations from one another for every land and city. The tropic circles do maintain the same distance from the equator at every latitude,[26] but the tropic circles do not keep the same separation from the arctic [circles] for all horizons; rather, the separation is less for some [horizons] and greater for others. 48 Similarly, the arctic [circles] do not maintain a distance from the poles that is equal for every latitude; rather, it is less for some and greater for others. However, all the spheres are inscribed for the horizon in Greece.

---

inhabitants at the equator when it is on either tropic. (3) In yet a different sense, at the Earth's equator all the parallel circles may be considered equators. The horizon bisects all the parallel circles; thus, at the Earth's equator, there is perpetual equinox.

[24] the circle that is closest to the region would be said to be the summer tropic. A parallel circle on the celestial sphere is "closer" to the region in question, the more nearly they have the same angular distance above the equator. For example, the summer tropic circle is 24° north of the equator. This circle is closest to the *klima* with latitude 24°, for there the tropic circle passes through the zenith.

[25] 6 sixtieths from the pole. Let the meridian circle be divided into 60 parts. Each part therefore corresponds to 6°. The arctic circle is then 36° from the pole, the summer tropic is 30° from the arctic circle, and the equator is 24° from the tropic. The figure of 24° for the distance from the equator to either tropic is a good round value. But the distance between the celestial pole and the (local) arctic circle varies with latitude. Geminos's figure of 36° strictly applies only at a geographical latitude of 36°, roughly that of Rhodes. It also results in the nice proportion of 6:5:4 for the three consecutive intervals.

[26] at every latitude. Literally, "for every inclination" of the axis of the cosmos.

THE COLURES

V 49 [Passing] through the poles are the circles that some call the colures; it is their property to have the poles of the cosmos on their own perimeters. They are called colures because certain parts of them are not to be seen.[27] 50 [In our *klimata*,] the other circles are seen in their entirety during the rotation of the cosmos; but a certain portion of the colure circles is unseen, namely the part cut off beneath the horizon by the antarctic [circle]. These circles are drawn through the tropic and equinoctial points and divide into 4 equal parts the circle through the middles of the signs.

THE ZODIAC

V 51 The circle of the 12 signs is an oblique circle. It is itself composed of 3 parallel circles, two of which are said to define the width of the zodiac circle, while the other is called the circle through the middles of the signs.[28] 52 The latter circle is tangent to two equal parallel circles: the summer tropic, at the 1st degree of Cancer, and the winter tropic, at the 1st degree of Capricorn. It also cuts the equator in two at the 1st degree of Aries and the 1st degree of Libra.   53 The width of the zodiac circle is 12 degrees. The zodiac circle is called oblique because it cuts the parallel circles.

THE HORIZON

V 54 The horizon is the circle that separates, for us, the visible and the invisible parts of the cosmos and bisects the whole sphere of the cosmos, so that a hemisphere is cut off above the Earth, and a hemisphere below.

[27] called colures because certain parts of them are not to be seen. See comment 12 to chapter ii, p. 130.

[28] the circle through the middles of the signs. The Sun moves along the ecliptic, which Geminos calls "the circle through the middles of the signs." The Moon and planets stay near this circle, but wander north and south of it. The *zodiac* is a band wide enough to encompass the wanderings of the planets (Achilleus, *Introduction to the "Phenomena" of Aratos* 23; Maass 1898, 53). The three circles that Geminos mentions are the ecliptic and the two circles 6° away from it, which are the northern and southern boundaries of the zodiacal band, as on the Farnese globe (fig. I.2). Geminos's band is wide enough to encompass the latitudinal wanderings of the Moon (which can be 5° north or south of the ecliptic) as well as those of Mercury, Jupiter, and Saturn (whose latitudes range up to 6°, 2° and 3°, respectively). However, Mars is sometimes nearly 7° away from the ecliptic, and Venus nearly 9°, so Geminos's band is not really wide enough.

V 55 There are two horizons—one perceptible [to our sight] and the other perceivable by reason.[29, e] 56 The perceptible horizon is the one described by our vision in accordance with the limitations of sight; thus, it has a diameter no larger than 2,000 stades.[30] 57 The horizon perceivable by reason is the one that extends all the way to the sphere of the fixed stars and bisects the whole cosmos.

V 58 The horizon is not the same for every land and city. However, the horizon remains sensibly the same for about 400 stades,[31] with the result that the lengths of the days, the *klima*,[32] and all the phenomena remain the same. 59 If the number of stades between the places is larger, the horizon begins to be different, in accordance with the changing *klima*, and all the phenomena change. It is, of course, necessary that the difference in excess of 400 stades between the places be taken in connection with the displacement[33] toward the north or south. 60 For people dwelling on the same parallel, even those who are a myriad stades[34] away, the horizon is different, but the *klima* is the same and all the phenomena are very similar. Of course, the beginnings and endings of the days will not be at the same time for all who dwell on the same parallel. 61 Strictly speaking, when there is the slightest displacement toward

[29] two horizons—one perceptible [to our sight] and the other perceivable by reason. The perceptible horizon is, for example, the edge of the sea that we observe. The horizon perceivable by reason is a great circle on the celestial sphere, defined by the intersection of the sphere with the horizon plane.

[30] a diameter no larger than 2,000 stades. If the perceptible horizon has a maximum diameter of 2,000 stades, then we can see for 1,000 stades (a bit more than 100 miles), at least in some circumstances. A simple geometrical calculation shows that the rim of the horizon is 1,000 stades away for an observer standing on a mountain 12.5 stades high. (We assume Eratosthenes' figure of 252,000 stades for the circumference of the Earth, given by Geminos at xvi 6.) Since Geminos puts the heights of some remarkable mountains at 10 to 15 stades (xvii 5), these geometrical considerations probably motivated his choice of 1,000 stades for the maximum possible distance of the perceptible horizon. Alternatively, Aujac (1975, 134) reads this passage as reflecting Greek experience with long-range signals, and refers to Diels's (1924, 71–90) discussion of a possible signal by fire from Mt. Athos to Euboea, a distance of a bit more than 100 miles, or roughly 1,000 stades.

[31] the horizon remains sensibly the same for about 400 stades. Geminos's estimate is reasonable, at least for solar phenomena. Given Geminos's figure of 252,000 stades for the circumference of the Earth (xvi 6), 400 stades amounts to 0.57° of the meridian or of the equator. Since the angular diameter of the Sun is about 0.5°, the fuzziness of shadows should, indeed, prevent detection of any change in the solar phenomena for about 400 stades. Kleomēdēs (*Meteōra* i 7.71–76) says that there are no shadows at noon at Syēnē on the summer solstice, and that the diameter of the shadowless area is 300 stades.

[32] the *klima*. This could also be translated "the inclination" [of the pole], or even "the latitude."

[33] displacement. *parados* could more literally be translated as a "passing."

[34] a myriad stades. A myriad is ten thousand, but the word is also used to denote any large, indefinite number.

whatever part of the cosmos, the horizon and the inclination[f] [of the axis] change and all the phenomena are different.[35]

V 62 The horizon is not engraved on the spheres, for the simple reason that, while the cosmos moves along from east to west, all the other circles turn with it, they too, with same motion as the cosmos; but the horizon is by nature motionless and preserves exactly the same position forever. 63 Thus, if horizons were engraved on the spheres, it would result that, when they turned, the horizon would move and would sometimes be at the zenith, which is absurd and incompatible with the theory of the sphere. The position of the horizon is, of course, perceived with the aid of the sphere's stand.[36]

THE MERIDIAN CIRCLE

V 64 The meridian circle is the circle traced through the poles of the cosmos and the zenith point. It is on this circle that the Sun produces midday and midnight. 65 This circle, too, is motionless in the cosmos, preserving the same position during the whole[g] rotation of the cosmos. This circle is not inscribed on the spheres that are figured with stars, because it is motionless and admits of no change whatsoever.

V 66 The meridian is not the same for every land and city. But as far as perception is concerned, the meridian remains the same for about 400 stades, although, in strict logic, with any displacement toward the east or toward the west the meridian becomes different. 67 For in a displacement toward the north or toward the south—even if there are a myriad stades between [the places]—the meridian remains the same, but in a displacement from east to west there are differences of meridian.

THE MILKY WAY

V 68 The Milky Way[37] also is an oblique circle. This [circle], rather great in width, slants <through> the tropic circles.[h] It is composed of a

---

[35] all the phenomena are different. This sentence is imprecise. Geminos has just pointed out that the *klima* only changes when the observer is displaced in the north–south direction.

[36] the sphere's stand. The stands of celestial globes and armillary spheres are commonly made in such a way that the horizontal ring of the stand represents the observer's horizon, as in fig. I.6. Geminos's comment shows that this feature of the design was incorporated in Antiquity.

[37] The Milky Way. The Greek expression is "the circle of milk (*galaktos*)," from which comes our *galaxy*. Aristotle (*Meteorology* 345a11–346b15) mentions several theories of the Milky Way advanced by his predecessors, which he refutes in turn. Aristotle's own

cloudlike mass of small parts and is the only [circle] in the cosmos that is visible. **69** The width of this circle is not well defined; rather, it is wider in certain parts and narrower in others. For this reason, the Milky Way is not inscribed on most spheres.[38]

This also is one of the great circles. **70** Circles having the same center as the sphere are called great circles on the sphere. There are 7 great circles [that we have mentioned]: the equator, the zodiac with the [circle] through the middles of the signs, the [circles] through the poles,[39] the horizon for each place, the meridian, the Milky Way.

---

opinion is that the Milky Way consists of the halos around many individual stars. These halos arise in the following way. Surrounding the Earth, at the upper limit of the air, is a warm, dry exhalation, which is carried around by the rotation of the heavens. This exhalation bursts into flame in the vicinity of bright stars. Aristotle rightly points out that the stars are brighter and more numerous in the vicinity of the Milky Way than in other parts of the sky. One might object that the exhalation ought also to be inflamed by the Sun, Moon, and planets, which are brighter than any of the stars. But according to Aristotle, the Sun, Moon, and planets dissipate the exhalation before it has a chance to accumulate sufficiently to burst into flame. Thus, only the stars along the Milky Way have halos. Note that for Aristotle the Milky Way is an atmospheric, and not a celestial, phenomenon: it is produced at the outer boundary of the air. For other theories of the Milky Way, including Poseidōnios's, see Macrobius, *Commentary on the Dream of Scipio* i 15.3–7. Geminos does not offer any theory of the Milky Way.

[38] the Milky Way is not inscribed on most spheres. Ptolemy, however, in his directions for constructing a celestial globe (*Almagest* viii 3), includes the Milky Way among the objects to be represented. In the preceding chapter, he gives a detailed description of the location of the Milky Way among the constellations. Leontios the mechanic is somewhat vague about whether one should try to draw the Milky Way. He points out that Aratos doesn't give a good description of this circle, but that Ptolemy does (Halma 1821, 68). One of the three preserved ancient spheres (see figs. I.3 and I.4, in our Introduction) clearly shows the Milky Way.

[39] the [circles] through the poles. That is, the colures.

VI. On Day and Night

VI 1 Day is spoken of in two senses. In one way, it is the time from the rising of the Sun until the setting, but in the other way a day is the time from a rising of the Sun until the [next] rising of the Sun.

VI 2 The day in the second sense[1] is [equal to] a rotation of the cosmos plus [the time required for] the rising of the arc that the Sun traverses, in

[1] day in the second sense. Refer to fig. 6.1. In the second sense, a day is the time from one sunrise to the next. At the first sunrise, let the Sun be at *B*. Simultaneously with the Sun, point *A* of the celestial equator rises. After one rotation of the cosmos (that is, one sidereal day), *A* will again be on the horizon. However, the Sun will not be rising, for in the course of the sidereal day, the Sun will have shifted eastward on the ecliptic approximately 1° from *B* to *C*. Therefore, as Geminos says, the day is equal to the time required for one rotation of the cosmos plus the time required for the rising of the small ecliptic arc (*BC*) that the Sun traverses in the course of a sidereal day. This statement is not perfectly precise, for it supposes the Sun to remain at *C* while the cosmos turns enough to cause *BC* to rise. But the error associated with Geminos's explanation is negligible for all practical purposes.

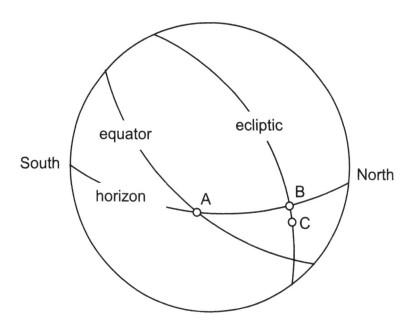

Fig. 6.1. A "day" as the time elapsed from one sunrise to the next.

the course of its motion opposite to that of the cosmos, during the rotation of the cosmos. 3 For this reason, the combined night and day is not exactly equal to every [other] night and day. However, as far as perception is concerned, the lengths are equal; but in strict reckoning, there is a certain small and imperceptible variation. 4 For the rotations of the cosmos are of equal duration, but the risings of the arcs that the Sun traverses during a rotation of the cosmos are not of equal duration.[2] It is for this reason that every combined night and day is not equal to every [other] combined night and day.

VI 5 In the second way of defining days, we say that the month is 30 days, the year 365¼ days. 6 The combined night and day is a time of 24 equinoctial hours. An equinoctial hour is the 24th part of the total time of a night and day.

LENGTH OF THE LONGEST DAY

VI 7 The lengths of the [longest] days are not the same for every land and city. Rather, the [longest] days are longer for those who live toward the north, and shorter for those who live toward the south. 8 In Rhodes, the longest day is 14½ equinoctial hours; around Rome, the longest day is 15 equinoctial hours; for those living north of the Propontis the longest day is 16 equinoctial hours; and for those living still farther north, the longest day is 17 or even 18 equinoctial hours.

VI 9 And it seems that Pytheas of Massalia[3] reached these places. At any rate, he says in his treatise *On the Ocean* that "the barbarians showed us where the Sun goes to sleep; for around these places it happens that the night becomes very short, 2 hours for some, 3 for others, so that, a little while after setting, the Sun rises straightaway."

---

[2] the risings of the arcs . . . are not of equal duration. In fig. 6.1, the Sun moves from *B* to *C* in the course of the sidereal day. However, since the Sun does not move at constant angular speed on the ecliptic, the length of arc *BC* depends upon the day of the year. Moreover, even if all the daily arcs *BC* were of the same length, the times required for them to rise above the horizon would not be equal, as some sections of the ecliptic make shallow angles with the horizon, while others make steep angles with the horizon. These are the two reasons for the variation in the length of the solar day.

[3] Pytheas of Massalia (Marseille). Pytheas was a navigator who, around 330 B.C., sailed north along the Atlantic coast of Europe and claimed to have gone as far as Thoulē, variously identified with the Shetland Islands, Iceland, or Norway. He recorded his experiences in a book, *On the Ocean (Peri tou Ōkeanou)*, of which some fragments survive. Opinion about his trustworthiness was divided in Antiquity. Strabo (*Geography* ii 3.5 and ii 4.1) considered Pytheas to be unreliable and even a great liar. For the fragments, see Mette 1952. For an English translation and commentary, see Roseman 1994, who discusses this passage at pp. 139–43. On Pytheas, see also Cunliffe 2002.

VI 10 Also, Kratēs the grammarian[4] says that Homer mentioned these places in the passage where Odysseus says:

Tēlepylos in Laistrygonia, where a shepherd hails while driving in
  his flock
and another answers while driving his out.
There a sleepless man might earn double wages,
one tending cattle, and the other pasturing white sheep:
For the roads of night and day are close together.[5]

11 For indeed, since around these places the longest day is 23 equinoctial hours, the night lacks only one hour of being shortened to nothing, so that the setting [point] draws near the rising [point and is separated from it only] by the very short arc of the summer tropic that is cut off beneath the horizon. 12 So if, he says, someone were able to stay awake during such long days, he would earn double wages, "one tending cattle, the other pasturing white sheep." And then he brings up the cause, which is mathematical and in conformity with the theory of the sphere: "for the roads of night and day are close together," that is, that the setting [point] lies near the rising point.

VI 13 If we go even farther north, the whole summer tropic circle is above the Earth,[6] so that at summer solstice the day at this place is 24 equinoctial hours. 14 For those who live still farther north, a certain portion of the zodiac circle is always above the Earth. In places where a length of one zodiac sign is cut off above the horizon, the day is a month long.[7] In places where two signs are cut off above the Earth, the longest

---

[4] Kratēs the grammarian. Kratēs (or Crates) of Mallos (c. 150 B.C.), Stoic philosopher and literary critic, wrote a commentary on Homer that is now lost. Kratēs interpreted Homer in light of the scientific knowledge of the second century B.C. and was generous in his appraisal of Homer's astronomy and geography. Other critics, notably Eratosthenēs, were much less generous. To those (Kratēs among them) who tried to puzzle out the route of Odysseus from clues in Homer, Eratosthenēs had this to say: "You will find the scene of the wanderings of Odysseus when you find the cobbler who sewed up the bag of winds" (Strabo, Geography i 2.15, H.L. Jones trans., p. 87). The debate over Homer's reliability, which continues to the present day, thus began with the Greeks themselves. Strabo (Geography i and ii) gives a survey of the ancient opinions, but was himself inclined to give Homer the benefit of the doubt. On Kratēs, see Pfeiffer 1968, 235, 238–46.

[5] Odyssey x 82–86.

[6] the whole summer tropic circle is above the Earth. Assume the modern value of 23.4° for the obliquity of the ecliptic. Then at a latitude of 66.6° N (i.e., on the terrestrial arctic circle), the celestial tropic of Cancer is tangent to the horizon. Thus, at this latitude, on the day of summer solstice, the Sun skims the horizon in the north but does not set: the longest day lasts 24 hours.

[7] the day is a month long. North of the arctic circle, there are portions of the ecliptic that are perpetually above the horizon. For example, at a latitude of approximately 67.4° N, the northernmost sign of the ecliptic, comprising the last 15° of Gemini and the first 15°

day turns out to be two months long.ª 15 The limit is a certain land lying in the extreme north, where the pole is at the zenith,[8] and where 6 signs of the zodiac circle are cut off above the horizon and 6 are cut off below the horizon: there the longest day is six months long, and similarly the night.

VI 16 As Kratēs the grammarian remarks, Homer seems to mention these places when he says, concerning the region of the Kimmerians:

> There lie the country and city of Kimmerian men,
> covered with mist and clouds. Never does the
> radiant Sun regard them with its rays, neither
> when it climbs toward starry heaven, nor when it
> turns again from heaven toward the Earth; but
> deadly night lies on these wretched mortals.[9]

17 Indeed, if the pole is at the zenith, it follows that both the night and the day are six months long. For it is in three months' time that the Sun travels up from the equator circle, which occupies the position of the horizon, to the summer tropic circle; and it is in another three months that it goes down from the summer tropic to the horizon. During all this time it moves along parallel circles above the Earth. 18 And since this region happens to be in the middle of the frigid, uninhabited zone, of necessity the place is always covered by clouds, the clouds are piled together in very thick mists, and the rays of the Sun cannot break through the clouds. 19 And so, logically, there is always night upon them, and darkness. For whenever the Sun is above the Earth, it is dark there because of the thickness of the clouds; and whenever the Sun is below the horizon, it is night there by physical necessity; thus their region is always without light. 20 So this, he says,[10] is what was meant by the poet: "Never does the radiant Sun regard them with its rays." Whether Homer really has this in mind must be another question. 21 But that there are certain places on the spherical Earth where the relative lengths of the days are as mentioned above is clear on the globe itself.[11] And, indeed,

---

of Cancer, stays always above the horizon. Thus, during the month it takes the Sun to travel these 30° of the ecliptic (in late June and early July), the Sun never sets. The longest day at this latitude therefore lasts a month.

[8] where the pole is at the zenith. This occurs at the north pole of the Earth. There, the northern half of the ecliptic is always above the horizon, and the southern half is always below.

[9] *Odyssey* xi 14–19.

[10] he says. That is, Kratēs. It is apparent from the shifts in style that most of this paragraph (at least vi 18–20) is a close paraphrase of Kratēs. The same is true of the earlier paragraph (vi 10–12) involving Tēlepylos.

[11] is clear on the globe itself. Geminos again refers his readers to a concrete model (celestial globe or armillary sphere) for the simplest and most convincing demonstration of the proposition.

these places are uninhabited on account of the excessive cold: they lie in the middle of the frigid zone.

VI 22 And, contrariwise, for those who live toward the south, the [longest] days become always shorter and shorter. For some, the longest day is 14 equinoctial hours; for others, 13. 23 The limit is a certain land, lying to the south of us, [the land] called "beneath the equator,"[12] where the poles lie on the horizon, and the sphere of the cosmos is right.[13] All the parallel circles described by the Sun during the rotation of the cosmos are cut in half <by the horizon>.[b] For this reason, it is always equinox for them.[14]

VI 24 The inequality of the days does not arise for any other reason than the altitude of the pole, which is called the inclination of the cosmos.[15] 25 Because of the elevation of the pole it happens that, for the [parallel] circles traced between the equator and the summer tropic, the larger sections are above the Earth and the smaller are below the Earth; but for the circles traced between the equator and the winter tropic, the smaller sections are above the Earth and the larger are below the Earth. 26 But when the poles fall on the horizon,[16] the cause of the inequality of the days—which was the inclination—is removed and, logically enough, it is always equinox for them. 27 For the Sun revolves around all the circles, both the larger and the smaller ones, in equal times, since the rotation of the cosmos is [made] about certain fixed points, the poles. 28 Thus, the inequality of the days is not due to the sizes of the circles, but rather to the inequality of the sections that the Sun travels below the Earth and above the Earth.

Seasonal Variation in the Length of the Day

VI 29 The increases [in the lengths] of the days and nights are not equal in all the zodiac signs. On the contrary, around the solstitial points

[12] [the land] called "beneath the equator." This is the equatorial region of the Earth. These parts of the Earth lie directly beneath the celestial equator. That is, the celestial equator passes through the zenith.

[13] the sphere of the cosmos is right (literally, "stands upright"). At the Earth's equator, the celestial sphere is said to be *right* because the parallel circles traced out by the Sun in its diurnal motion (including the tropics and the equator) are perpendicular to the horizon. In medieval Latin astronomy, this arrangement is called *sphaera recta*, "the right sphere."

[14] it is always equinox for them. At the Earth's equator, all the parallel circles are bisected by the horizon. Thus, on every day of the year, the Sun is up for 12 hours.

[15] the altitude of the pole, which is called the inclination of the cosmos. The altitude of the pole is the angular distance of the celestial pole above the observer's horizon. The Greek astronomers regarded the cosmos as upright when the poles are on the horizon. Thus the angle by which the cosmos is tilted, or inclined, is the same as the altitude of the pole.

[16] when the poles fall on the horizon. That is, at the Earth's equator.

they [the changes] are very small and imperceptible, so that for about 40 days the lengths of the days and nights remain the same. **30** For both in going toward and again in coming back from the solstitial points, it [the Sun] makes insensible movements in declination; thus, reasonably enough, during the number of days indicated above, there is an apparent tarrying by the Sun around the [same] place. **31** This is why[c] the greatest heats and the greatest colds occur after the solstices. For it[d] [the Sun] runs over the same place twice in succession, both while advancing insensibly and while retreating; logically enough, from the tarrying at one place, it produces an intensification of the heat at one time and of the cold at another. **32** This is clear, too, from the sundials[17]: for the tip of the gnomon's shadow remains on the tropic curves for about 40 days.

VI 33 But around each of the equinoxes the increases [in the lengths] of the days are large, so that the following day differs sensibly from the preceding one. It is for this reason that, on the sundials, the tip of the gnomon's shadow makes a perceptible departure from the equator circle in the course of a day.

VI 34 The cause of the inequality in the lengthening of the days is the obliquity of the zodiac circle.[e] For it is tangent to the tropic circles, and the contact extends for a great length, so that over a large space the separation from the summer tropic is small. **35** From this it follows that the difference between the sections that the Sun runs above the Earth [on one day and the next] is small and imperceptible. **36** But on the equator circle, there is an intersection with the zodiac circle at the equinox; and in either direction from the intersection, the inclination requires a large separation [of the zodiac circle] from the equator. **37** From this it follows that the difference between the days is large, owing to the difference between the sections that the Sun runs above the Earth [on one day and the next]. **38** It is for this reason that around the tropic circles the increases [in the lengths] of the days and the nights[f] are small and imperceptible; moreover, the increases have nearly a constant difference,[g] so that the daily increase around the equinox is about ninety times the daily increase around the solstices.[18] **39** The same reasoning [applies] both to the

---

[17] sundials. See fig. I.9 (in our Introduction) for an illustration of an ancient horizontal, plane dial. On the day of summer solstice, the tip of the shadow traces out the summer solstitial curve. As Geminos remarks, for roughly 20 days before and 20 days after the summer solstice, the tip of the shadow departs insensibly from this curve. On the day of the equinox, the shadow track is a straight line. However, the Sun's declination changes so rapidly around the equinox that, as Geminos remarks, the tip of the shadow moves noticeably away from the line in a single day. The same phenomena could be noted on other types of sundial, such as the spherical and conical dials shown in figs. I.7 and I.8.

[18] the daily increase around the equinox is about ninety times the daily increase around the solstices. At 40° latitude, at the equinox, the length of daylight changes by a little less than 3 minutes from one day to the next; but the day following the solstitial day is shorter

days and to the nights. For the night always diminishes by as much as the day increases.

VI 40 The days are longer than the nights in six signs: Aries, Taurus, Gemini, Cancer, Leo, Virgo, which are a semicircle of the zodiac circle, from the 1st degree of Aries to the 30th degree of Virgo. They are [the] northern [signs]. 41 Contrariwise, the nights are longer than the days in the remaining signs: Libra, Scorpio, Sagittarius, Capricorn, Aquarius, Pisces, which, again, are a semicircle of the zodiac circle, from the 1st degree of Libra to the 30th degree of Pisces.[h]

VI 42 The increase [in the lengths] of the days occurs from the 1st degree of Capricorn to the 30th degree of Gemini, which is a semicircle of the zodiac circle, from winter solstice to summer solstice. 43 But the increase [in the lengths] of the nights occurs from the 1st degree of Cancer to the 30th degree of Sagittarius, which is, again, a semicircle of the zodiac circle, from summer solstice to winter solstice.

ON SIGNS IN SYZYGY

VI 44 Some used to maintain that days are longest in Cancer, since the summer solstice is in that sign, and that nights are longest in Capricorn, since the winter solstice is in Capricorn, thus committing an error similar to the one concerning the syzygies.[19] 45 For if it were solstice in the whole sign, the aforementioned [claim] would be true. But in fact the solstitial points are perceivable [only] by reason.[20] The whole sign of

---

than the solstitial day itself by only about 2 seconds. The ratio is indeed about 90. But a change in the length of daylight of only 3 minutes was too small to be measured in Antiquity, let alone a change of 2 seconds. Geminos's statement is therefore not based on observation but on a theoretical calculation. This is the case with every statement of a similar nature in any ancient astronomical text. Geminos's discussion is based on the assumption that the changes in the length of the day form an *arithmetic progression*. Let the day of the winter solstice be called day 0; let the next day be called day 1; and so on. Then day 1 is longer than day 0 by a certain amount of time, x; day 2 is longer than day 1 by $2x$; day 3 is longer than day 2 by $3x$. This is what Geminos means when he says that the changes have nearly a constant difference. That is, the lengths of successive days are not themselves separated by constant differences; rather, the *changes* in the lengths of the days have constant differences. In modern terms, this is a case of constant second differences. It then obviously results that the daily change in the length of the day around the equinox is 90 times the daily change around the solstice. For more detail, see sec. 12 of the Introduction.

[19] an error similar to the one concerning the syzygies. Geminos is referring to his earlier discussion at ii 27–44.

[20] perceivable [only] by reason. We might say that the solstices are *theoretical* or mathematical points, i.e., true points with no extension. Geminos is insisting that the solstice occurs at a point, rather than throughout an entire sign of the zodiac. He uses the same expression ("perceivable by reason") to characterize the circles of the sphere, such as the

Cancer has the same distance from the summer solstitial point as does Gemini; it is enclosed by the same parallel circles; it makes its rising from the same place [on the horizon], and its setting in the same place: for they [Cancer and Gemini] are in syzygy with one another. 46 Therefore, the lengths of the days and of the nights are equal in Gemini and Cancer. For on the sundials the tip of the gnomon's shadow traces the same lines [when the Sun is] in the aforementioned signs.

VI 47 The same reasoning [applies] also to the winter solstice. For in this case one must not suppose that nights are longest in the whole of Capricorn; rather, there is one certain point perceivable by reason, [the point] that is in common to Capricorn and Sagittarius. 48 The whole signs of Capricorn and Sagittarius have the same distance from the winter solstitial point; they are enclosed by the same parallel circles: for they are in syzygy with one another. Thus, too, the lengths of the days and of the nights are equal in Sagittarius and Capricorn.

VI 49 In general, if signs are in syzygy with one another, those signs contain equal days and nights. 50 The days will be equal in Gemini and Cancer; in Taurus and Leo; Aries and Virgo; Pisces and Libra; Aquarius and Scorpio; Sagittarius and Capricorn.

---

tropics and the equator, which are without thickness and cannot therefore be perceived by the senses (v 11).

VII. <On the Risings of the 12 Signs>[1]

WHERE ON THE HORIZON DOES THE ZODIAC RISE?

VII 1 Since the cosmos is of spherical form and moves with a circular motion from east to west, it results that all the points on the sphere are carried on parallel circles. 2 From this it is evident that all the stars also make their motion on parallel circles. Because of this, too, all the fixed stars[2] rise from the same place and set in the same place. 3 Similarly, the parallel circles also rise from the same place and set in the same place.

VII 4 But the circle of the signs, being slanted in position with respect to the parallels, does not have all its parts rising from the same place and setting in the same place. 5 For this reason, the 12 signs do not rise from the same place or set in the same place.[a] For the zodiac circle makes its risings and settings over a breadth [of the horizon]. 6 The breadth of its rising is from the rising [point] of the 1st degree of Cancer to the rising [point] of the 1st degree of Capricorn.[3] For, however great is the arc[b] on the horizon between these degrees,[c] just so great is the displacement in breadth upon the horizon for the zodiac circle.

VII 7 And Aratos appears in agreement with these things when he speaks as follows:

> It passes over as much water of Ocean
> as rolls from the rising Capricorn

---

[1] <On the Risings of the 12 Signs>. Manitius suggests the title "On the Rising Times of the 12 Signs." Rising times are an important topic in Greek astronomy and do take up most of the chapter. However, vii 1–8 are concerned with rising directions, rather than times.

[2] the fixed stars. Literally, "the nonwandering stars," i.e., excluding the planets. Each fixed star always comes up at the same point on the eastern horizon and goes down at the same point on the western horizon.

[3] The breadth of its rising is from the rising [point] of the 1st degree of Cancer to the rising [point] of the 1st degree of Capricorn. This is the arc of the horizon between the rising positions of the Sun at summer and at winter solstice. The Sun's rising direction θ (measured northward or southward from due east) is given by $\sin \theta = \sin \delta / \cos \varphi$, where δ is the declination of the Sun and φ is the latitude of the observer. (Historians of astronomy call θ the *ortive amplitude*.) Thus, at the equator, the rising direction at summer solstice is 24° north of east; but, at latitude 36° (Rhodes), about 30° north of east. This problem is treated ingeniously by Ptolemy in *Almagest* ii 2. Let d denote the difference in length between the equinoctial day and the day in question, this difference being expressed in degrees rather than hours. Then Ptolemy's result is $\cos \theta = \cos \delta \cos (d/2)$, where the information about the observer's latitude is contained in the length of the day. See Neugebauer 1975, 37–39, and Pedersen 1974, 101–102.

to the rising Cancer. As much as it occupies in rising,
so much it occupies on the other side in setting.[4]

8 In these [verses], he delimits the displacement of the zodiac circle,
which it makes in breadth in the rising and the setting, [and he does it] in
harmony with mathematics and the phenomena.

THE RISING TIMES OF THE SIGNS

VII 9 The obliquity of the zodiac circle being such as this, it results
that the twelfth-parts, which are equal in size, make their risings and settings in unequal times. 10 For those [signs] that make their rising
when the zodiac circle is right[5] make their rising[d] in the most time. For
they meet the horizon perpendicularly, so that the rising of the sign is
one point at a time.[6] Because of this, <it results that>[e] a great deal of
time is used up by the rising.[f]

---

[4] Aratos, *Phenomena* 537–40.

[5] right. That is, upright or perpendicular to the horizon.

[6] one point at a time. Geminos argues (1) that a sign that rises when the ecliptic is perpendicular to the horizon will take a long time to come up. All 30 degrees of the sign must
be dragged successively, as it were, across the horizon. Geminos also says (2) that a sign
that rises when the ecliptic makes a shallow angle with the horizon takes a smaller amount
of time to rise. His justification of these statements is crude and, mathematically speaking,
nonsensical: when the ecliptic makes a shallow angle with the horizon, many parts of the
sign are able "to rise at the same time," rather than one after another. It should be noted,
however, that at the Earth's arctic circle the ecliptic can coincide with the horizon. In this
situation, the slightest rotation of the sphere causes an entire sign to rise instantaneously.
Thus Geminos's expression, that "many parts of the sign rise at the same time," is not as
far-fetched as it seems.

Geminos's statements (1) and (2) here are well suited for use with an armillary sphere in
convincing a student that some signs take longer to rise than others. Statement (2) is even
accurate; for, in every northern latitude below the arctic circle, Aries and Pisces are the
signs most shallowly inclined to the horizon, and do indeed take the least time to rise. But
statement (1) needs correction. For example, at the Earth's tropic of Cancer, the ecliptic is
perpendicular to the horizon when the first point of Libra is rising. According to Geminos's
prescription, Libra and Virgo should take the most time to rise. But, in fact, Leo and Scorpio take the most time to rise at this latitude.

The rising time of a sign depends upon *two* factors. One of these—the inclination of the
ecliptic to the horizon when the sign is rising—Geminos has pointed out. The other factor
is the sign's extent in right ascension. Signs near the solstices stretch over longer intervals
of right ascension than do signs near the equinoxes. This factor stretches out the rising
times of signs near the solstices and diminishes the rising times of signs near the equinoxes.
Thus, there is no way to estimate the rising time of Libra without a more detailed analysis
than Geminos undertakes. Indeed, the best that Euclid could do for rising times—in his much
more mathematical (but nontrigonometric) treatment of the problem in the *Phenomena*—
was (in Prop. 13) to make a comparative ordering of the rising times for the six signs following Capricorn. For a full discussion, see Berggren and Thomas 1996, 4–6.

VII **11** But those [signs] that make their rising when the zodiac circle is slanted with respect to the horizon rise in less time. For the signs meet the horizon slantingly, so that many parts make the rising at the same time as one another.[g] Because of this, it results that the rising is rapid.

VII **12** Whence arises a question, also posed in Aratos: How is it that, in the longest nights as well as in the shortest, six twelfth-parts [of the zodiac] rise and six set, even though there is a large difference in the nights? For the longest night exceeds the shortest by six equinoctial hours.[7] **13** Aratos speaks as follows:

> always, in every night,
> six twelfths of the circle set,
> and as many rise. Each night always stretches
> for as long as a semicircle takes to rise
> above the Earth from the beginning of the night.[8]

VII **14** So the question is how, in the longest as well as in the shortest nights, a semicircle of the zodiac circle both rises and sets. This happens because of the inclination of the zodiac circle. For, because of the slant, the semicircles of the zodiac <circle> rise and set in unequal times.[9] **15** When the zodiac circle is lowest on the horizon, which occurs when the first degree of Capricorn is culminating, then the semicircle from the 1st degree of Capricorn to the 30th degree of Gemini[h] makes a rapid rising; for it meets the horizon slantingly, and many parts make their rising at the same time. **16** But when the zodiac circle is most upright, which occurs when the first degree of Cancer is culminating, then the semicircle from the 1st degree of Cancer to the 30th degree of Sagittarius[i] rises perpendicularly; thus, it makes the rising in a long time. **17** So it is logical that, in winter as well as in summer nights, 6 signs rise and 6 signs set: for the rising times of the signs, which are equal in size, are unequal with respect to times. In the winter nights, those [signs] rise which make a long-protracted rising; but in the summer nights those [signs] rise which make a rapid rising.

---

[7] the longest night exceeds the shortest by six equinoctial hours. This implies longest and shortest nights of 15 and 9 hours, i.e., the *klima* of 15 hours, which Geminos associates with Rome (see vi 8).

[8] Aratos, *Phenomena* 554–58.

[9] Euclid (*Phenomena*, Propositions 14–18) discusses the time of passage of semicircles (and, more generally, of equal arcs) of the ecliptic across the visible and invisible hemispheres. See Berggren and Thomas 1996, 97–113. It is characteristic of Geminos's sense of his audience that he invokes Aratos rather than Euclid.

CORRECTING A COMMON ERROR ABOUT RISING TIMES[10]

VII **18** Now, just as they went wrong on the signs in syzygy,[11] so too the ancients committed an error in the rising times of the signs. **19** For, having supposed the zodiac circle to be most upright when the first degree of Cancer is culminating, [then] since at this time Libra rises and Aries sets, they asserted that Libra rises in the most time and that Aries sets in the most time. **20** Again, since the circle of the signs is lowest when the first degree of Capricorn is culminating, and [since] at this time Aries rises and Libra sets, they asserted that Aries rises most rapidly, and that Libra sets in the least time **21** And again, since the circle of the signs takes an intermediate inclination to the horizon when the first degree of Aries is culminating (or when the first degree of Libra is culminating[j]) <and at this time>[k] Cancer rises and Capricorn sets, they asserted[l] that Cancer has an intermediate time of rising, and Capricorn an intermediate time of setting. **22** And again, when the Claws[12] are culminating, Capricorn will have an intermediate time of rising and Cancer an intermediate time of setting.

VII **23** But such an exposition of the rising times by the ancients is completely mistaken. For since Gemini and Cancer have in the zodiac circle the same inclination to the horizon, and the same distance from the summer solstitial point, and equal lengths of days; **24** and [since] when Gemini is culminating, Virgo rises,[m] but when Cancer is culminating, Libra rises; [therefore] Virgo and Libra rise in equal times. **25** And since the circle of the signs is most upright when Gemini and Cancer are culminating, Virgo and Libra rise in the most time, and not the Claws alone, as the ancients supposed.

VII **26** Again, since Sagittarius and Capricorn have the same inclination to the horizon; and when Sagittarius is culminating, Pisces rises and

---

[10] a Common Error about Rising Times. In vii 18–32, Geminos corrects those who foolishly believe that Libra takes more time to rise than any other sign, and who do not realize that Libra and Virgo take exactly the same amount of time to rise. Perhaps the "ancients" made this mistake because they had the wrong idea about the pattern of signs in syzygy, which Geminos criticized at ii 27–43. (See comment 16 to Geminos's chapter ii.) Or, they may have simply fooled themselves by thinking about zodiac points rather than whole signs. Thus, as Geminos says, the ecliptic is most upright when the first point of Libra is rising. The "ancients" therefore wrongly concluded that Libra alone takes the most time to rise. But as Geminos points out, Virgo and Libra take the same amount of time to rise. The reader will find it helpful to work through this section of the text with a celestial globe at hand.

[11] as they went wrong on the signs in syzygy. Geminos is referring to his discussion at ii 27–43.

[12] Claws. This was the name used by "the ancients" for Libra. See comment 12 to chapter i, p. 117.

Virgo sets; but when Capricorn is culminating, Aries rises and Libra sets; 27 and <since> the circle of the signs is lowest when Sagittarius and Capricorn are culminating; [therefore] Virgo and Libra set in the least amount of time.

VII 28 And again, since when Pisces is culminating, Gemini rises; and when Aries is culminating, Cancer rises; 29 [and since] when Pisces and Aries are culminating, the zodiac circle has an intermediate inclination; therefore, Gemini and Cancer have an intermediate time of rising; Sagittarius and Capricorn, [an intermediate time] of setting.

VII 30 In the same way, since when both Virgo and Claws are culminating, the zodiac circle is said to have an intermediate inclination; and [since] when Virgo is culminating, Sagittarius rises and Gemini sets, 31 and when the Claws are culminating, Capricorn rises and Cancer sets; [therefore] Sagittarius and Capricorn will have an intermediate time of rising; Gemini and Cancer, of setting.

VII 32 From these things it is clear that [signs] that are equally distant from the solstitial and equinoctial points rise and also set in equal time.[13, n]

The Sum of the Rising and Setting Times

VII 33 And since in every night 6 signs rise in 12 [seasonal] hours, and in every day 6 signs rise in 12 [seasonal] hours, it is clear that, in a day and a night, 12 signs rise in 24 hours, and that one sign rises in 2 hours, according to the average time of the risings. 34 And again, since in every night 6 signs set in 12 [seasonal] hours, 35 it is clear

---

[13] [signs] that are equally distant from the solstitial and equinoctial points rise and also set in equal time. The text is sloppy here. A correct assertion would be: "signs that are equally distant from an equinoctial point rise in equal times and set in equal times." The same cannot be said of signs that are equally distant from a solstitial point. Euclid (*Phenomena*, Props. 12 and 13) gives this result correctly.

Table 7.1 gives the rising and setting times of the signs computed from Ptolemy's table of ascensions (*Almagest* ii 8) for the *klima* of 15 hours (north latitude 40°56').

Signs situated symmetrically with respect to an equinoctial point (e.g., Virgo and Libra) rise in equal times and set in equal times. But this is not true of signs situated symmetrically with respect to a solstitial point (e.g., Gemini and Cancer). Numerical comparisons could also be made using a celestial globe.

Geminos has stated, correctly, that Virgo and Libra have equal rising times (vii 24). But although he asserts that both Gemini and Cancer have an "intermediate time of rising" (vii 28), he does not explicitly claim that their rising times are equal. Again (vii 30), although Geminos says that Sagittarius and Capricorn have an "intermediate time of rising," he does not actually say that they rise in the same time. It would be easy for a reader mistakenly to infer these equalities; so the claim in vii 32, that the rising times are equal for signs situated symmetrically with respect to a solstice, may be a corruption of the text. In any

from the same [considerations] that one sign sets in 2 hours as well as rises in 2 hours, so that in 4 equinoctial hours it rises and sets. **36** And for all of the signs, the total time of the rising and the setting is equal to 4 equinoctial hours.[14] **37** Those signs that rise in the most time set in the least time; while those that rise in the least time set in the most time.

TABLE 7.1

Rising and Setting Times of the Signs in the *Klima* of 15^h, Computed Trigonometrically

Sign	Rising Time	Setting Time	Total
Aries	1^h10^m	2^h33^m	3^h43^m
Taurus	1 25	2 34	3 59
Gemini	1 55	2 24	4 19
Cancer	2 24	1 55	4 19
Leo	2 34	1 25	3 59
Virgo	2 33	1 10	3 43
Libra	2 33	1 10	3 43
Scorpio	2 34	1 25	3 59
Sagittarius	2 24	1 55	4 19
Capricorn	1 55	2 24	4 19
Aquarius	1 25	2 34	3 59
Pisces	1 10	2 33	3 43

---

case, this passage shows the danger in arguing qualitatively when precise rules can only be obtained from mathematical analysis.

[14] the total time of the rising and the setting is equal to 4 equinoctial hours. This is approximately but not exactly true. Table 7.1 in comment 13 gives the sum of the rising and setting times for each sign for the *klima* of 15 hours, as obtained from trigonometric calculation. (Interestingly, a form of the 4-hour rule continues to hold north of the arctic circle: for a sign that rises and sets north of the arctic circle, the *difference* between the rising and setting times is approximately 4 hours.) Geminos's "proof" of the 4-hour rule depends on hand-waving (and a sly substitution of equinoctial for seasonal hours). But the approximate validity of the rule could be demonstrated on a globe.

As we have seen (vi 38), Geminos is familiar with the use of arithmetic progressions to obtain approximate solutions to problems of spherical astronomy. If the rising times of the signs form an arithmetic progression (Introduction, sec. 12), *the 4-hour rule is satisfied exactly*. Geminos's discussion gives the impression that the rule of 4 hours is exactly true, rather than a good approximation. This seems to imply that Geminos was more familiar with arithmetic methods than with trigonometry.

VIII. On Months

THE SYNODIC MONTH

VIII 1 A month is the time from conjunction to conjunction, or from full Moon to full Moon. There is a conjunction whenever the Sun and the Moon are at the same degree [of the zodiac]; this is around the thirtieth day of the Moon. It is said to be full Moon whenever the Moon is diametrically opposite the Sun; this is around the midmonth.

VIII 2 The monthly period is 29 + 1/2 + 1/33 days.[1,a] In the monthly period the Moon passes through the circle of the signs plus the arc that the Sun passes over[2] in a month's time in the direction of the following

---

[1] The monthly period is 29 + 1/2 + 1/33 days. Geminos uses "month" (*mēn*, as at viii 1) and "monthly period" (*mēniaios chronos*, viii 2) interchangeably. For Geminos this always means the synodic month, i.e., the time from new Moon to new Moon (or from full Moon to full Moon). Geminos's length for the synodic month is 29 + 1/2 + 1/33, or 29.53030 days, a value that is not used by any other ancient writer (Neugebauer 1975, 584). The "+" signs have been added by the translators for clarity: in Greek mss. the symbols for 29 and for the unit fractions 1/2 and 1/33 are written right next to each other. (See textual note a.) On Greek numerals and unit fractions, see Cajori 1993, vol. 1, 21–29 or Heath 1921, vol. 1, 31–42.

[2] plus the arc that the Sun passes over. In fig. 8.1, O is the Earth. Let the Moon be at M_1 and the Sun at S_1, so that the Moon is new. When the Moon has run all the way around its circle and returned to M_1, it will not quite be new again, for in the meantime the Sun will have advanced on the ecliptic. New Moon will not again occur until the Moon has reached M_2 and is again in conjunction with the Sun at S_2. Thus, as Geminos says, a synodic month lasts for the time it takes the Moon to make a complete circuit of the zodiac plus the time

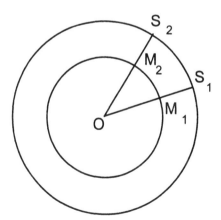

Fig. 8.1. Two successive new Moons M_1 and M_2 occuring about 29° apart, because the Sun moves from S_1 to S_2 in the course of the synodic month.

signs: this is approximately [the distance of] a sign. Thus the Moon moves through approximately 13 signs in the monthly period.

GREEK CALENDRICAL PRACTICE

VIII 3 The precise monthly period is, as has been said, 29 + 1/2 + 1/33 days, but the accepted [length] for civil practice is the more rounded [value] of 29½ days, so that the double month[3] is 59 days. It is for this reason that the civil months are reckoned alternately full and hollow,[4] since the double month according to the Moon is 59 days.

VIII 4 From these things it follows that the lunar year is 354 days; for if we multiply the 29½ days of the month by twelve, the 354 days of the lunar year are produced. 5 For the year is one thing by the Sun and another by the Moon. For the solar [year] comprises a circuit of the Sun around 12 signs, which is 365¼ days; while the lunar [year] comprises the time of 12 lunar months, which is 354 days.

VIII 6 Since neither the month nor the solar year is composed of [a] whole [number of] days, a time [interval] was therefore sought by the astronomers that will contain [a] whole [number of] days, whole months, and whole years. The goal for the ancients was to reckon the months by the Moon and the years by the Sun. 7 For the command, by the laws as well as the oracles, to sacrifice in the manner of the fathers,[5] was taken by all the Greeks to mean reckoning the years in accordance with the Sun and the days and months with the Moon.

VIII 8 Reckoning the years by the Sun means for the same sacrifices

---

required for the Moon to run the arc (M_1M_2) that the Sun traverses in a synodic month. If we take 29.53 days for the length of the synodic month and 365.25 days for the length of the year, then in one synodic month the Sun travels, on average, $360° \times (29.53 /365.25) = 29.1°$, approximately a sign, as Geminos says.

[3] the double month. An astronomical cycle need not contain a whole number of days, as is apparent in Geminos's length for the synodic month. But a *calendar* month must contain a whole number of days. Since calendar months vary in length, it is convenient to define a longer unit of time that may be invariable in length but still contain a whole number of days: thus the *dimēnos*, or "double month," consisting of a month of 29 days and a month of 30 days in succession.

[4] full and hollow. A full month is a month of 30 days; a hollow month, one of 29 days.

[5] to sacrifice in the manner of the fathers. That correct calendrical practice was important for religious observances is clear in a famous passage of Aristophanēs (*Clouds* 615–26). The Moon complains that although she renders the Athenians many benefits—saving them a drachma in lighting costs each month through moonlight—nevertheless they do not reckon the days correctly. Consequently, the gods threaten her whenever they are cheated of their dinners because the sacrifices have not been held on the right days.

to the gods to be performed in the same seasons of the year, the spring sacrifice always to be performed in the spring, and the summer in the summer, and, in the same way, for the same feasts to fall at the remaining proper times of the year;   9 for this they took to be suitable and pleasing to the gods. And this could not come about in any other way if the solstices and the equinoxes were not in the same months.

VIII 10 Reckoning the days by the Moon means for the names of the days to be in conformity with the phases of the Moon,   11 for the names of the days were named for the phases the Moon. Thus, on the day the Moon appears new it [the day] was named, by contraction, *noumēnia*⁶; on the day it makes its second appearance they called it "second"; and they called the phase of the Moon occurring at the middle of the month *dichomēnia* from this circumstance.   12 And in general they named all the days for the phases of the Moon; thus they also called the 30th day of the month, which is last, *triakas*, from this very circumstance.

VIII 13 And Aratos declaims in conformity with these things when he speaks thus about the names of the days:

> Do you not see? When the little horned Moon
> appears from the west, it teaches of the waxing
> month. When this first sheen spreads
> and is great enough to throw a shadow, it is coming to the fourth
> day.
> It is the eighth at half-Moon, midmonth at whole face.
> Always turning from one face to another
> it tells on which day of the month the morning rises.⁷

For in these [verses] he says clearly that the names of the days are named after the phases of the Moon.

VIII 14 The proof that the days are strictly reckoned according to the Moon is that eclipses of the Sun occur on the thirtieth, for then the Moon is in conjunction with the Sun and is at the same degree; but the eclipses of the Moon occur during the night leading to the midmonth, for then the Moon is diametrically opposite the Sun and falls into the shadow of the Earth.

VIII 15 And, therefore, when the years are reckoned strictly by the Sun and the months and days by the Moon, then the Greeks keep to the custom of sacrificing in the manner of the fathers, which is for the same sacrifices to the gods to be performed at the same times of the year.

---

⁶ *noumēnia, dichomēnia, triakas.* Literally, "new month," "dividing the month," and "thirty," respectively.

⁷ Aratos, *Phenomena* 733–39.

THE EGYPTIAN CALENDAR[8]

VIII 16 The Egyptians have held an opinion and a goal opposite to those of the Greeks. For they do not reckon the years by the Sun, nor the months and days by the Moon; but they have need of a certain plan of their own. 17 For they wish the sacrifices to the gods to occur not at the same times of the year, but to pass through all the seasons of the year, and the summer festival to be also a winter,[9] and a fall,[b] and a spring [festival].[c]

VIII 18 They reckon the year to be 365 days, for they count 12 thirty-day months and 5 epagomenal days. They do not add the ¼ [day] for the reason already mentioned, in order that, for them, the festivals might step backward. 19 For in 4 years they fall behind by one day with respect to the Sun, and in 40 years they will be behind by 10 days with respect to the solar year, so that the festivals will step backwards by just so many days in comparison with coming at the same seasons of the year. In 120 years the difference will be a month, both with respect to the solar year and with respect to the seasons of the year.

VIII 20 This, moreover, is the reason why the widespread error, which has been accepted among the Greeks through a tradition of long standing, has been believed until our own time. For most of the Greeks suppose the winter solstice according to Eudoxos to be at the same time as the feasts of Isis [reckoned] according to the Egyptians,[10, d] which is

---

[8] The Egyptian Calendar. As Geminos explains, the Egyptian calendar year is always 365 days long. It consists of 12 months, of 30 days each, followed by five additional days, which the Greeks called *epagomenal days* (*epagomenai*, "added on"). The advantage of the Egyptian calendar was its perfect uniformity: every year and every month were the same. The Egyptian calendar therefore stood in stark contrast to the chaotic lunisolar calendars of the Greeks and was adopted by the Greek astronomers of Alexandria as the most suitable for time reckoning in astronomical work. Ptolemy, in the *Almagest*, uses the Egyptian calendar as a matter of course. For an introduction to the Egyptian calendar, see Bickerton 1980, 40–43; Evans 1998, 175–81.

[9] and the summer festival to be also a winter. Geminos's language seems to echo the Canopus Decree set up by Ptolemaios III Euergetēs in 238 B.C., in his unsuccessful attempt to reform the Egyptian calendar by adding a sixth epagomenal day once every four years: "It came about that the festivals which were celebrated in winter fell in the summer, and that those celebrated in summer were instead in the winter" (trans. Bickerman 1980, 41). On the Canopus Decree, see Hölbl 2001, p. 105–11, with references to the published inscriptions at p. 121n186.

[10] most of the Greeks suppose the winter solstice according to Eudoxos to be at the same time as the feasts of Isis [reckoned] according to the Egyptians. "The winter solstice according to Eudoxos" almost certainly represents a traditional bit of astronomical lore, as expressed in a *parapēgma* or similar document. For example, in the Geminos *parapēgma*, at day 4 of Capricorn, the winter solstice according to Eudoxos is said to come 3 days later than the winter solstice according to Euktēmōn. Even more to the point is a mention near

completely false. For the feasts of Isis miss the winter solstice by an entire month.

VIII **21** The error originated from the aforementioned cause. For 120 years ago the feasts of Isis happened to be celebrated at the winter solstice itself.[11] But in 4 years a shift of one day arose; this of course did not involve a perceptible difference with respect to the seasons of the year. And in 40 years a shift of 10 days arose; it happens that [this] difference is not so very noticeable. **22** But now, when the difference is a month in 120 years, those who take the winter solstice according to Eudoxus to be during the feasts of Isis [reckoned] according to the Egyptians[12] are not lacking an excess of ignorance. For to be tripped up by one or two days is possible, but a shift of a month cannot escape notice. **23** For the lengths of the days can disprove [the proposition], since they have a great difference in comparison to the winter solstice. Also, the engravings of the sundials make the true solstice quite plain,[13] most of all among the Egyptians, who are engaged in observation. **24** Thus, the feasts of Isis were earlier celebrated at the winter solstice and, earlier still, at the summer solstice, as Eratosthenēs mentions in [his] treatise, *On the Octaetēris*.[14] And they will be celebrated in the fall, on the summer

---

the end of the *Celestial Teaching* of Leptinēs: "According to Eudoxos and Dēmokritos, the winter solstice falls sometimes on the 20th, sometimes on the 19th, of Athyr (Tannery 1893, 294). Athyr is a month of the Egyptian year. When we combine this with Plutarch's testimony (on p. 18, sec. 6 of the Introduction) that the feasts of Isis fall on Athyr 17–20, we can see the origin of the calendrical confusion that Geminos is striving to correct.

It is grammatically possible to translate the sentence, "most of the Greeks suppose the winter solstice to be at the same time as the feasts of Isis [reckoned] according to the Egyptians as well as according to Eudoxos" (e.g., Aujac 1975, 51; Manitius 1898, 109). But we have no reason to think that Eudoxos was an authority on the feasts of Isis, while we do know he had something to say about the date of the solstice. Our interpretation of the passage is similar to that of Jones 1999a and Neugebauer 1975, 579. We should point out, however, that the expression "according to Eudoxos and the Egyptians" already occurs on the second Miletus *parapēgma* (Diels and Rhem 1904, 107). On this expression, see Lasserre 1966, 214ff.

[11] 120 years ago the feasts of Isis happened to be celebrated at the winter solstice itself. This passage permits an attempt at dating Geminos's composition. See sec. 6. of our Introduction.

[12] those who take the winter solstice according to Eudoxus to be during the feasts of Isis [reckoned] according to the Egyptians. As with comment 10 above.

[13] the engravings of the sundials make the true solstice quite plain. Geminos appeals to an easily made observation. Since Greek sundials were ordinarily engraved with the shadow tracks for the equinoxes and for summer and winter solstice, a resident of Egypt could see at a glance that, on the feast of Isis, the tip of the shadow was not on the solstitial shadow track.

[14] the *octaetēris*. This is the 8-year lunisolar cycle, which contains, in principle, a whole number of tropical years and a whole number of synodic months. Eight years are set equal to 99 months. Thus, after 8 years, both the Sun and the Moon return, very nearly, to their original positions in the zodiac. See our Introduction, sec. 13. Eratosthenēs's treatise on the

solstice, in the spring, and again[e] at the winter solstice. For in 1,460 years[15] it is necessary that every feast pass through all the seasons of the year and return again to the same time of the year.

THE OCTAETĒRIS

VIII 25 The Egyptians, therefore, have achieved the proposed end according to their own plan; but the Greeks, having the opposite purpose, reckon the years by the Sun and the months and days by the Moon. 26 Now, the ancients reckoned thirty-day months, and embolismic months[16] every other year.[17] But, as the truth was quickly demonstrated in the phenomena, through the days and months not harmonizing with the Moon and the years not being in line with the Sun, they sought[f] a period that will be in harmony with the Sun in respect of the years, and with the

---

octaetēris has not come down to us. However, an *octaetēris* is discussed in the *Celestial Teaching* of Leptinēs, a papyrus of the second century B.C. (Blass 1887, 20; Tannery 1893, 290). On the *octaetēris*, see Neugebauer 1975, 620–21.

[15] 1,460 years. The Egyptian calendar year of 365 days falls short of the solar year by about ¼ day. Therefore, in 4 × 365 years (or 1,460 years), the Egyptian calendar will shift with respect to the Sun by an entire year. To express things more precisely, 1,461 Egyptian years contain the same number of days as 1,460 Julian years.

[16] embolismic months. The ordinary year of a lunisolar cycle contains 12 months. However, several years in each cycle must contain an embolismic (or *inserted*) month: these years therefore contain 13 months. *Intercalary* and *embolismic* are synonymous.

[17] embolismic months every other year. Geminos claims that the Greeks originally used a 2-year cycle, with months of 30 days, and embolismic months every other year. The 1st year in the cycle would have contained 12 months of 30 days each, or 360 days. The 2nd year of the cycle would have contained 13 months of 30 days each, or 390 days. It is best to read Geminos's brief remark on this cycle as pedagogical: he means to show his reader how one might begin to solve the problem of finding a good lunisolar cycle with a simple first step. The insertion of an embolismic month every other year is not such a bad procedure. But 30-day months would get out of step with the Moon so quickly that it is difficult to believe that this system was ever used.

In Herodotos (*Histories* i 32, also ii 4), Solon lectures Croesus: "Take 70 years as the span of a man's life: those 70 years contain 25,200 days, without counting the embolismic months. Add a month every other year, to make the seasons come round with proper regularity, and you will have 35 additional months, which will make 1,050 additional days. Thus the total of days for your 70 years is 26,250, and not a single one of them is like the next in what it brings" (trans. Sélincourt). This calculation is consistent with Geminos's remark and may have been its source. But it is clear that Herodotos is using round numbers to produce a rhetorical effect.

Censorinus (*On the Birth Day* 18.2) also claims that the ancient Greek cities used a cycle of 2 years, in which a year of 12 months alternated with a year of 13. This testimony is weakened by the rest of Censorinus's story, i.e., that the ancients recognized their error and produced a new cycle of 4 years by doubling the old one, and that they then produced the *octaetēris* by doubling the 4-year cycle. Nearly all modern writers agree on the unreality of the 2-year cycle. See Samuel 1972, 33–35; Bickerman 1980, 28.

Moon in respect of the months and days. The time of the period[g] contains [a] whole [number of] months, whole days, and whole years.

VIII 27 First they formed the period of the *octaetēris*, which contains 99 months (among which 3 are embolismic), 2,922 days, and 8 years. They formed the *octaetēris* in this way. 28 Since the solar year is 365¼ days and the lunar year is 354 days, they took the excess by which the solar year exceeds the lunar: it is 11¼ days. 29 If then we reckon by the Moon the months in the year, we will be 11¼ days short with respect to the year of the Sun. Therefore they inquired how many times these days must be multiplied to produce whole days and whole months: multiplying [11¼] by eight produces 90 days, or three months.

VIII 30 Since in the year we fall 11¼ days short in relation to the Sun, it is clear that in 8 years we will fall 90 days short in relation to the Sun, which is 3 months. 31 For this reason, 3 embolismic months are reckoned in each *octaetēris*, so that the deficiency occurring each year with respect to the Sun is made up, and after the 8 years have elapsed from the beginning, <the feasts>[h] are again in harmony with the seasons of the year. This being the case, the sacrifices to the gods will always be performed at the same seasons of the year.

VIII 32 Now they arranged the embolismic months so that they were, as much as possible, evenly distributed. For one ought not to wait until a difference of a month arises with respect to the phenomena, nor to anticipate a whole month with respect to the course of the Sun. 33 For this reason they prescribed that the embolismic months be reckoned in the 3rd, 5th, and 8th year—two [embolismic] months [each] falling after two years, and one reckoned after one year. It would make no difference if one were to make the same disposition of the embolismic months in other years.

VIII 34 The lunar year is reckoned at 354 days.[i] For such a reason, they took the lunar month to be 29½ days, and the double month to be 59 days. Therefore they reckon a hollow and a full month alternately, because the double month according to the Moon is 59 days. 35 There are, then, 6 full and 6 hollow [months] in the year: together they total 354 days. For this reason they reckon alternate months full and hollow.

The Defect of the Octaetēris and Attempts to Correct It

VIII 36 If, then, it were necessary to keep ourselves in harmony only with the years according to the Sun,[18] use of the aforementioned period

---

[18] in harmony only with the years according to the Sun. Geminos refers to the fact that the *octaetēris* treats the length of the solar year accurately, but is based upon an inexact length for the lunar month.

would suffice to maintain harmony with respect to the phenomena. But since it is necessary to reckon not only the years by the Sun, but also the months and days by the Moon, they examined how they might reach this goal too.

VIII 37 Since, therefore, the lunar month is assumed accurately to be $29 + 1/2 + 1/33$ days; and [since], including the embolismic months, there are 99 months in the *octaetēris*, they multiplied the $29 + 1/2 + 1/33$ days of the month by the 99 months: there are then 2,923½ days. Therefore, in the 8 solar years, 2,923½ days must be reckoned according to the Moon. 38 But since the solar year is 365¼ days, the 8 solar years contain 2,922 days; for the days of the year, when multiplied by eight, produce just such a number. Since, therefore, the days according to the Moon in the 8 years were 2,923½, we will fall one and a half days short with respect to the Moon each *octaetēris*. 39 Then in 16 years we will fall 3 days short with respect to the Moon. For this reason, 3 days are added to the Moon's course each sixteen-year period, in order that we may reckon the years by the Sun and the months and days by the Moon.

VIII 40 But if such a correction is [introduced], another error results. For the three days added in the 16 years on account of the Moon make in 10 sixteen-year periods an excess with respect to the Sun of 30 days,[j] which is a month. 41 For this reason, in the course of 160 years, one of the embolismic months is removed from <one> of the *octaetērides*. For, instead of the 3 months due to be reckoned in the 8 years, only 2 are put in, so as to be in harmony with the Moon in respect of the months and days, and with the Sun in respect of the years, again from the beginning [of the 160-year cycle] after the month has been removed.

VIII 42 But even if such a correction is made, harmony with respect to the phenomena does not result in this way. For it has turned out that *octaetēris* completely fails with respect to the months, with respect to the days, and with respect to the embolismic [months] as well. 43 For the monthly period has not been taken accurately.[19, k] 44 Thus, one must not in any period reckon the hollow [months] equal [in number] to the full; rather, the full must exceed the hollow. For if the monthly period were only 29½ days, it would be necessary to reckon equal [numbers of] full and hollow months. 45 But in fact there is a perceptible fraction in the monthly period [in excess of 29½ days], which [with time] fills up the length of a day. For this reason it will be necessary for the full to exceed the hollow months.

---

[19] *For the monthly period has not been taken accurately.* Immediately after these words, the mss. include a spurious passage mentioning the more exact figure of 29;31,50,8,20 days for the length of the month. This is apparently a gloss drawn from Ptolemy's *Almagest*, or some source dependent on it. See textual note k.

46 Nor, indeed, are there really 3 embolismic months in the 8 years. For if the lunar year were 354 days, the excess of the solar year would be 11¼ days; and when these were multiplied by 8, they would fill up the 3 embolismic months. **47** But in fact the lunar year is very nearly 354⅓ days.[20] If, therefore, we take the 354⅓ from the 365¼ [days], 10 + 1/2 + 1/3 + 1/12 will be left; and this when multiplied by eight yields 87⅓ days, very nearly.[21] These days do not fill up 3 months, for which reason one ought not reckon 3 embolismic months in the 8 years.

VIII 48 This is also [made] clear through the nineteen-year period.[22] For in the 19 years, 7 embolismic months are reckoned; and the nineteen-year period will be in harmony for a longer time, as far as concerns the reckoning of the months. In 8 nineteen-year periods, therefore, 56 embolismic months will be reckoned. But in the *octaetēris* 3 embolismic months are reckoned; thus, in 19 *octaetērides*, which is 152 years, 57 embolismic [months] are reckoned. **49** But according to the nineteen-year period, which is in harmony with the phenomena, 56 embolismic months are reckoned in the same [amount of] time, so that the *octaetēris* is in excess by one embolismic month. Therefore the *octaetēris* does not [really] have 3 embolismic months; rather, the period has failed completely in this respect.

The Nineteen-Year Period

VIII 50 Therefore, since it turned out that the *octaetēris* is a failure in every respect, the astronomers around Euktēmōn, Philippos, and Kallip-

---

[20] the lunar year is very nearly 354⅓ days. Geminos here introduces a new parameter, as an improvement over the nominal value of 354 days for the length of 12 synodic months. The new figure is roughly, but not exactly, consistent with Geminos's value for the length of the month: note that 12 (29 + 1/2 + 1/33 days) = 354 12/33, which is a bit larger than 354⅓.

[21] very nearly. The result of the multiplication is exact: 8 (10 + 1/2 + 1/3 + 1/12) = 87⅓. Perhaps Geminos means that this value for the defect accumulated over 8 years is only approximate because the length (354⅓ days) used for the lunar year is only approximate.

[22] the nineteen-year period. This is called the Metonic cycle by most modern writers. Geminos explains it at viii 50–58. In this paragraph, he uses it to estimate the error committed by the *octaetēris*. His conclusion is that the *octaetēris* requires 1 embolismic month too many every 152 solar years. This is in good agreement with his earlier argument, based upon different numerical parameters, that in 20 *octaetērides* (160 years), one embolismic month ought to be omitted. See viii 40–41 and our Introduction, sec. 13.

[23] the astronomers around Euktēmōn, Philippos, and Kallippos. *Peri* can signal the persons "associated with," or "around" one, e.g., family, friends, or colleagues. Sometimes *peri* is translated, as by Manitius, to indicate a "school": "the astronomers of the school of Euktēmōn, Philippos, and Kallippos." When this is too strong, it might be better to write, "Euktēmōn, Philippos, and Kallippos and the astronomers who follow them." But it is also

pos[23] constructed another period, that of nineteen years. 51 For they observed that in the 19 years 6,940 days are contained, and 235 months, including the embolismic months; and in the 19 years, 7 embolismic months are reckoned. The year according to them is therefore $365\frac{5}{19}$ days.[1] 52 And in the 235 months they designated 110 hollow and 125 full, so that hollow and full are not reckoned alternately, but sometimes two full [months] are reckoned in a row. For the nature of the phenomena admits of this, as far as regards the treatment of the Moon, which was not the case with the *octaetēris*.

VIII 53 In the 235 months they designated 110 hollow for the following reason. Since 235 months are reckoned in the 19 years, they [temporarily] supposed them all to be of thirty days: and the days total 7,050.[m] 54 Thus, if all the months are reckoned to be thirty days, the 7,050 days exceed the 6,940 days: there are <110 days> [in the difference] <and> they therefore reckon[n] 110 months as hollow, so that the 6,940 days of the nineteen-year period may be completed in the 235 months.

VIII 55 In order that the treatment of the removed days may be as uniform as possible, they divided the 6,940 days into 110 [parts]: there are then 63 days [in each part]. It is therefore necessary to reckon a removed day every 63 days in this period. 56 The thirtieth [day of the month] is not always the removed day; rather, the [day] falling once every 63 days is selected for removal.

VIII 57 In this period the months seem to have been taken rightly, and the embolismic [months] to have been arranged in harmony with the phenomena, but the annual period has <not> been taken in harmony with the phenomena. 58 For the annual period, carefully observed for many years, has been in agreement with $365\frac{1}{4}$ days, but the year inferred from the nineteen-year period is $365\frac{5}{19}$ days, which exceeds the $365\frac{1}{4}$ days by 1/76 day.

THE SEVENTY-SIX-YEAR PERIOD[24]

59 For this reason the astronomers around Kallippos set right the excess [portion] of the day and constructed the seventy-six-year period, put

---

possible that Geminos really means only Euktēmōn, Philippos, and Kallippos themselves. We print simply "around," so as not to impose too strict an interpretation on a flexible Greek preposition. In some ancient sources, Metōn and Euktēmōn are cited as a pair in connection with the 19-year cycle or with the determination of the summer solstice of 432 B.C. (Ptolemy, *Almagest* iii 1). But sometimes, as here in Geminos, Metōn's name does not appear (see Neugebauer 1975, 623n12).

[24] The Seventy-Six-Year Period. Modern writers often refer to this as the Kallippic Cycle. See the Introduction, sec. 13. There is a large modern literature on the 19- and 76-year

together from 4 nineteen-year periods, which contain 940 months, of which 28 are embolismic, and 27,759 days. **60** They treated the arrangement of the embolismic [months] in the same way. And this period, most of all, seems to agree with the phenomena.

---

cycles. For an introduction, see Neugebauer 1975, 615–24; Bowen and Goldstein 1989; Jones 2000.

IX. On Phases of the Moon

That the Moon Is Illuminated by the Sun

IX 1 The Moon is illuminated by the Sun. For it always keeps the bright [side] turned toward the Sun. And when it rises before the Sun, the bright [side] of it faces toward the east.[a] And when it sets before the Sun, or sets after the Sun, the bright [side] is turned toward the Sun. 2 On some days,[1] although rarely, the Moon has been observed setting after the Sun and holding the bright face toward the west; then, having passed by the Sun during the night, and having risen before the Sun, it is seen holding the bright [side] toward the east. From this it is clear that the Moon is illuminated by the Sun.

IX 3 The following things have also been observed.[2] When the Sun rises at the [time of] winter solstice, then the middle of the illuminated [portion] faces toward the Sun, so that the straight line joining the horns of the Moon is cut in half and at right angles by the straight line leading from the center of the Sun to the [line of] bisection of the Moon.

---

[1] On some days. This sentence is clearly corrupt, for the astronomy is wrong: the order in time of the two observations is backwards. Probably, Geminos originally wrote something like this: "On some days, although rarely, the Moon has been observed rising before the Sun and holding the bright face toward the east; then, having passed by the Sun during the day, and having set after the Sun, it is seen holding the bright side toward the west." (Cf. Manitius 1898, 269–71.) The error is so elementary that Geminos can hardly be responsible for it.

[2] The following things have also been observed. The situation is shown in fig. 9.1 The Sun is rising and may be seen at the same time as the crescent Moon. Line *AB* through the horns of the Moon has center *C*. The line from the center *D* of the Sun to *C* meets *AB* at right angles. So, the middle *F* of the illuminated portion of the Moon must face toward the Sun. Since, as Geminos says, the same stituation applies at all times of year, the Moon must be illuminated by the Sun. That the Moon gets its light from the Sun was old knowledge, which had been attributed by Plato to Anaxagoras (fifth century B.C.) as a recent discovery (*Cratylus* 409A). Even earlier, Parmenidēs (sixth century B.C.) had described the Moon

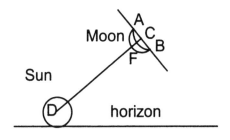

Fig. 9.1. An argument that the Moon gets its light from the Sun.

4 And when the Sun rises at the [time of] summer solstice, again the middle of the illuminated [portion] is turned toward the middle of the Sun, so that, in the same way, the aforementioned straight line is bisected and cut at right angles. The same thing happens also at the setting. And so it is inferred from this evidence that the Moon is illuminated by the Sun.

EXPLANATION OF THE PHASES

IX 5 Its illuminated [portion] is of course always equal to a hemisphere; <but> its illuminated [portion] does not always appear equal to our sight because of the [varying] distances away from the Sun. 6 For when, on the thirtieth [day of the month], the Sun and the Moon are in the same degree, then the hemisphere that is illuminated is the one facing toward the Sun and turned away from our sight; for the Moon moves below the Sun. 7 And when the Moon passes by the Sun about the first of the month, then the Moon is seen crescent-shaped, for a small <part> of the illuminated hemisphere is turned toward our sight. 8 And when the Moon stands away from the Sun in the following days, the illuminated [portion] is always more and more seen by us. When the Moon is a fourth part of the zodiac circle away, it is seen bisected, for then half of the hemisphere illuminated by the Sun has turned toward us. 9 The greater <the distance of the Moon>[b] from the Sun, the greater also [appears] the illuminated [portion]. And when it is diametrically opposite the Sun, the illuminated hemisphere faces toward our sight. And, in general, the sizes of the illuminations are seen in relation to the separation. 10 And at the last [of the month], when the Moon runs under the Sun, it is seen deprived of light, for its illuminated hemisphere has turned upward toward the Sun. And so, logically, the illuminated part of the Moon is unobservable for us. It is clear from this that the Moon is illuminated by the Sun.

THE FOUR PHASES

IX 11 The Moon takes on all 4 of its phases in the monthly period, producing [each of] them twice. The phases are these: crescent, half Moon, gibbous Moon, full Moon. 12 The crescent occurs around the

---

poetically as "always fixing its gaze on the beams of the Sun" (Parmenidēs, frag. 15, trans. Heath 1932, 20). Geminos is attempting to convince his reader that simple and precise observations can demonstrate the truth of the common saying that the Moon gets its light from the Sun. Of course, these are "textbook observations"—more likely to be described than performed.

beginnings of the months, half Moon around the 8th of the month, gibbous Moon around the 12th, and full Moon around the midmonth; and again gibbous after the midmonth, half Moon around the 23rd, crescent around the ends of the months.

VARIABILITY IN THE TIMES OF THE PHASES

IX 13 The Moon does not always produce the corresponding phases on the days that are named for them, but rather on different days, in accordance with the anomaly of the [Moon's] movement.[3] 14 For the Moon appears crescent-shaped at the earliest on the first of the month and at the latest on the 3rd; it remains a crescent sometimes until the 5th, and sometimes, at the latest,[c] until the 7th. It becomes a half Moon at the earliest around the 6th and at the latest around the 8th. It becomes gibbous at the earliest around the 10th and at the latest around the 13th. It becomes full at the earliest around the 13th and at the latest around the 17th. 15 It becomes[d] gibbous for the second time at the earliest around the 18th and at the latest around the 22nd. It becomes a half Moon for the second time at the earliest around the 21st and at the latest around the 23rd. It becomes a crescent for the second time at the earliest around the 25th and at the latest around the 26th.

IX 16 The entire monthly period[e] is 29 + 1/2 + 1/33 days. It is the time from conjunction to conjunction, <or> from full Moon to full Moon. A conjunction is the time at which the Sun and the Moon are at the same degree, which happens on the 30th [of the month].

---

[3] the anomaly of the [Moon's] movement. The Moon does not move about the Earth at a uniform angular speed. From the modern point of view, the Moon's motion on its elliptical orbit, according to Kepler's laws, is responsible for this anomaly. (Additional nonuniformities in the Moon's motion are due to gravitational perturbations by the Sun.) In Hipparchos's lunar theory, the variable rate of motion is produced by an epicycle. In system B of the Babylonian lunar theory, the Moon's daily shift along the zodiac increases by constant increments each day from least motion to greatest motion. (Geminos takes up the Babylonian lunar theory in chapter xviii). Regardless of the theoretical view one adopts, it follows that the four principal phases (new Moon, first quarter, full Moon, third quarter) are not separated by intervals of exactly one-fourth of the synodic month.

X. On the Eclipse of the Sun

X 1 The eclipses of the Sun[1] occur because of covering by the Moon. Since the Sun moves higher, and the Moon lower, when the Sun and the Moon are at the same degree,[2] the Moon, having run in under the Sun, blocks the rays leading from the Sun toward us. **2** Therefore, one must not speak of them as *eclipses* in the proper sense, but rather as

[1] eclipses of the Sun. Ms. C contains a diagram, squeezed into the lower margin of fol. 28, illustrating a conjunction of the Moon and Sun. Since the text makes no mention of this figure, it is probably a later addition, and we print it here in the commentary, rather than in the body of the translation. A similar diagram in C is squeezed into the lefthand margin of fol. 28v to illustrate a lunar eclipse. (See comment 1 to chapter xi.) There is an interesting difference between the two diagrams. In the diagram for chapter xi, the "shadow of the Earth" is shown and labeled. But in the diagram for chapter x, there is no shadow of the Moon. This difference parallels a textual difference in Geminos's treatment of the two kinds of eclipses. Thus (xi 1), "eclipses of the Moon occur because of the Moon's falling into the shadow of the Earth." But (x 1) "eclipses of the Sun occur because of covering by the Moon." Thus, Geminos has no need to mention the Moon's shadow: he does not say that during a solar eclipse we fall into the shadow of the Moon, but rather

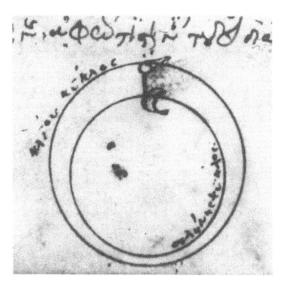

Diagram 7. An illustration of a solar eclipse from Vaticanus gr. 318, folio 28 (early fourteenth century). The outer circle is labeled "circle of the Sun"; and the inner one, "circle of the Moon." The crescent symbol for the Moon and the cone-shaped symbol for the Sun (probably meant to suggest its rays) have ancient roots and are attested on papyri from the fourth century A.D.

coverings.[3] For not one part of the Sun will ever be eclipsed [i.e., "fail"]: it [simply] becomes invisible to us through the covering of the Moon.

X 3 For this reason the eclipses are not equal for all [observers][4]; rather, there are great variations in the magnitudes of the eclipses, in accordance with the differences of the *klimata*. 4 For at the same time the whole Sun is eclipsed for some, half [is eclipsed] for others, less than half for others, while for still others no part of the Sun is observed to have been eclipsed at all. 5 For those dwelling vertically beneath the covering, the whole Sun is invisible; for those dwelling partly outside the covering, a certain part of the Sun is seen to have been eclipsed; and for those dwelling wholly outside the covering not one part of the Sun is observed to have been eclipsed.

X 6 The greatest proof that the Sun is eclipsed because of covering by the Moon is that the eclipses do not occur on another day, but only on the 30th, when the Moon is in conjunction with the Sun, and also from the fact that the magnitudes of the eclipses are relative to the locations [of the observers].

---

that the Moon covers the Sun. This, of course, reflects what we actually see during eclipses. The diagrams in chapters x and xi were probably not originally part of Geminos's text. But the person who added the diagrams thought about eclipses in the same way as Geminos did, and the diagrams conform to the spirit of the text.

[2] whenever the Sun and the Moon are at the same degree. Geminos here ignores the inclination of the Moon's path to the plane of the ecliptic. An eclipse of the Sun does not occur every time there is a new Moon, but only when the new Moon happens to be near one of the two nodes of the Moon's orbit. Thus, roughly speaking, solar eclipses are possible at two new Moons approximately 6 months apart.

[3] not . . . *eclipses* . . . but *coverings*. An eclipse (*ekleipsis*) is literally a "failing," or "cessation," from which it is clear that this word was applied to the astronomical event long before its cause was understood. Geminos therefore prefers to speak of a "covering" (*epiprosthēsis*).

[4] eclipses are not equal for all [observers]. Such a statement need not be based on observation. It follows from the geometry of the shadow diagrams drawn to illustrate the theory of eclipses, e.g., in Aristarchos's *On the Sizes and Distances of the Sun and Moon* (propositions 1, 2, and 13). Geminos asserts (x 6) that the dependence of eclipse magnitude on the observer's location constitutes evidence that solar eclipses are due to the covering of the Sun by the Moon. In fact, this is an example of the backward science so characteristic of textbook writing—in our own day, no less than in Antiquity. The variation of eclipse magnitude with the observer's location is a consequence deduced from the theory, used then as if it were a fact of observation to support the theory. It is characteristic that Geminos does not cite observations of any particular eclipse.

Recorded simultaneous observations of an eclipse from different locations were very rare in Antiquity. Hipparchos used information about a solar eclipse (probably of March 14, 189 B.C.) to determine the parallax of the Moon. The Sun was totally eclipsed near the Hellespont, but only four-fifths eclipsed at Alexandria. Hipparchos's efforts are mentioned by Ptolemy (*Almagest* v 14, 19). But it is to Pappos's *Commentary on the Almagest* that we owe the information about the circumstances of the solar eclipse (Rome 1931, vol. i, 68.) On Hipparchos's method, see Swerdlow 1969; Toomer 1975; Neugebauer 1975, 322–29; Van Helden 1985, 10–14. The same eclipse circumstances are mentioned by Kleomedēs (*Meteōra* ii 3.1).

XI. On the Eclipse of the Moon

XI 1 The eclipses of the Moon[1] occur because of the Moon's falling into the shadow of the Earth. Just as other bodies illuminated by the Sun cast shadows, so too the Earth, illuminated by the Sun, casts a shadow. And, of course, because of the size of the Earth, it results that the shadow is pure and deep.[2] 2 When the Moon is diametrically opposite the Sun, then too [as always] the shadow of the Earth is diametrically opposite the Sun; therefore the Moon, moving along lower than the shadow,[3] logically falls into the shadow of the Earth. 3 The part of it that falls into the shadow of the Earth always becomes unilluminated by the Sun because of the covering by the Earth. For at that time the Sun, the Earth, the shadow of the Earth, and the Moon are on the same straight line. 4 For this reason eclipses of the Moon do not occur on another day, but rather on the midmonth, for then the Moon is diametrically opposite the Sun.

[1] eclipses of the Moon. Ms. C contains a diagram, squeezed into the left margin of fol. 28v, illustrating the eclipse of the Moon. Because we doubt that it was part of Geminos's original work, we print it here in the Commentary, rather than in the translation.

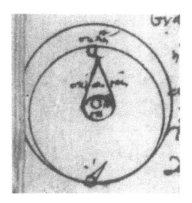

Diagram 8. An illustration of a lunar eclipse from Vaticanus gr. 318, folio 28. The three labels in the diagram read, from top to bottom, "Moon," "shadow of Earth," and "Earth."

[2] pure and deep (*eilikrinē . . . batheian*). The first adjective refers to the intensity, or absoluteness, of the shadow. The second refers primarily to the shadow's extension in space. However, *bathus*, like English *deep*, can also be used of colors.

[3] the Moon, moving along lower than the shadow. The Earth's cone-shaped shadow reaches out a certain distance from the Earth. The Moon is nearer to the Earth (i.e., "lower") than is the tip of the Earth's shadow.

XI 5 The eclipses of the Moon are, however, equal for all [observers]. For the coverings that occur in the eclipses of the Sun are different on account of the locations [of the observers], for which reason the magnitudes of the eclipses are different. But the Moon's falling into the shadow is equal for all during the same eclipse.

XI 6 However, an equal [part] of the Moon is certainly not always eclipsed. For when the Moon makes its passage through the middle of the eclipse zone,[4] the whole falls into the shadow of the Earth so that, of necessity, the whole of it is eclipsed.   7 But when it grazes the shadow,

---

[4] eclipse zone. By *to ekleiptikon* here and in the following sentence, Geminos means a zone into which the Moon must enter in order to be eclipsed, hence the translation "eclipse zone." (*Ekleiptikon* must not be confused with what we call the "ecliptic," known to Geminos as "the circle through the middles of the signs," as at ii 21 or v 52.) In xi 7, Geminos goes on to say that the *ekleiptikon* or eclipse zone is 2 degrees [wide]. That is, Geminos regards the eclipse zone of the Moon as a belt extending ±2° in latitude from the ecliptic. This figure is almost certainly a borrowing from system A of the Babylonian lunar theory. In Babylonian lunar ephemerides, there is a column, called column E by modern scholars, which represents the latitude of the Moon (probably the Moon's center), and which varies between the extremal values +6° and −6° (Neugebauer 1975, 514–17). E is used in calculating eclipse magnitudes. And, indeed, eclipses are deemed to be possible if the value of E falls within the range ±2° and the other necessary conditions apply (Neugebauer 1975, 522). So, in this short but fascinating chapter, Geminos employs a parameter borrowed from the Babylonians to illuminate a Greek, geometrical explanation of eclipses. In the Babylonian ephemerides, E is tabulated, not in terms of the degree, but in terms of the "barleycorn" (*she*): 72 *she* = 1° (Neugebauer 1975, 514–15). The extremal value of the lunar latitude for which an eclipse is possible is $E = 144$ *she*, or 2° exactly. Geminos has heard enough about Babylonian eclipse theory to express the range of latitude in units comprehensible to a Greek reader. On the other hand, he is vague about just what the *ekleiptikon* is. For example, he does not specify that 2° is the maximum latitude *of the Moon's center* for which a grazing eclipse is possible. And Geminos gives no sign that he understands how this parameter is used in practice. In the Babylonian ephemerides, E is not of interest in itself, but is used to calculate values for column Ψ, which represent eclipse magnitudes. Cuneiform tablet ACT 60 is a table of lunar eclipses for the years 175–152 B.C., computed theoretically in system A. Its eclipse magnitudes compare well with those obtained from modern tables. See the graph in Neugebauer 1955, 108 and the table in Neugebauer 1955, vol. 3, plate 38.

There does not appear to be any way to explain Geminos's *ekleiptikon* of ±2° in the context of elementary Greek astronomy, since any plausible argument would result in a considerably narrower band. A simple analysis in the spirit of elementary Greek astronomy might go like this: According to Hipparchos (Ptolemy, *Almagest* iv 9), the angular diameter of the Moon is 1/650 of a great circle, and the angular diameter of the Earth's shadow is 2½ times the angular diameter of the Moon. (Aristarchos said only twice as large.) Both of Hipparchos's figures apply to the Moon at mean distance in the syzygies. From these figures, the diameter of the Earth's shadow is 2.5 × 360°/650 = 1.38° (not a bad value). Half of this is 0.69°, which is the maximum possible geocentric latitude of the eclipsed part of the Moon. Adding 0.28° (a semi-diameter of Hipparchos's Moon), we get ±0.97° for the extremal latitude of the Moon's center during a grazing eclipse. So there appears to be no way for Geminos to arrive at an *ekleiptikon* that is twice as large, following this line of thought. A more sophisticated analysis, taking the Moon's parallax into account, *will*

[only] a certain part of the Moon is eclipsed. Its eclipse zone is 2 degrees [wide], for all the eclipses of the Moon occur in this space.[a]

---

broaden the eclipse zone, but would also make it asymmetric. Thus, in our view, there is no way that Geminos's 2° figure could have arisen from a purely Greek context. Finally, in our view, attempts to explain the 2° figure as the difference in apparent positions of the eclipsed Moon viewed by two widely separated observers (e.g., Aujac 1975, 145n3) have little probability, for it seems clear that Geminos is describing the range in the latitudes of the eclipsed Moon that might be noted by a single observer at a fixed location.

How well does the Babylonian parameter of ±2° correspond to the facts? To answer this question, we start from Aristarchos's eclipse diagram, in Fig. 11.1. σ is the angular radius of the Sun, as observed from the Earth's center C. τ is the geocentric angular radius of the Earth's shadow, at the distance of the Moon. Finally, p_S and p_M are the horizontal parallax of the Sun and the Moon, respectively. Now, $p_S + p_M = 180° - \chi$. Also, $\sigma + \tau = 180° - \chi$. Therefore, $p_S + p_M = \sigma + \tau$. We can use this relation to obtain a value for the maximum angular radius of the Earth's shadow. However, absorption effects in the Earth's atmosphere increase the theoretical radii by about 2% (Smart 1977, 380). Thus we use $\tau = (p_S + p_M - \sigma)51/50$. Letting the Moon be as close to the Earth as possible maximizes p_M. Letting the Sun be as far way as possible minimizes σ. (p_S is so small that its change in value scarcely matters.) Thus, we take $p_M = 1.006°$ (the largest possible value, corresponding to a geocen-

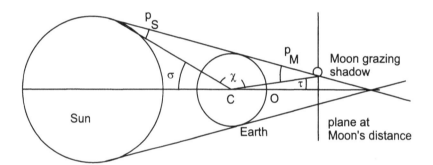

Fig. 11.1. Aristarchos's eclipse diagram, used as a starting point for investigating the maximum possible latitude of the Moon during a grazing lunar eclipse.

tric lunar distance of about 57 Earth radii), $\sigma = 0.262°$ (Sun at apogee), and $p_S = 9''$. With these figures, we obtain $\tau = 0.761°$. An observer at O is one Earth radius closer to the Moon than is C, and the apparent angular radius of the shadow, viewed from O, becomes $\tau_O = \tau(57/56) = 0.775°$. Without consideration of parallax effects, this is the maximum latitude of the edge of Earth's shadow. Now, since Babylonians tabulate the latitude of the Moon's *center*, when the edge of the Moon is eclipsed, we should add 0.274° (the maximum angular radius of the Moon), so we get a theoretical range in E (again, before the consideration of parallax effects) of about ±1.05°. This is close to the figure of ±0.97° obtained above from elementary considerations.

At the latitude of Babylon (32.5°), when the Moon is eclipsed at the meridian it may be as high as about 82° above the horizon (eclipse in midwinter, Moon slightly above ecliptic), in which case parallax shifts the Moon only about 0.14° lower in the sky. (The parallax may be calculated from $\sin p = (1/57) \sin z$, where p is the parallax, z is the Moon's ap-

XI 8 It is clear from these things that the eclipses of the Moon are due to the [Moon's] falling into the shadow of the Earth, for the magnitudes of the eclipses are in due relation to the Moon's motion in latitude.[b] And the eclipses of the Moon do not occur on other days than on the mid-month.

---

parent zenith distance, and the Moon is assumed to be at a geocentric distance of 57 Earth radii.) The eclipsed Moon on the meridian may be as low as about 32° above the horizon (eclipse in midsummer, Moon slightly below ecliptic), in which case parallax makes the Moon appear 0.85° lower. To conclude: for eclipses that take place at the meridian, at Babylon, the apparent latitude E of the Moon's center ought to range from about +0.91° to about −1.90°. (Note that 0.91 = 1.05 − 0.14 and 1.90 = 1.05 + 0.85.) The Babylonian convention, to which Geminos alludes, of putting the range at ±2° somewhat overstates the actual range (roughly +0.91° to −1.90°), but is actually pretty good. It should be noted that considerations of parallax were outside the scope of Babylonian astronomy.

XII. That the Planets Make the Movement Opposite to [That of] the Cosmos

XII 1 The cosmos moves in a circular motion from east to west. For all the stars that are observed in the east after sunset are observed, as the night advances, rising always higher and higher; then they are seen at the meridian. 2 As the night advances, the same stars are observed declining toward the west; and at last they are seen setting. And this happens every day to all the stars. Thus it is clear that the whole cosmos, in all its parts, moves from east to west.

XII 3 That it makes a circular motion is immediately clear from the fact that all the stars rise from the same place and set in the same place. 4 Moreover, all the stars observed through the *dioptras*[1] are seen to be making a circular motion during the whole rotation of the *dioptra*s.

THE EASTWARD MOTION OF THE SUN AND MOON

XII 5 The Sun, however, moves [also] from west to east, oppositely to the cosmos. This is clear from the stars that rise before the Sun: for those stars that are observed before sunrise to have risen before the Sun are observed on the following nights to have risen earlier in advance [of the Sun]. And this happens every night, one after another. 6 From this it is clear that the Sun passes toward the following signs,[2] moving from west to east, oppositely to the cosmos.

XII 7 If, instead, the Sun were moving from the east to the west, it would always happen that the stars rising in advance [of the Sun] become invisible. For, passing toward the preceding parts, it would be bound to hide them with its own rays. For the stars near the Sun are always invisible, as they are outshined by the Sun. 8 In fact, this does

---

[1] stars observed through the *dioptras*. *Dioptra* is the general term for a sighting instrument, and several different instruments were called by this name. Geminos here refers to a sighting tube (or pair of sights attached to a rod) that could be aimed at a star. After the sights were aimed at the star, the instrument could be rotated around the polar axis. In this way, the sights could be kept on the given star during the whole course of the night. This circular rotation of the instrument was then a demonstration of the circular motion of the star. See fig. I.13 and sec. 8 of our Introduction.

[2] toward the following signs. That is, toward the signs that follow behind the Sun during the diurnal rotation of the cosmos. Thus, "toward the following signs" is toward the east; and "toward the preceding signs" is toward the west.

not happen; rather, the stars rising in advance [of the Sun] are observed on the following nights to be always at a greater and greater distance from the rising [point], so that in the monthly period a whole sign rises in advance of the Sun—the one that was formerly in the rays of the Sun. For the sign following the Sun is always invisible because of the rays of the Sun, while the one preceding it is visible.[a]   9 In the monthly period the following sign always becomes invisible by the Sun's passing over it, while the preceding sign is seen to have stood away an interval of two signs.[b] And this always happens to the 12 signs.   10 From this it is clear that the Sun makes its passage moving oppositely to the cosmos, toward the following signs and not toward the preceding ones.

XII 11 The motion is observed more plainly in the case of the Moon, for this too is observed moving oppositely to the cosmos from west to east. This can be grasped in one night because vision bears witness to the phenomenon.   12 For when the Moon is observed near a certain one of the fixed stars, as the night advances it departs toward the east from the carefully noted star; and <the star, toward the> west <from the Moon>.[c] And often, during the whole night, it separates by 8 degrees toward the east from the carefully noted star.   13 Thus in one night the motion opposite to [that of] the cosmos is observed, for it does not pass toward the preceding stars, but toward the following ones.

Refutation of an Erroneous Opinion

XII 14 Some say that the displacement is toward the following signs for the Sun and the Moon, but that they do not actually move oppositely to the cosmos. Rather, they are left behind the sphere of the fixed stars because of their sizes. But the displacement seems to us to be toward the following signs in contrast to the opposite motion.   15 But this [they say] is not true; rather, Sun and Moon move from east to west, but, being outstripped by the cosmos before they run around the circle, they are observed in the following signs.

XII 16 Some also use this comparison: for if, they say, one supposes 12 runners[3] possessing equal speed and performing the motion in a circle,

---

[3] 12 runners. The twelve runners represent the signs of the zodiac. One should imagine them as equally spaced around a circular race track and all running at the same speed. The thirteenth and slower runner is the Sun. Vitruvius (*On Architecture* ix 1.15) asks his reader to imagine seven ants on a potter's wheel. The ants, representing the Sun, Moon, and planets, travel eastward at their own individual rates, while being carried westward by the rotation of the wheel. The ants on a wheel metaphor is much better, since it at least allows for two physically distinct motions—the westward motion of the wheel and the eastward march of the individual ants. However, Geminos would probably have disapproved of this metaphor, too,

and then another, slower one moving among them and performing a motion in a circle similar to theirs, he will seem, while being overtaken, to be passing toward the following [runners]. **17** This will not be the truth, but while he moves in the same fashion [as the other runners] he will seem, on account of the slowness, to be moving in the opposite direction.

XII 18 And this [they suppose] to have happened both in the case of the Sun and in the case of the Moon. For, while moving toward the same parts as the cosmos,[d] because of the slowness they are carried toward the following [signs], just as ships descending rivers, and being outrun by the current, seem to move backwards.[4] And this, they say, happens both in the case of the Sun and in the case of the Moon.

XII 19 But this opinion, maintained by many philosophers,[5] is not in harmony with the phenomena. For if they moved by a falling behind— the bodies slipping behind because of their sizes—then the falling behind would be bound to occur along parallel circles, just as all the fixed stars move on parallel circles because of the motion of the cosmos being circular from east to west. **20** But, in fact, they are not left behind along parallels. Rather, the Sun, moving on the circle through the middles of the signs, also makes at the same time its passage in declination,[6] from tropic to tropic, as if, I suppose, the motion proper to it were [along the ecliptic] from west to east.[e] **21** The Moon makes its passage over the whole width of the zodiac circle.[7] But nothing[f] that moves backwards in

---

since it fails to account for the Sun's (and planets') motion in declination. The ants are also used by Kleomēdēs (*Meteōra* i 2.17–19) as well as by Achilleus (see Maass 1898, 48.16–18).

[4] just as ships descending rivers, and being outrun by the current, seem to move backward. For another cosmic analogy involving ships and relative motion, see Kleomēdēs, *Meteōra* i 2.13–17.

[5] this opinion, maintained by many philosophers. Lucretius (*On the Nature of Things* v 621–36) maintains that the Sun and the Moon are left behind during the rotation of the cosmos and therefore *appear* to have a movement toward the east. He tells us, too, that Dēmokritos was of a similar opinion. In Geminos's day, the idea that the apparent eastward motions of the Sun and Moon might be due to a simple falling behind was certainly not maintained by anyone who was well versed in astronomy. This is perhaps why Geminos refers to the "philosophers," some of whom may have taught the mistaken view as a simplification suitable for beginning students. In fragment 2, Geminos makes a famous distinction between the methods of the astronomer and those of the physicist (or philosopher of nature). Kleomēdēs (*Meteōra* i 5.16–19), too, distinguishes between the "mathematicians" (we might say, "scientists") and the philosophical schools.

[6] in declination. *kata platos*, literally "in width." Like most Greek astronomical writers, Geminos uses this expression to describe a motion perpendicular to an astronomical circle of reference. Here, the motion is from north to south, perpendicular to the celestial equator. But Geminos uses the same expression for a motion in celestial latitude, i.e., a motion perpendicular to the ecliptic. (See comment 7).

[7] The Moon makes its passage over the whole width of the zodiac circle. Here Geminos uses *platei* ("width") for celestial latitude, i.e., for a displacement perpendicular to the

the manner of a falling behind can at the same time move in width; rather, it would be bound to perform the falling behind in accordance with the motion of the cosmos.

The Motion of the Planets

XII 22 But, most of all, the motion in the case of the five planets proves the opinion to be false. For those are sometimes left behind the fixed stars, sometimes take the lead, and sometimes remain by the same stars, which [events] are called "stations."[8]   23 As the motion in their case is of such a kind, it is clear that the shift toward the following [signs] does not occur by a falling behind, for they would always be falling behind. But, in fact, there is a certain spherical construction[9] proper for each, in accordance with which they pass sometimes toward the following [signs], sometimes toward the preceding, and they sometimes stand still.   24 Thus too in the case of the Sun and the Moon, the motion in declination is proper to them, separate [for each], and natural, in accordance with which they make the passage in declination while moving from west to east.

XII 25 That the shift cannot be produced by a falling behind toward the following signs is clear also from the fact that the shifts are in proportion neither to the sizes nor to the distances[10] [of the planets].   26 For if

---

ecliptic. The Moon, moving eastward around the zodiac once each month, also wanders north and south of the ecliptic by up to 5°. Thus, as Geminos says, the Moon, in making its eastward passage, makes use of the full width of the zodiac circle.

   [8] "stations." When a planet goes into or comes out of retrograde motion, it slows and then apparently stands still with respect to the background stars for a few days or a few weeks, depending upon the planet. The Greek term for a station is *stērigmos*.

   [9] spherical construction. *sphairopoiïa*. This is probably a reference to deferents and epicycles, a form of planetary theory that goes back at least to Apollōnios of Pergē (c. 200 B.C.). Geminos's description of it as a *spherical* construction seems to indicate that, already by his time, deferent-and-epicycle planetary theory had been worked into a three-dimensional system. In the second century A.D., the accommodation of deferent-and-epicycle planetary theory with nested-sphere cosmology was described in detail by Theōn of Smyrna and, most notably, by Ptolemy in his *Planetary Hypotheses*. Geminos's remark seems to indicate that Theōn and Ptolemy elaborated an already commonplace idea. Here, as elsewhere, Geminos also seems to take the *spherical construction* (a model of planetary motion) as actually representing nature: it is not merely a mathematical device for saving the phenomena. See sec. 10 of our Introduction and Evans 2003.

   [10] the shifts are in proportion neither to the sizes nor to the distances. If a planet fell behind the fixed stars through a failure to keep pace with them, the rate at which it fell behind might conceivably depend on the size of the planet and on its distance from the Earth. Presumably, the larger planets should fall behind more quickly, as should those that are closer to the Earth and therefore farther from the fixed stars. That there is no relation between the

they moved backwards because of the sizes of the bodies, through their having a slower motion than the fixed stars, it would be necessary that the fallings behind be in proportion to the sizes and the distances. 27 As this is not the case, one is forced to say that the opposite motion is natural for the planets. It has resulted, then, from the individual spherical construction[11] of each [planet] that the shifts are different.

---

planets' sizes and the rate at which they fall behind may be taken as a fair assertion. Of the two brightest (and apparently largest) planets, Venus and Jupiter, one falls behind very rapidly and one very slowly. By contrast, Geminos's case based on the distances suffers from circularity, for the planets' periods were often used as a criterion for judging their distances from the Earth. In fact, since most writers (including Geminos: see i 23–30) put the planets with the longest tropical periods farthest from the Earth, the traditional ordering would lead to a correlation of exactly the kind that Geminos denies. But as there was no independent method of determining the distances, no valid argument could really be based on them.

[11] spherical construction, *sphairopoiïan*.

XIII. On Risings and Settings[1]

DIURNAL RISINGS AND SETTINGS

XIII 1 The cosmos, moving from east to west in a day and a night, returns [to its starting position] from rising to rising. During the rotation of the cosmos all the stars both rise and set each day. 2 The appearance[2] at the horizon that takes place each day is a rising,[3] and the disappearance[4] below the horizon that takes place each day is a setting.[5]

HELIACAL RISINGS AND SETTINGS

XIII 3 In another sense, one speaks of [heliacal] risings and settings, which some ignorant people suppose to be used for the same concept. But there is a great difference between rising and [heliacal] rising.[6] For a rising is what has already been said, but a [heliacal] rising is the appearance taking place at the horizon that is considered in connection with the distance to the Sun. 4 The same principle [applies] also to the setting. For, in one sense, the disappearance below the horizon that takes place each day is called a setting,[7] but, in a different sense, the [disappearance] that

---

[1] On Risings and Settings. The title of the chapter is that of a book by Autolykos of Pitanē, written about 320 B.C. (See sec. 11 of the Introduction for a discussion of the central concepts and technical vocabulary.) The text of chapter xiii contains a large number of elementary errors, and is the most defective of Geminos's work. But the subject was elucidated clearly by Autolykos, whom Geminos follows in all significant details. It is unlikely that Geminos himself could have made all the elementary errors we shall encounter in chapter xiii. Many must be due to an early copyist or editor who tried to "improve" the text of a rather tricky chapter.

[2] appearance. *phasis*. The plural, *phaseis*, is the title of one of Ptolemy's works, which is devoted to the heliacal risings and settings of the fixed stars.

[3] rising. *anatolē*.

[4] disappearance. *krupsis*. This could also be translated as "hiding."

[5] setting. *dusis*.

[6] a great difference between rising and [heliacal] rising. In Greek astronomy, *anatolē* is used for a diurnal rising, and *epitolē* for a rising with respect to the Sun. We always translate the former as "rising." There is no English noun equivalent to *epitolē*; therefore, this is also usually translated as "rising." But whenever Geminos explicitly contrasts the two terms (as in this paragraph), we shall indicate *epitolē* by "[heliacal] rising."

[7] a setting. In contrast with the situation for diurnal risings and heliacal risings (comment 6), Greek does not offer a vocabulary distinction for the two kinds of setting. For both the diurnal setting and the heliacal setting one word, *dusis*, is used.

occurs with respect to both the horizon and the Sun at the same time [is also called a setting].

KINDS OF HELIACAL RISINGS

XIII 5 For each of the stars there are two [heliacal] risings: some of them are called morning and some are called evening. There is a morning rising whenever, at the same time as the Sun is rising, a certain star rises with it and is on the horizon at the same time. But there is an evening rising whenever, while the Sun is setting, a certain star rises and is on the horizon at the same time.

XIII 6 And of the morning and evening risings there are two different kinds: some of them are called true and some [are called] visible. They are true [morning risings] whenever a certain star, being truly on the horizon at the same time as the Sun is rising, rises with it.   7 But this rising is unobservable because of the rays of the Sun. On the following day the Sun, moving oppositely [to the cosmos] passes toward the following [signs], and the star rises before the Sun by just so much as the Sun shifts toward the following [signs] in the daily period.   8 The rising of the star cannot yet be seen, and it is still outshined by the Sun. But again on the following day the Sun passed toward the following [signs], and the star rises before the Sun by just so much as the Sun shifted in the two days.   9 During the following days, the star rises always earlier and earlier in advance [of the Sun]; and when it rises in advance by just so much that the rising of the star is visible, [the star] having escaped from the rays of the Sun [for the first time], then this star is said to have made its visible morning rising.   10 For this reason[a] the visible risings of the stars are announced in the public decrees.[8] For the true [risings] are unobserved and unobservable, while the visible are both announced and observed.

XIII 11 The same principle [applies] again[b] to the evening risings. For there are also two different kinds of these: some of them are called true, some visible. They are true whenever, at the same time as the Sun is truly setting, a certain star rises and is <on> the horizon at the same time, in

---

[8] the visible risings of the stars are announced in the public decrees (*psēphismasin*). Manitius translates *psēphisma* by "Kalender." The *parapēgmata*, or star calendars, were the most prominent public use of heliacal risings and settings, and Geminos probably does have them at least partly in mind. But *psēphisma* usually means a "decree" or "act." Heliacal risings and settings were sometimes used in legal contexts. For the use of the visible morning rising of Arcturus in a contract that wound up in court, and that was discussed in an oration of Dēmosthenēs, see "Androcles against Lacritos in Reply to the Latter's Special Plea," in Murray 1936–39, vol. 1.

strict reckoning. **12** These risings are also unobservable because of the rays of the Sun. And in the following days, as the distance to the star diminishes because of the motion of the Sun, [the star] rises before sunset; but it is still unobservable, being outshined <by> the Sun. **13** When it is observed after sunset,[9] having escaped for the last time[c] from the rays of the Sun, then it is said to have made its visible evening rising. On the following nights it appears always higher and higher.

[9] When it is observed after sunset. Geminos's discussion in this paragraph may seem confusing unless it is noted that his sequence of statements does not follow the sequence of events in time. Refer to fig. 13.1. Let the star be rising at S. Statement xiii 11, the definition of the true evening rising (TER), applies to the sunset that occurs when the Sun is at B: the star S rises exactly as the Sun B sets. Statement 12 applies to the situation "in the following days," when the Sun has reached C. Then, as Geminos says, the distance between the Sun and the star has diminished, and the star rises before sunset. But statement 13, the definition of the visible evening rising (VER), applies to the sunset that occurs when the Sun is at A. Then the Sun is just far enough below the western horizon for the rising of the star to be seen for the last time. The Sun moves eastward on the ecliptic and thus passes through A before B. In this paragraph, Geminos aims merely to introduce the basic concepts. He makes the time sequence clear below, at xiii 19.

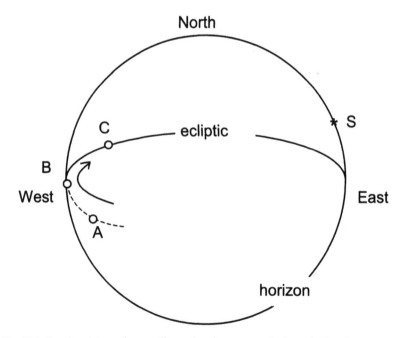

Fig. 13.1. Evening risings of a star, illustrating the sequence in time of sidereal events discussed by Geminos in xiii 11–19. The arrow indicates the direction of the Sun's motion on the ecliptic.

KINDS OF HELIACAL SETTINGS

XIII 14 In the same way there are also said to be two different settings: some of them are morning, some evening. They are called morning settings whenever a certain star sets while the Sun is rising. It is called an evening setting whenever a certain star goes down while the Sun is setting and goes under the horizon at the same time.

XIII 15 Of the morning settings there are two different kinds: some are true, some visible. They are true whenever both the Sun and the star are <exactly>[d] on the horizon, the Sun rising and the star setting. These settings are unobservable because of the rays of the Sun. **16** It is a visible morning setting whenever the star is seen setting for the first time[e] before sunrise.

XIII 17 In the same way in the case of the evening settings, there are two different kinds: some of them are true, some visible. They are true whenever both the Sun and the star are exactly on the horizon, both setting. These settings are also unobservable because of the rays of the Sun. **18** They are visible evening settings whenever, after sunset, a certain star goes down after the Sun and is seen by us.

PROPOSITIONS

XIII 19 Of the morning risings and settings, the true come first, and the visible come later.[10] But of the evening risings and settings, the visible come first and the true come later.[11]

XIII 20 From morning rising to morning rising, or from evening rising to evening rising, or in general from the same kind [of phase] to the same kind, is a year for all the stars.[12] For the Sun, passing through the zodiac circle in a year, is again next to the same stars.

XIII 21 For the <stars>[13, f] lying on the zodiac circle, [the time] from evening rising to morning rising is six months, and also [the time] from evening setting to morning setting.

---

[10] Of the morning risings and settings, the true come first, and the visible come later. This is proved by Autolykos, *On Risings and Settings* i 1.

[11] of the evening risings and settings, the visible come first and the true come later. Autolykos, *On Risings and Settings* i 1.

[12] from the same kind [of phase] to the same kind is a year for all the stars. Autolykos, *On Risings and Settings* i 6 and ii 5.

[13] For the <stars>. From here to the end of xiii 27, the text is badly corrupted. Every proposition in xiii 21–24 and 27 is either wrong or problematical as stated. See textual note f for a full discussion and proposed emendation, pp. 265–67.

XIII 22 But for the stars lying north of the zodiac circle, from evening rising to morning rising is a greater time than six months.

XIII 23 And for those lying south of the zodiac circle, from evening rising to morning rising is less than six months' time.

XIII 24 The time in excess of the six months is not fixed for all the stars, but is greater for some and less for others. 25 For those [stars] lying farther and farther north, the time becomes always greater and greater, because greater segments [of the stars' diurnal circles] extend above the Earth.[g] 26 For those lying farther south, the time becomes always less and less, because smaller segments [of the star's diurnal circles] extend above the Earth.[h]

XIII 27 Contrariwise, for the [stars] lying north of the zodiac circle, the time from morning setting[i] till evening rising is less than six months, while for those toward the south the time from morning setting[j] till evening rising is more than six months.[k] 28 The variation in the times depends upon the distances from the zodiac circle, in accordance with the variation in the segments cut off above the Earth by the horizon.

XIII 29 For the [stars] lying on the zodiac <circle>, morning rising and evening setting occur at the same time[14] [i.e., on the same day]; and, again, morning setting and evening rising [occur] at the same time. But for the remaining stars the aforementioned kinds [of phases] do not take place at the same time; rather, there is a difference in the times.

---

[14] morning rising and evening setting occur at the same time. That is, *on the same day*. This (correct) proposition applies only to the true phases. For a star on the zodiac circle, the true morning rising occurs when the Sun is located at the star itself. The true evening setting will obviously occur on the same day. Geminos's next proposition also applies to the true phases. For a star on the zodiac, true morning setting and true evening rising occur on the same day. In this case, the Sun is diametrically opposite the star on the zodiac, so that the Sun rises while the star sets and vice versa.

XIV. [On the Paths of the Fixed Stars]

DIVISION OF THE STARS' PATHS BY THE HORIZON

XIV 1 As the stars move in a circular motion from east to west, those that lie on the equator circle run equal paths above the Earth and below the Earth. For the equator circle is bisected by the horizon. 2 Those stars that lie north of the equator circle move a longer time above the Earth and a shorter [time] below the Earth.[1] For greater segments of all the circles on which the [northern] fixed stars[a] move are cut off above the Earth by the horizon, and smaller ones below the Earth, because of the elevation of the pole. 3 But those stars that lie south of the equator circle run a shorter path above the Earth, and a longer one below the Earth. For, contrariwise, of the circles on which the fixed stars move, those lying in the south[b] have smaller segments above the Earth and greater ones below the Earth.

SIMULTANEOUS RISINGS AND SETTINGS

XIV 4 Such paths for the fixed stars being granted, it follows that not all those that rise at the same time also set <at the same time>. Rather, of those that rise <at the same time>, the ones lying farther south always set first,[2] because they run shorter segments above the Earth. 5 In the same way, those that set at the same time do not also rise at the same time, for the ones lying farther north rise first, because they run shorter segments below the Earth.

XIV 6 And, again, [all] those <rising> first do not also set <first>.[3] Rather, certain ones of those rising before [another star] set at the same time [as it], and certain ones [set] later. 7 In the same way, of those set-

---

[1] stars that lie north of the equator circle move a longer time above the Earth and a shorter [time] below the Earth. Geminos naturally assumes that the phenomena are observed at a locality north of the equator.

[2] of those that rise <at the same time>, the ones lying farther south always set first. Geminos's xiv 4 and xiv 5 together are equivalent to Autolykos, *On the Moving Sphere* 9.

[3] [all] those <rising> first do not also set <first>. The points made in this paragraph (xiv 6–7) were given a more precise formulation in Euclid's *Phenomena*. Propositions 4 and 5 of the *Phenomena* may be paraphrased as follows. *Phenomena* 4: If two stars lie on a great circle that does not cut or touch the local arctic circle, the one rising first also sets first; and the one setting first also rises first. *Phenomena* 5: But if two stars lie on a great circle that cuts the local arctic circle, the one closer to the arctic pole will rise first and set last.

ting before [another star], certain ones do not rise before [it does];
rather, some rise at the same time, and some later.ᶜ

XIV 8 Aratos mentions these things to a certain extent when he says:

> But always is Taurus borne before the Charioteer
> to descend on the other side, though they come up in companion-
> ship.⁴

For in these [lines] he says that Taurus rises at the same time as the Charioteer but sets earlier. This occurs because of the difference of the segments that the fixed stars run above the Earth and below the Earth.

TYPES OF BEHAVIOR

XIV 9 Because of such a spherical arrangement,⁵ not all the stars both
rise and set each night. Rather, certain ones rise and set, certain ones rise
but do not set, while some neither rise nor set. 10 Indeed, the ones lying farther north that begin high [above the ground] after sunset are seen
still higher before sunrise.⁶ And those lying farther south are observed
neither rising nor setting, but move below the Earth the entire nighttime.

XIV 11 So, too, certain of the stars are called "doubly visible,"⁷ such
as Arcturus; for frequently it is observed setting after sunset and, on the
same night, it is seen rising before the Sun. For this reason it is called

---

⁴ Aratos, *Phenomena* 177–78. That is, Taurus and the Charioteer (Auriga) rise at the
same time, but Taurus sets first, since Auriga is farther north on the celestial sphere than
Taurus. This description (for the latitude of Greece) was reasonably accurate in Antiquity
and remains so today, as the reader may confirm on a celestial globe. This passage, however, was criticized by Hipparchos as being out of agreement with the phenomena (*Commentary on the Phenomena of Eudoxos and Aratos* i 5.14–18). Hipparchos pointed out
that only the feet of the Charioteer rose with Taurus, while the remaining parts of the body
rose with Pisces and Aries.

⁵ spherical arrangement, *sphairopoiian*.

⁶ the ones lying farther north that begin high [above the ground] after sunset are seen
still higher before sunrise. This is the kind of behavior called "dock-pathed" by Ptolemy:
neither the star's rising nor its setting may be observed, even though the star is in the sky
for the entire night. Thus, the two ends of the star's path across the visible hemisphere are
"docked," or cut off. See the Introduction, sec. 11.

⁷ "doubly visible." A star is said to be doubly visible (*amphiphanēs*, literally, "seen on
both sides") if it can be seen setting in the west after sunset and then rising in the east before sunrise on the same night. This behavior is only possible for stars located well to the
north. Ptolemy uses *amphiphanēs* in the same way as Geminos to characterize these northern stars (Ptolemy, *Phaseis* 6). But Ptolemy also calls such a star *eniautophanēs*, "visible all
year," because it has no period of invisibility. See the Introduction. Autolykos (*On Risings
and Settings* ii 15) points out that a star far enough north has no period of complete invisibility, but may be seen at some time of night during each night of the year.

"doubly visible," since it is observed both setting in the evening and rising in the same night.

XIV 12 Quite opposite is the arrangement for those that set before sunset and rise after sunrise, so that they are not seen the whole night long, either rising or setting. Some call these [stars] "night-escapers."[8]

XIV 13 These[d] properties do not exist for all times for the same stars. Rather, because of the displacement of the Sun, they change—both those involving the risings and those involving the settings.

---

[8] "night-escapers," *nuktidiexoda*. Manitius translates this by *Nachtpassanten*; Aujac, by *évadées de la nuit*. If a star sets before sunset and rises after sunrise, it will be above the horizon only during the daytime and will therefore be completely invisible. This is possible for any star that is not too far north. See the Introduction, sec. 11, p. 69.

Ptolemy uses this term in a completely different sense (*Phaseis* 6). For Ptolemy, a star is *nuktidiexodos* (which we translate as "night-pathed") if it is visible for the whole of its path above the horizon. This is possible only for stars far enough south of the ecliptic. Ptolemy uses *nuktidiexodos* in contrast to *kolobodiexodos*, "dock-pathed." For Ptolemy, then, any star that rises and sets falls into one of three categories: night-pathed, dock-pathed, or doubly visible, according as the star is located well south of the ecliptic, near the ecliptic, or well north of the ecliptic.

Geminos appears to use *nuktidiexodos* in contrast to *amphiphanēs*; i.e., the distinction is between stars that have a period of invisibility and those that do not. Thus, we must assume either that a shift in the meaning of *nuktidiexodos* occurred between the time of Geminos and that of Ptolemy (favored by Aujac, p. 148n4), or that Geminos has simply misused the word (favored by LSJ, s.v. *nuktidiexodos*). We have argued (in sec. 11 of the Introduction) that the technical terms for the three star classes were not standard before Ptolemy's time. Thus, the boundary between a mistake and an alternative use of the term is a fuzzy one.

XV. Concerning the Zones on Earth

XV 1 The surface of the whole Earth is spherical in shape and is divided into 5 zones.[1] Of these, the two around the poles, lying farthest from the path of the Sun, are called frigid and are uninhabited because of the cold. They are bounded by the arctic [circles and extend] to the poles. 2 The next ones, lying in an intermediate relation to the path of the Sun, are called temperate. These are bounded by the celestial arctic circles and the tropic circles, and lie between them. 3 The remaining [zone], in the middle of the aforementioned ones, lying under the very path of the Sun, is called torrid. It is bisected by the terrestrial equator circle, which lies beneath the celestial equator circle.

---

[1] The surface of the whole Earth . . . is divided into five zones. *Zonē* means a "belt," in this case a belt around the spherical Earth, lying parallel to the equator. Although various Greek writers attribute the zones to Parmenidēs or Pythagoras, we find the first clear system in Aristotle (*Meteorology* 362a32–b30). Aristotle divides the Earth into regions (although he does not call them *zones*) by means of the tropic, arctic, and antarctic circles, and points out that there are therefore two habitable regions—the one where we live and the corresponding one in the southern hemisphere. The lands beyond the summer tropic (where the noon shadow ceases to fall to the north) are uninhabitable because of the heat, and "the lands beneath the Bear" are uninhabitable because of the cold. Thus the system described by Geminos was in place by the middle of the fourth century B.C.

The arctic circle for Aristotle and Geminos is not the fixed circle of modern convention. Rather, it is the *local arctic circle*, i.e., the boundary between the always-visible stars for a given locality and those that rise and set. Aristotle refers to this as "the always-visible" circle. (For Geminos's discussion, see v 2.) The local arctic circle is by definition a circle on the celestial sphere: its radius, expressed as an angle, is equal to the latitude of the place of observation. The projection of this arctic circle onto the Earth serves as the boundary between the frigid and temperate zones. Since the size of the arctic circle varies with the latitude of the observer (v 29–39), it makes little sense to use this circle to define zones of habitability. However, as Geminos says (v 45–48), the convention was to use the arctic circle for the latitude of Greece, which results in a frigid zone that extends 36° from the pole. (See also xvi 7–12.)

The most informative source on the later history of terrestrial zones is Strabo (*Geography* ii 2.1–3.3), who summarizes and criticizes the views of several geographical writers, notably Poseidōnios and Polybios. According to Strabo, Poseidōnios asked how one could fix the limits of the temperate zones, which are nonvariable, by means of arctic circles that vary with the place of observation. Poseidōnios fixed the boundaries of the arctic and antarctic zones by means of celestial phenomena: in these regions the shadow of a vertical gnomon is *periskian*—that is, there are days in the year when the tip of the shadow traces a path all the way around (*peri*) the gnomon. For Poseidōnios, then, the boundary of the arctic zone is the same as that favored by modern convention. Similarly, the temperate zones are called *heteroskian*, because the noon shadow always falls in the same direction (always north in the north temperate zone). The zone between the tropics is called *amphiskian*, because the noon

XV 4 Of the two temperate zones, it happens that the northern one is inhabited by those in our *oikoumenē*,[2] which is approximately 100,000 stades in length and about half that in width.[3, a]

---

shadow falls both north and south in the course of the year. These terms are also used by Kleomēdēs (*Meteōra* i 4.132–33) and Ptolemy (*Almagest* ii 6).

Although Poseidōnios adopted five zones on the basis of celestial phenomena, he also claimed that, for human interest, two others could be distinguished. These are narrow bands straddling the two tropics. In these regions, the Sun is directly overhead for about half a month each year. Thus, they are more scorched than places either north or south of them. Polybios distinguished six zones, since he divided the tropical region into two zones by the equator, but this view was not very popular. On zones, see Dilke 1985 and Aujac 1976. (For Poseidōnios on the zones, see Edelstein and Kidd 1972, fragments 49 and 208–11, with translation in Kidd 1999, 108–15, 272–78, and commentary in Kidd 1988.)

[2] *oikoumenē*. The known inhabited part of the Earth. See comment 3 to chapter v.

[3] our *oikoumenē*, which is approximately 100,000 stades in length and about half that in width. The view that the latitudinal width of the *oikoumenē* is half its longitudinal extent was attributed to Eudoxos (Aujac et al. 1987a, 143) and was common in early Greek geography. Geminos asserts it again at xvi 4. But other writers maintained that the length is more than twice the width, and Strabo (*Geography* i 4.5) says that this position was taken, not only by the later writers, but also by the best of the early ones. Eratosthenēs put the length at 78,000 stades and the width at 38,000 stades, for a ratio of only slightly more than 2:1 (Aujac et al. 1987b, 156). (Much later, Ptolemy was also to argue for a ratio greater than 2:1.) There is nothing problematic about Geminos putting the ratio at 2:1, for, in his day, this was a traditional round value that avoided argument with competing authorities. Geminos's figure of 100,000 stades for the length of the *oikoumenē* may have arisen in the following way. At xvi 6, Geminos uses Eratosthenēs' figure of 252,000 stades for the circumference of the Earth. Since this is the circumference of the equator, the circumference of the latitude circle for latitude 37.5° (roughly that of Athens) would be 200,000 stades (cf. Strabo, *Geography* i, 4.6). Thus a 100,000-stade *oikoumenē* at that latitude would stretch halfway around the globe, for an extent of 180°. This is, in fact, just the longitudinal extent that Ptolemy was later to argue for (Berggren and Jones 2000, 110), against his predecessor, Marinos of Tyre, who had put the longitudinal extent of the *oikoumenē* at 225° (Berggren and Jones 2000, 71). To sum up: Geminos's remark seems to be based on (1) the old round figure of 2:1 for the ratio of the length to the width of the *oikoumenē*, (2) Eratosthenēs' value of 252,000 stades for the circumference of the Earth, and (3) an assumption that the *oikoumenē* is bounded by two meridians 180° apart. The remark is thrown in so casually, without explanation, that one could reasonably suspect that a gloss has worked its way into the text. But there is no compelling historical reason why the remark could not be Geninos's own.

XVI. On Geographical Regions[1]

XVI 1 Of those who dwell on Earth, some are called *synoikoi*, some *perioikoi*, some *antoikoi*, and some *antipodes*.[2] *Synoikoi* are those who dwell around the same place in the same zone [as we do]. *Perioikoi* are those who dwell in the same zone [as we] but around the circle. *Antoikoi* are those who dwell in the southern zone in the same hemisphere. *Antipodes* are those who dwell in the southern zone in the other hemisphere, lying on the same diameter as our *oikoumenē*,[3] which is why they have been called "with feet opposite" [*antipodes*]. 2 For all heavy bodies tend together toward the center, because the motion of the bodies is toward the middle[4]; thus, if from a certain region in our *oikoumenē* a straight line is joined to the center of the Earth, and extended, those who

---

[1] On Geographical Regions (*Peri Oikēseōn*). An *oikēsis* is a place of residence, an inhabited district. But as the residents of several *oikēseis* were purely hypothetical (see xvi 19–20), we have given the title of this chapter a more neutral translation. Throughout the body of the chapter, *oikēsis* is translated by "region." The Greek title of this chapter is the same as that of a book by Theodosios of Bithynia (c. 100 B.C.), and the subject matter of the chapter represents a common genre. The title of Theodosios's book is often rendered "On Habitations."

[2] some are called *synoikoi*, some *perioikoi*, some *antoikoi*, and some *antipodes*. It seems best to leave these terms untranslated. But if translations were desired, these might suffice: *synoikoi* = co-dwellers; *perioikoi* = near-dwellers; *antoikoi* = opposite-dwellers; *antipodes* = those with feet opposite. Inhabitants of North America could classify other North Americans as *synoikoi*, the Chinese as *perioikoi*, South Americans as *antoikoi*, and Australians as *antipodes*. Thus the two temperate zones are divided into four habitable regions. Kleomēdēs (*Meteōra* i 1.209–34) uses these terms in the same way, although he does not mention the *synoikoi*, and adds that the *antoikoi* are also called *antōmoi* ("shoulder to shoulder"). Achilleus uses a different scheme (see Maass 1898, 65):

Geminos	Achilleus	Geminos	Achilleus
synoikoi	*perioikoi*	*perioikoi*	*antoikoi*
antoikoi	*antichthones*	*antipodes*	*antipodes*

[3] *oikoumenē*. The inhabited part of the Earth. The term varies in its inclusiveness from writer to writer, but generally denotes the Mediterranean lands known to the Greeks: nearer Asia, Europe, and North Africa.

[4] the motion of the bodies is toward the middle. This is a reference to the doctrines of natural place and natural motion. Aristotle (*Physics* 230a19–231a2, 253b33–35) asserts

are at the end of the diameter in the southern zone are with feet opposite to those dwelling in the northern zone.

THE OIKOUMENĒ

XVI 3 Our *oikoumenē* is divided into three parts: Asia, Europe, and Libya. The length of the *oikoumenē* is approximately double the width. 4 For this reason, those who draw world maps[a] in proportion[b] draw them on oblong panels so that the length is double the width. Those who draw circular world maps have wandered far from the truth,[5] for the length [of a circular map] is equal to the width, which is not the case in nature. 5 Of necessity, the proportions of the distances are not preserved in the circular world maps, for the inhabited part of the Earth is a certain segment of a sphere having the length the double of the width, which cannot be bounded by a circle.

DIMENSIONS OF THE TERRESTRIAL ZONES[6]

XVI 6 The terrestrial great circle <that lies>[c] opposite the celestial meridian has been measured and found to be 252,000 stades,[7] with a

---

that the natural place of elemental earth is the center of the cosmos. Any particle of earth, if let go, naturally moves radially downward toward the center. From this doctrine, Aristotle (*On the Heavens* 296b7–297b22) deduces that the shape of the Earth is spherical and that the Earth is located at the center of the cosmos. Geminos probably mentions the center as the place to which all heavy matter tends, in order to reassure the reader that our *antipodes* are no more likely to fall off the Earth than we are.

[5] Those who draw circular world maps have wandered far from the truth. This was a traditional complaint, already made by Aristotle (*Meteorology* 362b12–15).

[6] *Dimensions of the Terrestrial Zones.* For a detailed discussion of this section of the text, see Aujac et al. 1987c, 170–71.

[7] 252,000 stades. This is Eratosthenēs' figure for the circumference of the Earth, attributed to him by many ancient writers, including Hero (*Dioptra* 35), Strabo (*Geography* ii 5.34), and Theōn of Smyrna (*Mathematical Knowledge* iii 3). The most detailed account of Eratosthenēs' procedure is that of Kleomēdēs (*Meteōra* i 7.49–110), from which it appears that Eratosthenēs' original value for the circumference of the Earth was 250,000 stades. It is likely that Eratosthenēs amended his figure to 252,000 in order to have a circumference evenly divisible by 60. Eratosthenēs' sixty-part division of the circle is attested by Strabo (*Geography* ii 5.7). The figure of 252,000 stades also results in an even number (700) of stades to the degree; however, the 360-part division of the circle was not used by the Greeks until about a century after Eratosthenēs (Neugebauer 1975, 590). In Geminos's day both the 60-part and the 360-part divisions were in common use, and Geminos himself makes use of both. Stades of several different lengths were used in Antiquity. On the

diameter of 84,000 stades. Then, the meridian circle being divided into 60 parts, one part is called a sixtieth, which is 4,200 stades. For if 252,000 stades be divided into 60 parts, the sixtieth [part] is 4,200 stades.

XVI 7 The distances between the zones are fixed in the following way. The width of each of the two frigid zones is 6 sixtieths [of the meridian circle], which is 25,200 stades. The width of each of the two temperate zones is 5 sixtieths, which is 21,000 stades. The width of the torrid zone is 8 sixtieths, so that from the equator to each of the tropics is 4 sixtieths, which is 16,800 stades.

XVI 8 From the terrestrial pole, which lies opposite the celestial pole, to the terrestrial arctic [circle] there are therefore 25,200 stades. From the terrestrial arctic [circle], which lies opposite the celestial arctic [circle], to the terrestrial <summer> tropic, which lies opposite the celestial summer tropic, there are 21,000 stades. From the summer tropic to the terrestrial equator, which lies opposite the celestial equator, there are 16,800 stades. 9 And, again, from the equator to the other tropic there are 16,800 <stades>; from the tropic to the antarctic[d] [circle], 21,000 stades; from the antarctic [circle] to the other pole, 25,200 stades. Thus the distance between the poles totals 126,000 stades, which is half of the perimeter of the Earth; for from pole to pole is a semicircle.

DIVISION OF THE CELESTIAL SPHERE

XVI 10 The distribution of these sixtieths is also the same on the armillary spheres.[8] For the armillary spheres are constructed in the following fashion. The arctic [circle] is separated from the pole by 36 degrees, which is 6 sixtieths, for 6 sixtieths are 36 degrees. The arctic [circle] is separated from the summer tropic by 30 degrees, which is 5 sixtieths. The summer tropic is separated from the equator by 24 degrees, which is 4 sixtieths. 11 The equator is separated from the winter tropic by an equal 24 degrees. The winter tropic is separated <from> the antarctic [circle] by 30 degrees. The antarctic [circle] is separated

---

problem of determining the length of Eratosthenēs' stade, see Dicks 1960, 42–46; Dilke 1985, 32–33; Aujac et al. 1987b, 148n3; and especially Engels 1985. A rough equivalence is about 9 stades to the mile, or 6 stades per kilometer. But some authors favor a shorter stade (see Fischer 1975 and Rawlins 1982).

[8] on the armillary spheres. On an ancient armillary sphere, the arctic and antarctic circles were represented by rings, which had, therefore, to be of fixed size. But the size of the local arctic circle varies with latitude. Therefore, as Geminos says, the fixed ring of the armillary sphere could really only represent the local arctic circle for one single *klima*. According to Geminos, it was customary to build the sphere so that the arctic circle was 36° from the celestial pole, which corresponds to the *klima* of Rhodes (latitude 36°).

from the southern pole by 36 degrees. And thus, again, from pole to pole the degrees total to 180, or 30 sixtieths. **12** For both the armillary spheres and the solid [globes] are constructed for this one *klima*, as only the arctic [circles] change in their distances [from the other circles] in some regions. In any case, the terrestrial zones receive their division in accordance with the one aforementioned *klima*.

VARIATION OF THE PHENOMENA WITH LOCALITY

XVI **13** For those inhabitants of the Earth who live on the same parallel, the [celestial] phenomena that depend on locality are the same: the lengths of the days are equal, as are the magnitudes of eclipses, and the lines inscribed on the sundials are the same. **14** And, in general, everything is the same for regions lying on the same parallel; for the inclination of the cosmos remains the same, and the phenomena become different because of the inclination of the cosmos.

XVI **15** Of course, the beginnings and ends of the days do not occur at the same time for all, but earlier for some and later for others. It is the first hour for some, when it is the middle of the day for others, and sunset for still others. **16** However, as far as perception is concerned, the horizon remains the same for approximately 400 stades[9] from east to west, so that, as far as perception is concerned, the rising occurs at the same time for them, and also the setting. But whenever the distance is greater than 400 stades, one rising occurs before the other, and one setting before the other.[e]

XVI **17** For those who dwell on the same meridian, the change of *klima* is imperceptible up to 400 stades. But with a shift of a greater distance toward the north or toward the south, there is another inclination [of the cosmos], so that all the phenomena become different. **18** For the lengths of the days, the magnitudes of eclipses, and the lines inscribed on the sundials all vary with locality for those who dwell on the same meridian. For the inclination changes if there is a displacement toward the north or toward the south. However, the middles of the days and the middles of the nights occur at the same time for all those who dwell on the same meridian.

IS THE SOUTHERN ZONE INHABITED?

XVI **19** When we speak of the southern zone and of those dwelling in it, as well as the so-called *antipodes* in it, we should be understood in

---

[9] the horizon remains the same for approximately 400 stades. Geminos's estimate is reasonably good. See comment 31 to chapter v, p. 158.

this way: that we have received no account of the southern zone nor [any report of] whether people live in it, but rather that, because of the whole spherical construction [of the cosmos], and the shape of the Earth, and the path of the Sun between the tropics, there exists a certain other zone, lying toward the south and having the same temperate character as the northern zone in which we live. **20** In the same way, we speak of the *antipodes*, not in the sense that people positively dwell diametrically opposite us, but rather that there is on the Earth a certain habitable place diametrically opposite us.

On the Torrid Zone

XVI 21 Certain of the ancients, and Kleanthēs[10] the Stoic philosopher among them, declared that Ocean spreads over the torrid zone between the tropics. **22** In conformity with them, Kratēs[11] the grammarian, in setting out the wanderings of Odysseus and in inscribing the whole sphere of the Earth with the defining circles, just as we have said, makes Ocean lie between the tropics, saying that the whole arrangement is made in conformity with the mathematicians.

XVI 23 But such an arrangement is alien both to mathematical and to physical thought, and has not been adopted by a single one of the ancient mathematicians, as Kratēs claims. **24** For now in our time [the zone] between the tropics has been reconnoitered and has been found habitable for the most part and not surrounded on every side by sea. And, of the 16,800 <stades'> distance between the summer tropic and the equator, approximately 8,800 stades have been traveled over, and the account concerning these places has been recorded, [the question] having been investigated through [the agency of] the kings in Alexandria.[12] And therefore those who hold that the Ocean spreads between the tropics are quite mistaken.

---

[10] Kleanthēs. Kleanthēs of Assos (third century B.C.), successor to and disciple of Zēnōn (Zeno of Citium). Among historians of astronomy, he is best known for his suggestion that Aristarchos of Samos be indicted "on the charge of impiety for putting in motion the hearth of the universe" (Plutarch, *On the Face in the Moon* vi, 922F–923A). Diogenēs Laertios (*Lives and Opinions* vii 174) says that Kleanthēs wrote a book *Against Aristarchos*, which was perhaps where Kleanthēs made this attack.

[11] Kratēs. See comment 4 to chapter vi. Kratēs asserted that the middle of the torrid zone is occupied by Ocean, and that there are temperate lands, inhabited by *Aithiopians*, both north and south of it. Kratēs' theory of the Ocean and the globe on which it was presumably represented are mentioned by Strabo (*Geography* i 1.7, 2.24; ii 5.10). See Aujac et al. 1987c, 162–64.

[12] investigated through [the agency of] the kings in Alexandria. Strabo (*Geography* xvii 1.5) says that the Ptolemaic kings of Egypt gradually acquired a knowledge of the upper Nile by sending men to those parts on elephant hunts and other business. According to Strabo, Ptolemaios Philadelphos (ruled 285–247 B.C.) was of an especially inquiring disposition.

XVI 25 From these things it is clear too that imagining that the land lying between the tropics is uninhabited because of the excessive heat, and especially [the land] around the middle of the torrid zone, is quite wrong. 26 For those inhabiting the boundaries of the torrid zone are *Aithiopians*, who have the Sun at the zenith on the [summer] solstice. Indeed, by the nature of things, one ought to suppose there are two *Aithiopias*, with *Aithiopians* dwelling both around our summer tropic circle[f] and around our winter tropic,[13] which is summer [tropic] for the *antipodes*.

XVI 27 Kratēs claims also that Homer means this in the verses where he says:

> *Aithiopians*, who are divided in two, the remotest of men,
> some at the setting of Hyperion, others at the rising.[14]

And so Kratēs, speaking in marvels, takes things said by Homer for his own purposes and in archaic fashion, and transfers them to the spherical construction that accords with reality.[g] 28 For Homer and the ancient poets,[15] nearly all of them, mean to say that the Earth is flat and [extends to and] meets the cosmos; that the Ocean lies around [the Earth] in a circle, occupying the place of the horizon; and that the risings [of the Sun] are from the Ocean and the settings are into the Ocean. Consequently, they supposed the *Aithiopians* near the rising and those near the setting to be burned by the Sun. 29 This notion is consistent with their proposed arrangement, but alien to the spherical construction in accord with nature.[h] For the Earth lies in the middle of the whole cosmos and has the status of a point. The risings and settings of the Sun are from the *aithēr*[16] and into the *aithēr*, since the Sun is always equally distant from the Earth.

XVI 30 Therefore, while the aforementioned *Aithiopians* [i.e., those of Homer] are incomprehensible, those lying under the celestial tropics,

---

[13] *Aithiopians . . .* around our summer tropic . . . and . . . winter tropic. Ptolemy (*Geography* i 9) also believes that similar people must live at equal distances to the north and south of the equator. He says, however, that *Aithiopians* do not live quite as far north as the summer tropic, from which he infers that they must not live quite as far south as the winter tropic. The *Aithiopians* of the ancient Greek writers should not be confused with the modern political entity of Ethiopia. An *Aithiopian* was someone "with burned face," that is, with dark skin.

[14] Homer, *Odyssey* i 23–24. These lines, and Kratēs' interpretation of them, are discussed by Strabo (*Geography* i 1.6, i 2.24–27).

[15] Homer and the ancient poets. For a discussion of the worldview of Homer and Hesiod, see Dicks 1970, 27–38, or Evans 1998, 3–5. That Ocean lies everywhere around the Earth is clear from Homer's description of Achilles' shield, which was decorated with an image of the cosmos (*Iliad* xviii 483–89, 607–8). The Sun rises from the Ocean and sets into Ocean (*Odyssey* xxii 197–98. *Iliad* viii 485; xviii 239–40).

[16] *aithēr*. The *aithēr* is the fifth element, of which the celestial bodies are made. See Aristotle, *On the Heavens* 268b11–270b26. Thus Geminos insists that when the Sun rises and sets, it does not at all descend into the Ocean, but remains in its place among the celestial spheres of elemental *aithēr*.

who are on[i] the boundaries of the torrid zones, are in conformity with nature. **31** One certainly must not suppose the torrid zone to be uninhabited, for people have now gone to many places of the torrid zone and have found them for the most part habitable.

XVI 32 Thus, too, it is asked by many whether the regions around the middle of the torrid [zone] are more habitable than those around the boundaries of the torrid zone. Polybios the historian has even composed a book that has the title "On the region about[j] the equator," which is in the middle of the torrid zone. **33** He asserts that the places are inhabited, and that the region has a more temperate character than those situated around the boundaries of the torrid zone.[17] First he cites accounts of those who have seen the regions and bear witness to the phenomena; then he reasons on the natural motion of the Sun.

XVI 34 For the Sun stays a long time near the tropic circles, both in the approach to them and in the retreat [from them], so that, as far as perception is concerned, it remains on the tropic circle for about 40 days. **35** For this reason the lengths of the days also remain the same for about 40 days.[k] And therefore, since the delay occurs at the regions lying under the [celestial] tropics, of necessity the region is completely burned and is uninhabited because of the excessiveness of the heat.

XVI 36 But the retreats [of the Sun] from the equator circle are rapid, which is why the lengths of the days receive large increases around the equinoxes.[l] It is quite logical, then, that the regions lying under the [celestial] equator are more temperate, as there is <no> delay[m] at the point overhead, and the Sun[n] quickly retreats.

XVI 37 For all those dwelling between the tropic circles lie in the same relation to the path of the Sun; but it tarries a longer time for those dwelling around the tropics. **38** For this reason it results that the regions around the equator, which lie in the middle of the torrid zone, are more temperate than those situated near the boundaries of the torrid zone, which lie under the tropic circles.

---

[17] [Polybios] asserts that [the equatorial] region has a more temperate character than those [places] situated around the boundaries of the torrid zone. Although Geminos mentions only Polybios, many others held the same opinion, including Eratosthenēs (see Strabo, *Geography* ii 3.2) and Poseidōnios (see comment 1 to chapter xv, pp. 208–9).

XVII. On Weather Signs from the Stars

XVII 1 The opinion concerning weather signs current among laymen involves a spurious belief, that the changes in the air occur because of the [heliacal] risings and settings of the stars. But mathematical and physical opinion has it otherwise.

ALL WEATHER OCCURS NEAR THE EARTH

XVII 2 And first it must be mentioned that what signs there are of rainstorms and wind occur near the Earth, and do not extend to a greater height. For exhalations from the Earth[1] are varied and irregular, and therefore cannot extend to the sphere of the fixed stars; rather, the clouds do not [even] reach up to ten stades[a] in height.

XVII 3 In any case, those who climb Kyllēnē,[2] the highest mountain in the Peloponnesos, and sacrifice to Hermēs, to whom offerings are made on the summit of the mountain, find, when they climb again and perform the sacrifices a year later, that the thigh bones and the ashes from the fire remain in the same condition in which they had left them, and that they have not been changed by winds or by rainstorms, because all the clouds and the formation of the winds occur below the summit of the mountain.

---

[1] exhalations from the Earth. In Aristotle's *Meteorology*, two kinds of exhalation are given off by the Earth when it is heated by the Sun. One, a hot, dry exhalation, is given off by earth itself; the other, cold and moist, is given off by water (*Meteorology* 340b26, even more explicit in Pseudo-Aristotle, *On the Cosmos* 394a9–22). Aristotle calls the cold, moist exhalation "vapor" (*atmis*) to distinguish it from the hot, dry "exhalation" proper (*anathumiasis*). The lower part of the air contains both exhalations and is therefore hot and moist. It is in this region that clouds form. The upper air contains only the hot, dry exhalation, which prevents the formation of clouds. According to Aristotle, a second reason why clouds cannot form above the tops of the highest mountains is that the air there is carried around by the celestial motion. Geminos's explanation is much simpler—that the cloud-forming exhalations (*anathumiaseis*) simply do not reach to the tops of the mountains. But this is in accord with Aristotle, who kept the cold, moist exhalation near the ground. In any case, Geminos here adopts a position opposed to the Stoic view that exhalations from the Earth reach to and nourish the stars (Cicero, *On the Nature of the Gods* ii 83; Kleomēdēs, *Meteōra* i 4.128–30 and 8.79–82; Kidd 1999, 174; Bouché-Leclercq 1899, 75).

[2] Kyllēnē. Mt. Kyllēnē or Cyllene (modern Killíni) is in northeastern Arcadia. According to Strabo (*Geography* viii 8.1), some assign it a height of 20 stades, and others 15 stades. Pausanias (*Description of Greece* viii 17.1) says that in his day (second century A.D.) the temple to Hermes at the summit was dilapidated. Pausanias gives the impression that there were, indeed, remnants of many sacrifices lying about. Today, Mt. Killíni is the site of the Kryonerion astronomical station of the National Observatory of Athens.

4 And many times, too, those who climb Atabyrion[3, b] make the ascent through the clouds and see the mass of clouds down below the summit of the mountain. 5 And the height of Kyllēnē is less than 15 stades, as Dikaiarchos[4] shows he has measured, while the vertical drop of Atabyrion is less than 10 stades.[c] For, as we have said, since all the clouds involve the exhalation from the Earth, they occur near the Earth.

Empirical Nature of the Parapēgmata

XVII 6 The predictions from the weather signs that occur in the *parapēgmata* are not [made] from particular, definite precepts, nor are they treated methodically by a particular science, nor do they involve a necessary result.[5] Rather, taking whatever was concordant from that which generally arises through daily observation, they [i.e., the old astronomers] inserted it in the *parapēgmata*. 7 The compilation and observation came about in this way: They took the beginning of the year and, having observed in which sign the Sun started at the beginning of the year, they recorded against the degree [of each zodiac sign], day by day and month by month, the important changes that occurred in the air, winds, rains, and hail; and they placed these beside the positions of the Sun reckoned by sign and by degree. 8 Having observed these things for more years, they recorded in the *parapēgmata* the changes that occur for the most part around the same places of the zodiac, not taking the record from any particular science or from a definite method, but rather taking from experience whatever accorded most closely.

XVII 9 Since they could not record a day, a month, or a definite year[6] in which a certain one of these [events] occurs, because the beginnings of

---

[3] Atabyrion. Mt. Atabyrion, or Atabyris (modern Attáviros), is the highest mountain in Rhodes, located in the southwestern interior of the island. Strabo (*Geography* xiv 2.12) says that the mountain is dedicated to Zeus. This was already the case in Pindar's day (fifth century B.C.) (Pindar, *Olympian Odes* vii 87).

[4] Dikaiarchos. Dikaiarchos of Messina in Sicily (fl. c. 320 B.C.), said to have been a pupil of Aristotle, wrote a geographical treatise that was known to Strabo. He also wrote a work on the measurement of the heights of mountains, perhaps the first of its kind. For the ancient citations of Dikaiarchos, see Bunbury 1879, i 616–18. On his measurement of mountains, see Lewis 2001, 158–62.

[5] result (*apotelesma*). As an astrological term, this word indicates the "result" of a particular position of the stars. Various astrological works were written under the titles *apotelesmata* or *apotelesmatika*. See Bouché-Leclercq 1899, 83, 328.

[6] they could not record a day, a month, or a definite year. Geminos refers to the fact that the civil calendars of the Greeks varied from city to city in the names of their months and the starting points of their years. Each city also inserted extra months and days at will. Often,

the years are not everywhere the same, and the names of the months are not everywhere the same, and the days are not reckoned in the same way, they wished to mark the changes of the air by means of certain fixed signs. 10 Thus the changes of the air occur <in connection with>ᵈ the risings of the stars, which define the seasons,ᵉ not in the sense that the stars have a power with respect to the changes in the winds and the rains, but in the sense that they are accepted for the sake of a sign for our foreknowledge of the conditions in the air. 11 And just as the signal fire is not itself a cause of a state of war, but is a sign of a season of war, in the same way the risings of the stars are not themselves causes of the changes in the air, but are established signs of such conditions.

XVII 12 Having observed carefully from the beginning and having compiled the *parapēgmata* [by] scrutinizing the places of the [Sun in the] zodiac circle at which the changes of the air generally occur, they observed at these times which one of the stars rises or sets with [each change]. And by means of the risings and settings of these [stars] they furnished themselves with signs for knowing the changes of the air in advance. 13 And therefore they used by preference the visible risings and settings in the prognostication of the aforementioned [changes]. For the true risings and settings are unobservable, while they were able to see the visible ones at the announced times. 14 They took it, then, that the Pleiades when setting have a certain such power, so as to engender a certain wetness in the air; or, again, when rising they signal the beginning of summer. Thus also Hesiod says:

When the Pleiades, daughters of Atlas, are rising,
begin the harvest; the plowing, when they set,[7]

not because of the power of the star, for that is completely stupid.[8]

---

days were inserted or deleted for purely administrative reasons, e.g., to delay a religious festival until the preparations were ready but still permit the festival to be held on the prescribed date. (For examples, see Pritchett and Neugebauer 1947, 20–22; Bickerman 1980, 35–36.) Even without arbitrary tampering, a particular civil date in a particular city's calendar could shift back and forth by a whole month with respect to the celestial phenomena. The heliacal risings and settings of the stars provided what the Greek civil calendars could not: fixed guideposts in the year of the seasons. Geminos suggests that the use of astronomical signs to tell the time of year (required because of the unsatisfactory nature of the Greek calendars) helped people leap to the mistaken conclusion that the stars cause the weather.

[7] Hesiod, *Works and Days* 383–84.

[8] completely stupid. *apoplēkton*, from which comes our "apoplectic." In its medical use, the word meant "struck senseless or paralyzed by a stroke." Geminos's use of this term signals the strength of his feeling on this subject.

The Stars Indicate but Do Not Cause the Weather

XVII 15 For whether the stars are fire or *aithēr*,[9] as some find pleasing, they all partake of the same substance and power and have no sympathy[10] with things occurring on the Earth. 16 The whole Earth has the relationship of a center to the sphere of the fixed <stars>,[11] and no efflux or emanation reaches from the fixed stars to the Earth. How can one suppose to be causes of rains, winds, and hail things from which no influence falls to us?

XVII 17 Now, the power from the Sun and Moon does reach as far as the Earth, and, in accordance with their motion, is either greater or smaller. Thus, logically, there is a sympathy involving these [two stars] that accords with the power of each of them. But the risings and settings of the fixed stars have [only] the status of a sign, as we said before.

[9] whether the stars are fire or *aithēr*. Aristotle says the stars are made of the fifth element, or *aithēr*, and refutes the earlier thinkers who said the stars were fire. (*On the Heavens* 289a11–35. See also Pseudo-Aristotle, *On the Cosmos* 392a6–9.) Earlier thinkers who asserted the fiery nature of the stars included Anaximandros, Anaximenēs, Xenophanēs, and Empedoklēs. For the preserved fragments, see Diels 1951, Heath 1932, or Kirk and Raven 1966. Nearer to Geminos's day, the fiery nature of the stars was maintained by the Stoics, including Kleanthēs (Cicero, *On the Nature of the Gods* ii 40) and Poseidōnios (fragment 127 in Kidd 199, 182). Geminos declines to be drawn into this philosophical debate.

[10] sympathy. Geminos's remark here could be translated, "and have no rapport with things occurring on the Earth." On sympathy, see comment 4 to chapter ii. *Sympatheia* is a technical term, which Geminos uses in two different ways. Here in chapter xvii he refers to a "sympathy" (which he rejects) between the stars and the Earth as the source of a purported influence of the stars and planets upon the weather. Geminos's argument here is squarely opposed to Stoic doctrine. In chapter ii, Geminos uses "sympathy" to refer to a possible rapport between people born under signs of the zodiac that stand in certain relations to one another. Geminos seems more disposed to accept this sort of sympathy, commonly used by the astrologers.

[11] The whole Earth has the relationship of a center to the sphere of the fixed <stars>. That is, the Earth is a mere point in comparison with the sphere of stars. Thus the distance of the sphere of stars from us is truly vast. Earlier authors who refer to the pointlike character of the Earth, relative to the cosmos, include Euclid, implicitly in *Phenomena*, prop. 1 (Berggren and Thomas 1996, 52), and Archimēdēs, explicitly, in his *Sand Reckoner* (Heath 1912, 222). Archimēdēs attributes this view to Aristarchos of Samos, but criticizes it on mathematical grounds, because a point has no magnitude. The pointlike character of the Earth posed a problem for the Stoics, who maintained that an emanation from the Earth nevertheless reaches and nourishes the stars. Kleomēdēs (*Meteōra* i 8.79–99) attempts to meet this difficulty by maintaining that the Earth is small, but powerful, since it contains most of the substance of the cosmos and its matter is almost indefinitely expansible. (Consider the huge expansion that takes place when wood burns and turns to smoke.) Moreover, according to Kleomēdēs, there is a sort of return flux from the stars to the Earth: thus, for the Stoics, the pointlike character of the Earth does not prevent stellar influences from reaching us. So the elementary proposition proven by the astronomers (that the Earth is a mere point in relation to the cosmos) served as a fulcrum for the arguments of the philosophers.

XVII **18** Thus one must not suppose the same weather signs [everywhere] to be produced from the same stars; rather, in accordance with a shift of *klima*, the risings and settings of the stars become different. **19** And it is necessary to have signs proper to each horizon for the changes in the air. For the same *parapēgma* cannot be in harmony [with the phenomena] in Rome, in Pontos, in Rhodes, and in Alexandria.[12] Rather, of necessity, the observations are different for different horizons; and for each city other stars are taken as producing weather signs. **20** From this it is clear that the risings and setting of the stars do not, of their own nature, generate conditions in the air; rather, for each horizon both the observations and the changes of the air are different.

XVII **21** This is why all the weather signs listed in the *parapēgmata* are not always in harmony [with the facts]. Rather, it is sometimes the case that they are completely out of accord: the risings and settings accompany the greatest storms while holding out a sign of fair weather. And sometimes there has been fair weather in the city but rain in the country. **22** Many times, three or four days after the fact, it [the weather change] has been "indicated" by the rising or the setting of the star; and sometimes one has taken the weather sign four days in advance. And thus they fail in the predictions from the weather signs, but have the excuse that they took the sign early, or that it came later.

XVII **23** From all this it is clear that, as far as the weather signs in the *parapēgmata* are concerned, they have been recorded in a general way, not treated by a particular science or compelling method, but rather recorded from continuous observation. This is why they are often wrong. Therefore one should not reproach the astronomers if they miss the mark with the weather signs. **24** If one is wrong in foretelling an

---

[12] the same *parapēgma* cannot be in harmony [with the phenomena] in Rome, in Pontos, in Rhodes, and in Alexandria. These places are all in different *klimata*, and thus the phenomena must be different. Ptolemy, in his own *parapēgma* (in his *Phaseis*), gives the dates of the heliacal risings and settings of the principal stars for five different *klimata* (the *klimata* of 13½, 14, 14½, 15, and 15½ hours). Moreover, he attributes each weather prediction to a particular authority (e.g., Euktēmōn, Eudoxos, Hipparchos). And at the end of the *parapēgma* he states the *klima* to which each authority's weather predictions apply. Ptolemy's scrupulous procedure is very unusual. Before his time, *parapēgmata* were often used indiscriminately for *klimata* other than those for which they were designed. This is clear from Geminos's complaint. Moreover, the surviving *parapēgmata* (apart from Ptolemy's) generally do not contain any information about the place for which they were made. Geminos objects principally to the misuse of the *astronomical* part of the *parapēgma*: the dates of the risings and settings vary with latitude. Geminos does not seem to make as strong an objection to the use of the weather predictions for the wrong *klimata*. There are two reasons for this. First, some weather phenomena really are universal, for example the onset of the summer heat 30 days after the solstice (see xvii 41). Second, weather phenomena are so variable that Geminos simply does not believe they can be reliably predicted by any *parapēgma* (xvii 21–25).

eclipse or the [heliacal] rising of a star, then both the practitioner and [his] practice will, with good reason, be deemed worthy of reproach. For all that is treated methodically by means of science is bound to have a decision free of error. 25 But matters connected with weather signs give grounds neither for praise when they succeed in a general way, nor for reproach when they miss the mark. For this particular part of astronomy is not scientific and does not deserve rebuke.

ON THE RISING OF THE DOG STAR

XVII 26 One must understand the same things also concerning the rising[13] of the Dog [i.e., the star Sirius]. For everyone supposes the star to possess an inherent power[14] and to be the cause of the intensification of the heat when it rises at the same time as the Sun. But this is not the way it is. Rather, since this star rose at the most fiery time of the year, they [the old astronomers] marked by means of its appearance the change of the air in regard to heat.

XVII 27 It is the Sun that is the cause of the intensification of the heat. For first after the winter, when we have been chilled, in measure as it draws near us[15] it begins to warm us. But it does not yet make the warmth evident, as the chill from the winter still remains. 28 After a delay, with the Sun drawing always nearer and nearer, there results a perception of the warmth. Then it happens that the Sun falls upon the same region twice in succession; for both in the approach to the summer tropic circle and in the retreat the Sun passes by the same regions. And hence there results from this cause the intensification of the heat. 29 Moreover, both the [daily] approaches toward the summer tropic and the [daily] retreats from it are extremely small and imperceptible;

[13] the rising. Throughout the following paragraphs, Geminos discusses the *visible morning rising* of Sirius. The star, after being invisible for nearly 3 months, is seen rising in the east just before sunrise. In Greek Antiquity, the morning rising of Sirius occurred in late July and marked the onset of the worst of the summer heat.

[14] everyone supposes the star to possess an inherent power. Ancient references to the influence of Sirius are legion. Among the earliest are Hesiod's in *Works and Days* 417, 587. Pliny is a good example of a complete believer in the effects of the Dog at its morning rising: "For who is not aware that the heat of the Sun increases at the rising of the Dog-star, whose effects are felt on Earth very widely? At its rise the seas are rough, wine in the cellars ripples in waves, pools of water are stirred. There is a wild animal in Egypt called the gazelle that according to the natives stands facing this Dog-star at its rise, and gazing at it as if in worship, after first giving a sneeze. It is indeed beyond doubt that dogs throughout the whole of that period are specially liable to rabies." (Pliny, *Natural History* ii 40. H. Rackham, trans.) This sort of credulity explains the pains that Geminos takes in refuting the popular belief in stellar influences on the weather.

[15] as it draws near us. That is, as the Sun is more nearly overhead at noon.

for <the Sun> stays on the summer tropic circle for about 40 days, and therefore the lengths of the days around the solstices involve imperceptible increases.

XVII 30 Since the Dog rose around this time, they marked by its appearance the time of the intensification of the heat—not that the star is the cause of it; the Sun, rather, is the cause.   31 If, then, someone takes the rising of the Dog as a sign of the season, he takes it rightly, just as Homer says about the Dog:

> It was made to be an evil sign,[16]

not in the sense that it has an inherent power in relation to the intensification of the heat, but rather that it was accepted for the sake of a sign.

XVII 32 However, those poets and philosophers who attribute to the Dog the power over the intensification of the heat have wandered far from the truth and from physical reasoning. For this star shares the same substance with all the stars.   33 For, whether the stars are fire or *aithēr*, all have the same power. And the emanation from the Dog ought to be overpowered by the multitude of the stars. For others are greater in size[17] than it, and [the stars are] countless in number.   34 If, then, the power from the whole of them does not reach to the Earth, and adds nothing to the power of the Sun, how is it plausible for the emanation from the one star to produce such a great intensification of the heat? 35 And if all the fixed stars, sharing the same power, do not work together at something, it is not possible for the warmth from the one star to manifest a sensible difference at its co-rising with the Sun.   36 It is the Sun itself that is the cause of the heat by its traveling <twice>[f] in succession over the same place in the region. But as they could not designate for all [people or localities] a common day on which the intensifications of the heat occur, since this star rises around this time, they marked the time by its appearance.

XVII 37 That the star is not the cause of the intensification of the heat is clear from what is about to be said. First, on many occasions, more stars and larger ones rise with the Sun and produce no perceptible

---

[16] Homer, *Iliad* xxii 30. Geminos cites Homer in defense of his own view that the stars are signs of the weather but do not cause it. However, in the very next line, Homer goes on to say that Sirius "brings much fever to wretched mortals."

[17] others are greater in size. Like all astronomers before the telescopic period, Geminos associates brightness with size. In asserting that some stars are larger than Sirius, Geminos means that there are stars that are brighter. In fact, Sirius is the brightest of all the stars. (This fact is mentioned by Ptolemy in *Almagest* viii 3.) Geminos does not appear to be thinking here of the planets, for he takes them up separately at xvii 38 below. Perhaps, as Aujac (1975, 156) has suggested, Geminos had in mind Canopus, the celebrated star of the Egyptians, which was visible only below latitude 36°. It is not brighter than Sirius, but its fame may have led Geminos to think so.

change at all; but, quite the contrary, there are times when storms occur and cold winds blow at their risings,[g] since they contribute nothing to the intensification of the heat.

XVII 38 And often in the same sign as the Sun are the largest of the five planets, Phaëthōn, Phōsphoros, and Pyroëis,[18] from which powers also fall upon the Earth, and nothing different occurs in the air because of them. From this it is clear that neither the fixed stars nor the planets contribute anything to the intensification of the heat.

XVII 39 For if the Dog exercised a certain power, the intensification of the heat ought to occur at the true[h] rising, for then it rises at the same time as the Sun. But this is not the case; rather, the greatest heat occurs at the visible rising. For around this time the Sun, for the aforementioned reasons, is the cause of the intensification of the heat.

XVII 40 In Rhodes the star rises 30 days after the <summer> solstice,[19] but at other places 40 days after the summer solstice, and at others 50 [days] after, so that it no longer makes its rising at the [time of the] intensification of the heat. 41 It is clear that for each [locality][i] the season embracing the intensification of the heat is one and the same, i.e., 30 days after the summer solstice; and wherever the Dog rises at this

[18] Phaëthōn, Phōsphoros, and Pyroëis. Jupiter, Venus and Mars, respectively. These three planets are brighter than Sirius, at least part of the time, so Geminos is on firmer ground here. Sirius is magnitude −1.6. The magnitudes of the planets vary in the course of their synodic cycles. Venus can be as bright as magnitude −4.3; Jupiter, −2.7; Mars, −2.5. Jupiter and Mars, however, are both sometimes dimmer than Sirius—Mars considerably so.

[19] In Rhodes the star rises 30 days after the <summer> solstice. This time for the visible morning rising of Sirius is mentioned by Hipparchos, *Commentary* ii 1.18. In the *parapēgma* that follows Geminos's *Introduction to the Phenomena*, Kallippos is said to have put the morning rising of the Dog on the 30th day of Cancer, which would be 30 days after the solstice (counting inclusively). According to the Geminos *parapēgma*, others put this event on various dates: Dositheus on day 23 of Cancer (for Egypt), Metōn on day 25, Eudoxos on day 27, and Euktēmōn either on day 27 of Cancer or on day 1 of Leo (there are two entries). Geminos's time of 30 days after the solstice thus appears to follow the tradition of Kallippos, perhaps filtered through Hipparchos.

Ptolemy's dates for the morning risings of Sirius, shown in table 17.1, give an idea of their dependence on the latitude of the observer. The dates come from Ptolemy's *Phaseis*, and the places associated with the various *klimata* are from the table of ascensions in the *Almagest*.

TABLE 17.1
Morning Risings of Sirius, According to Ptolemy

Klima of 14[h]	Lower Egypt	27 days after summer solstice
14½	Rhodes	33
15	Hellespont	38
15½	Middle of Pontos	43

time, it indicates the season; but elsewhere, it is some other one of the stars in [another] constellation.[j] For the [heliacal] risings and settings of the stars do not happen to occur at the same time for all.

XVII 42 The assertion made by most,[20] that around this time it rises with the Sun, is completely uninformed, for at this time the star stands at its greatest distance from the Sun. 43 For [at this time of year] the Sun makes its [diurnal] path wholly upon the summer tropic circle, while it [i.e., the Dog] lies on the winter tropic circle, so that they are separated from one another by the greatest distance. How then could it be the cause of the intensification of the heat? 44 If the star had a certain power, it would produce an intensification when it was together with the Sun at the winter solstice, when the star moved upon the same circle as the Sun. For at that time there ought to be, in accordance with the [celestial] phenomenon, some perceptible change in the air. But this is not the case; rather, quite the contrary, it is winter. Thus in the *parapēgmata* it is cited [only] as a sign.

XVII 45 Therefore it is clear from all this that neither this star nor another has any power so great as to work changes in the air. Rather, the commanding cause involves the Sun. Their risings and settings are cited for foreknowledge[k] of the changes in the air, for which reason they are not always in harmony [with the facts].

XVII 46 Thus one would better make use of signs given us by nature, which, too, Aratos has used. For he omitted as mistaken[l] the changes of the air [predicted] from the risings and settings of the stars, but inserted those arising naturally and from some cause in his treatise of the *Phenomena*, at the end of the whole work.[21] 47 For he takes the prognoses from the rising and setting of the Sun, from the rising and setting of the Moon, from the halo that occurs around the Moon, from shooting

[20] The assertion made by most. Manitius would delete this entire paragraph (xvii 42–44) as an addition by a later excerptor containing "nothing but nonsense." Manitius points out that in Hipparchos's time the declination of Sirius was only some −16°, so the star was not really on the winter tropic circle (declination −24°). But we see no reason to reject the paragraph. Geminos's argument is a physical one: if the summer heat were truly produced by Sirius reinforcing the effect of the Sun, then this ought to happen when Sirius and the Sun follow the same diurnal path through the sky. But the Sun and Sirius share a diurnal path in the winter, when the Sun is far enough south to be near the declination of Sirius. Geminos is simply emphasizing that the mere rising of Sirius at the same time as the Sun is not capable of producing a physical effect. For the Sun and Sirius are not in any significant sense "together" at that time.

[21] at the end of the whole work. The last half of Aratos's *Phenomena* is devoted to weather signs drawn from the behavior or appearance of animals and inanimate objects, as well as from rings around the Sun and Moon, etc. Examples: A single ring around the Moon promises wind or calm; a broken ring, wind. Two rings foretell storm; three, a greater storm. Ducks beating their wings on the shore indicate a gale. Dogs dig before a storm. Snuff gathering on the nozzle of a lamp is a sign of rain.

stars, and from [the behavior of] dumb animals. **48** For the prognoses from these things, arising from some natural cause, involve necessary results. Thus also Boēthos the philosopher,[22] in the fourth book of his commentary on Aratos, has given an account of the[m] natural causes of both winds and rains, and has indicated the prognoses [to be drawn] from the aforementioned conditions. **49** These signs are used also by Aristotle the philosopher, and by Eudoxos[23] and many other astronomers.

[22] Boēthos the philosopher. Boēthos of Sidon (second century B.C.), Stoic philosopher, was the author of a lost commentary on Aratos (see Maass 1898, 324).

[23] by Aristotle the philosopher, and by Eudoxos. Aristotle discusses the natural causes of winds and rains in the *Meteorology*. The reference to Eudoxos is puzzling, since he was a leading parapegmatist. His weather predictions from the heliacal risings and settings are quoted in one of the Miletus *parapēgmata*, in the *parapēgma* attached to Geminos's *Introduction to the Phenomena*, and in Ptolemy's *Phaseis*. Perhaps, in the lost work that served as the inspiration for the astronomical part of Aratos's *Phenomena*, Eudoxos also made some use of "natural signs."

XVIII. On the *Exeligmos*

XVIII 1 The *exeligmos* is the least time that contains whole [numbers of] months, whole days, and whole returns of the Moon [in anomaly].[2] For, since the monthly period[3] was observed and is very nearly 29 + 1/2 + 1/33 days, while the return of the Moon [in anomaly] is very nearly 27 + 1/2 + 1/18 days, the least time was sought which contains whole days, whole months, and whole returns. It is as follows.

XVIII 2 The Moon appears to travel through the zodiac circle nonuniformly. If the Moon[a] has run a certain <smallest>[b] arc,[4] on the following day it travels a greater one than this, and always a greater one on the succeeding days,[c] until it travels its greatest arc; <then, in turn,> always[d] a smaller one than the preceding, until it has returned to the original smallest arc. The time from the smallest motion to the smallest motion is called a return [in anomaly].

XVIII 3 The *exeligmos* has been observed[5] to contain 669 whole

---

[1] The *Exeligmos*. See sec. 13 of our Introduction for a discussion of this chapter. See also Neugebauer 1975, 585–87, 602–4, as well as Bowen and Goldstein 1996.

[2] returns of the moon [in anomaly]. This is what we today call an *anomalistic month*. Geminos refers to this period simply as a "return" or "restoration" (*apokatastasis*) of the Moon. Ptolemy (*Almagest* iv 2) speaks of "returns in anomaly" (*apokatastaseis anōmalias*), to distinguish them from, for example, "returns in latitude" (*apokatastaseis platous*), or, as we would say, to distinguish the anomalistic month from the draconitic month. "Anomaly" refers generally to a nonuniformity—in this case, the speeding up and slowing down of the Moon.

[3] the monthly period. By this, Geminos always means the synodic month (the time from one new Moon to the next), as at viii 2.

[4] <smallest> arc. Geminos is referring to the zodiacal arc that the Moon traverses in the course of 24 hours. As he says below (xviii 5), this can vary from a minimum of 11 or 12° to a maximum of 15 or 16°.

[5] has been observed. When Geminos invokes observation of the parameters of the *exeligmos* (as here), or of the Metonic cycle (viii 51), or of the length of the month (xviii 1), he means only that these phenomena have been established and are to be accepted as data for the basis of the argument he wishes to make. He has nothing to say about, and certainly did not know, what sorts of observations the Babylonians used to establish the *exeligmos*. This parallels the common practice of modern, elementary science textbooks of invoking "observation" without giving any details. For Geminos, *paratērein* ("to observe") can be used of a single, definite act of observation (of the Moon next to a star, as at xii 12, or of the crescent Moon, at ix 2). But it can also be used for the perusal of calendrical data that leads to the establishment of a lunisolar cycle (as with the Metonic cycle at viii 51). Finally, *paratērein* sometimes clearly only means "to note carefully" (as when the ancient parapegmatists noted which sign the Sun was in at the beginning of a particular year, at

months, and 19,756 days.[6] In this time the Moon makes 717 returns of anomaly in longitude, while in the same time passing through 723 zodiacal circles plus 32 degrees.

ELEMENTS OF THE BABYLONIAN LUNAR THEORY

XVIII **4** In the possession of these phenomena, which were investigated from ancient times, when it was necessary to establish the Moon's daily anomaly in longitude,[7] <they asked>[e] what its smallest motion is, what the greatest is, what the mean is, and what its daily increase or diminution is, **5** taking also this fact from the phenomena: that when it travels the smallest [daily arcs, it travels] more than 11 degrees but less than 12; and when it travels the greatest [arcs], it travels more than 15 degrees but less than 16.

XVIII **6** Moreover, since the Moon has been observed to pass in 19,756 days through 723 zodiac circles plus 32 degrees, and each of the circles contains 360 degrees, I reduced the number of circles to degrees and added the 32. The entire number of degrees is 260,312. Therefore, in 19,756 days the Moon passes through the aforementioned number of degrees. **7** <Then, dividing the> number <of degrees>[f] by the number of days, we will find the mean daily motion of the Moon. For whenever, without taking account of the increase or decrease of the motion from the average, we divide the number of degrees by the number of days, then the [quantity] found is called the mean motion. This is found to be 13 degrees, 10 minutes, 35 seconds. **8** The sixtieth of one degree[g] is called a first sixtieth, and the sixtieth of the first sixtieth is called a second sixtieth. In the same way, when the second sixtieth is divided into 60 parts, one part is called a third sixtieth. The same system applies also to the remaining sixtieths. **9** Such being the disposition of the numbers,

---

xvii 17). Nothing in Geminos's *Introduction* gives any insight into ancient methods of astronomical observation.

[6] 19,756 days. This figure for the length of the *exeligmos* is not quite consistent with the value for the length of the month that Geminos has just given (xviii 1), since $(29 + 1/2 + 1/33) \times 669 = 19,755.773$ days. However, it *is* consistent with the Babylonian value of 29;31,50,8,20 days. (On the Babylonian month length, see textual note k to chapter viii). For $29;31,50,8,20 \times 669 = 19,755.967$ days. The Babylonian value for the length of the month is rounded to 20 parts in the last sexagesimal place (i.e., a third of a unit in the preceding place), so we cannot expect better agreement. Geminos's figure for the length of the month is a reasonably accurate and simply expressed approximation to the Babylonian value. But if he had actually known the Babylonian paramter, it would have been odd for him not to mention it here.

[7] the Moon's daily anomaly in longitude. That is, the difference between the Moon's mean daily motion and the actual motion for a given day.

the mean [daily] motion[8] of the Moon has been found by the Chaldeans[9] to be 13;10,35°.

XVIII 10 And since the Moon makes 717 returns in 19,756 days, if we wish to discover in how many days the Moon makes one return, we will divide the number of days by the number of returns. There are 27;33,20 days per return.[10] In so many days, therefore, the Moon goes from least motion to least motion.

XVIII 11 And since in an entire return there are 4 equal time intervals, they took[11] the 4th part of the 27;33,20 days; and this is 6;53,20 days. In so many days, therefore, the Moon goes from least motion to mean [motion], and from mean to greatest, and again in the same way

---

[8] the mean [daily] motion. In printing base-sixty numbers, the convention we follow is that the integer part of the number stands to the left of the semicolon and the fractional part stands to the right, with successive sexagesimal places separated by commas. Thus 13;10,35° = 13°10'35''.

[9] found by the Chaldeans. Geminos is correct in attributing this figure for the Moon's mean daily motion to the Babylonians. Though Geminos does not say so explicitly, the other parameters of the lunar theory mentioned here are also Babylonian. The parameters of daily motion in the Babylonian lunar theory of system B are discussed in Neugebauer [1955], 76–77 and Neugebauer 1975, 480–81. The function giving the Moon's velocity in degrees per day is called F^* by Neugebauer. It is a linear zigzag function with the following parameters:

least value	11;06,35
mean value	13;10,35
maximum value	15;14,35
daily change	0;18.

These are exactly as in Geminos. The Babylonian parameters are deduced most easily from tablet ACT 190 (Neugebauer [1955], pp. 179–80 and plate 131). This tablet, of the Seleucid period, lists the Moon's daily motion for 248 consecutive days (248 days = 9 anomalistic months, very nearly). The actual values of the greatest, mean, and least motion parameters do not happen to occur on this tablet because of the "phase" of the linear zigzag wave. That is, one must find the extreme values by extrapolating the rising and falling sections of the graph to determine the values at which they intersect.

[10] There are 27;33,20 days per return. Geminos asserts that $19,756/717 = 27;33,20$. More accurate arithmetic gives a different value in the second sexagesimal place: 27;33,13 days. See sec. 13 of the Introduction for an explanation of Geminos's unusual rounding. In any case, Geminos's final answer of 27;33,20 days for the anomalistic month is equivalent to the value $27 + 1/2 + 1/18$ days that he gave at viii 1.

[11] they took. Throughout this chapter, Geminos is vague about just who "they" are. He uses verbs in the third-person plural without subjects (here, xviii 13, and 18), or verbs in the passive voice ("the least time was sought" at xviii 1). Perhaps he has in mind the Chaldeans mentioned in xviii 9, perhaps other unnamed ancient astronomers. Geminos knows that the *exeligmos* was of Babylonian origin, but he has no real knowledge of how it was originally constructed. His discussion in this chapter should be regarded as a pedagogical effort to show how one *might* arrive at this cycle, given certain data, rather than a historical account of what actually happened.

from greatest to mean, and from mean to least. For these 4 time intervals are equal to one another.[12]

XVIII **12** Next, if there are three numbers exceeding one another by equal amounts, the sum of the extremes is the double of the mean. In the motion of the Moon there are three numbers exceeding one another by equal amounts, i.e., the least motion, the mean, and the greatest. If, therefore, we add together the greatest and the least, they will be the double of the mean motion. **13** But the mean motion was 13;10,35°. They multiplied this by two: [the result] is 26;21,10°. Therefore the greatest motion of the Moon and the least, added together, are exactly[h] 26;21,10°.

XVIII **14** But the greatest and the least [motion] taken in a rough way from observation are, <added together,>[i] 26 degrees. Thus there is a remainder of 0;21,10° that escaped observation of the phenomena by means of instruments. **15** This, then, must be distributed between the least and the greatest [motion], so that the two motions added together produce 26;21,10°. But it is necessary to add the surplus in such a manner that the least motion be not greater than 12 degrees, and the greatest not greater than 16 degrees.

XVIII **16** We shall make the distribution in the following fashion. Since the Moon goes in 6;53,20 days from the least motion to the mean, and from the mean to the greatest, and since it proceeds always by equal increases and decreases, one must find a number[13] **17** which, when multiplied by the fourth part of the time for a return [in anomaly], will yield a certain number which when added to the mean motion will produce a number greater than 15 degrees but less than 16, and when subtracted from the mean motion will leave a number greater than 11 degrees but less than 12. And the [sum of the] excesses beyond the 15 degrees and the 11 [degrees] will be equal to the same 0;21;10°.

XVIII **18** And 0;18° is found to do this. For if this is multiplied by the 4th part of the [time for a] return [in anomaly], i.e., by 6;53,20 days, 2;4° are produced. This is added to the mean motion, 13;10,35°, and 15;14,35° is obtained. And they took the 2;4° from the mean motion, and 11;6,35° is left.

XVIII **19** Thus the least motion of the Moon has been found to be 11;6,35°; the mean motion, 13;10,35°; the greatest motion, 15;14,35°; and the daily augmentation, 0;18°.

---

[12] these 4 time intervals are equal to one another. This is strictly true of the Babylonian lunar theory that Geminos is describing, but not quite true of actual lunar phenomena, nor of an epicycle theory of the Moon's motion.

[13] one must find a number. The problem may be paraphrased as follows. The Moon's zodiacal motion for a day is assumed to change by equal increments each day, from the day of least motion to the day of greatest motion. The size of the daily change is the number x that "one must find." See sec. 13 of the Introduction, pp. 97–98.

Parapēgma

The manuscripts of Geminos's book conclude with the following *parapēgma*, which is effectively the nineteenth chapter of the *Introduction to the Phenomena*. (For a complete discussion of the Geminos *parapēgma*, including the scholarly debate over its authorship, see appendix 2.)

The notices at the beginning of each section (e.g., "The Sun passes through Cancer in 31 days") refer to the time that the Sun spends in each zodiacal *sign*. Throughout a sign, the date is indicated by the position of the Sun within the sign: 1st day of Cancer, 11th day of Cancer, etc. But the notices of risings and settings refer to *stars and constellations*. So, for example, according to Kallippos, on the 27th day of the sign of Cancer, the constellation Cancer finishes rising.

The star phases in the *parapēgma* are, of course, *visible* risings and settings. Throughout the *parapēgma*, a statement that a constellation simply "rises" or "sets" almost invariably indicates a *morning* rising or setting. Thus, there is in general no difference between saying "Cancer rises" and "Cancer rises in the morning." Evening risings and settings are usually explicitly identified as such by the ancient writer. The few exceptions to this rule are indicated in our comments. In contrast to Ptolemy, who is scrupulous in his use of technical vocabulary, the writer of our *parapēgma* tends to use *epitellei* and *anatellei* interchangeably. We translate both simply as "rises." (On the technical vocabulary of star phases, see sec. 11 of the Introduction.)

THE TEXT

[Parapēgma]

The times in which the Sun passes through each of the signs and, for each sign, the weather predictions, which are written underneath. We shall begin from summer solstice.

> The Sun passes through Cancer in 31 days.

<On the> 1st day, according to Kallippos, Cancer begins to rise; summer solstice; and it signifies.[1]

---

[1] it signifies. Throughout the *parapēgma*, this is our rendering of *episēmainei*. This probably means that the star signs indicate that a particular day's weather bears special watching, either because it is likely to undergo a sudden change, or because it can serve to indicate the

<On the> 9th day, according to Eudoxos, the south wind blows.

<On the> 11th day, according to Eudoxos, the whole of Orion rises in the morning.[2]

<On the> 13th day, according to Euktēmōn, the whole of Orion rises.

<On the> 16th, according to Dositheus, the Crown[3] begins to set in the morning.

<On the> 23rd, according to Dositheus, the Dog[4] becomes visible in Egypt.

<On the> 25th, according to Metōn, the Dog rises in the morning.

<On the> 27th, according to Euktēmōn, the Dog rises. According to Eudoxos, the Dog rises in the morning, and the etesian winds[5] blow for the following 55 days: the first 5 are called the "forerunners." According to Kallippos, Cancer finishes[a] rising; windy.

<On the> 28th, according to Euktēmōn, the Eagle[6] sets in the morning[b]; a storm at sea follows.

<On the> 30th, according to Kallippos, Leo begins to rise; the south wind blows; and the Dog is visible rising.

<On the> 31st, according to Eudoxos, the south wind blows.

---

weather for the immediate future. Many days throughout the year are deemed significant, but the basis on which they were singled out is not clear.

Lehoux (2000, 130–45) gives a valuable discussion of the possible meanings of "episēmainei" in ancient meteorology. He argues that in the parapēgmata, episēmainei is usually intransitive and impersonal and should just be translated "there is a change in the weather." Because most instances of "episēmainei" in the Geminos parapēgma are so bare, they could well conform to Lehoux's thesis. But we can see that in two cases something more complicated is going on. Thus, on day 17 of Virgo, the constellation Virgo is clearly doing the signifying. And on day 16 of Sagittarius, particular things (thunder, lightning, and rain) are being signified.

[2] the whole (holos) of Orion rises in the morning. In the Geminos parapēgma, holos is frequently used in this way, to indicate the completion of an extended rising. In this case, Orion has finished making its morning rising: the whole constellation may be seen above the eastern horizon just before dawn. The ancient compiler sometimes uses other expressions for the same purpose. Thus, under day 5 of Libra: according to Kallippos, Virgo "finishes rising" (legei anatellousa).

[3] The Crown. Our Corona Borealis.

[4] The Dog. The star Sirius, α CMa. The constellation Canis Major is also called "Dog" (Kuōn). But in the Greek parpēgmata, notices about the Dog almost invariably refer to the star itself.

[5] etesian winds. The etesian winds are annually recurring (hence their name) north winds that blow in the Mediterranean in the summer, giving some relief from the heat. Aristotle (Meteorology 361b36–362a2) says that the etesian winds blow after the summer solstice and the rising of the Dog star, and that they blow in the daytime but fall off at night. Aratos (Phenomena 150–54) says that the etesian winds begin when the Sun enters Leo. According to our parapēgma, Eudoxos allows 5 days for the forerunners, which begin on the 27th day of Cancer. If we suppose that Aratos is ignoring the forerunners, he appears to be in agreement with Eudoxos.

[6] the Eagle. Our Aquila.

The Sun passes through Leo in 31 days.

On the 1st day, according to Euktēmōn, the Dog is visible[7]; the stifling heat follows[8]; it signifies.

On the 5th, according to Eudoxos, the Eagle sets in the morning.

On the 10th day, according to Eudoxos, the Crown sets.[c]

On the 12th, according to Kallippos, Leo, rising to its middle, produces stifling heat in the greatest degree.

On the 14th, according to Euktēmōn, there is stifling heat in the greatest degree.

On the 16th day, according to Eudoxos, it signifies.

On the 17th, according to Euktēmōn, the Lyre[9] sets; and it rains; the etesian winds cease; and the Horse[10] rises <in the evening>.[d]

On the 18th, according to Eudoxos, the Dolphin[11] sets in the morning. According to Dositheus, the Harbinger of the Vintage[12] sets at nightfall.[e]

On the 22nd, according to Eudoxos, the Lyre sets in the morning; it signifies.

On the 29th, according to Eudoxos, it signifies. According to Kallippos, Virgo rises[13]; it signifies.

---

[7] according to Euktēmōn, the Dog is visible (*ekphanēs*). This is part of a peculiar double entry for the morning rising of the Dog star (*Kuōn*) according to Euktēmōn. For the 27th day of Cancer we have "according to Euktēmōn, the Dog rises (*epitellei*)." It seems that the Dog makes its morning rising on day 27 of Cancer, but it may be only fleetingly visible. By the first day of Leo, the star is easily seen. The word *ekphanēs* may be translated as "visible," but also more emphatically as "evident," or "manifest." Geminos uses this word in a similar way, in his discussion of Procyon: "In Rhodes it is hard to see, or is seen completely from high places; but in Alexandria it is completely evident." (*pantelōs ekphanēs*. *Introduction to the Phenomena* iii 15). We can rule out the possibility that Euktēmōn was referring to the extended rising of the whole constellation of the Dog: this would take more than two weeks. Moreover, a similar double entry, according to Euktēmōn, occurs for the morning rising of Arcturus (days 10 and 20 of Virgo).

[8] the stifling heat follows. The worst of the summer heat comes with the morning rising of Sirius. Ancient references to this connection are legion. See Hesiod, *Works and Days* 587; Aratos, *Phenomena* 331–35. Hipparchos (*Commentary* ii 1.18) concurs: "For the heat is at its greatest around the rising of the Dog, which occurs very nearly 30 days after the summer solstice."

[9] The Lyre. Our Lyra. But the bright star of this small constellation, Vega (α Lyr), is also called the Lyre.

[10] the Horse. Our Pegasus.

[11] the Dolphin. Our Delphinus.

[12] the Harbinger of the Vintage. The star Vindemiatrix, ε Vir.

[13] Virgo rises. Probably the text originally said "begins to rise." Three other notices due to Kallippos track the progress of the morning rising of Virgo at days 5 and 17 of Virgo and day 5 of Libra.

The Sun passes through Virgo in 30 days.

On the 5th day, according to Eudoxos, a great wind blows and it thunders. According to Kallippos, the shoulders of Virgo rise; and the etesian winds cease.

On the 10th day, according to Euktēmōn, the Harbinger of the Vintage appears,[14] Arcturus rises, and the Bird[15, f] sets at dawn[g]; a storm at sea; south wind. According to Eudoxos, rain, thunder; a great wind blows.

On the 17th, according to Kallippos, Virgo, risen to the waist, signifies; and Arcturus is visible rising.

On the 19th, according to Eudoxos, Arcturus rises in the morning; <winds> blow for the following 7 days; fair weather for the most part; at the end of this time there is wind from the east.

On the 20th, according to Euktēmōn, Arcturus is visible: beginning of autumn.[16] The Goat,[17] great star in the Charioteer,[h] rises <in the evening>[i]; and afterwards, it signifies; a storm at sea.

On the 24th day, according to Kallippos, the Wheat-Ear[18] of Virgo rises; it rains.

The Sun passes through Libra in 30 days.

On the 1st day, according to Euktēmōn, autumnal equinox; and it signifies. According to Kallippos, Aries begins to set; autumnal equinox.

On the 3rd, according to Euktēmōn, the Kids[19] rise in the evening; it is stormy.

On the 4th, according to Eudoxos, the Goat rises at nightfall.

On the 5th, according to Euktēmōn, the Pleiades appear in the evening[j]; it signifies. According to Kallippos, Virgo finishes rising.

On the 7th day, according to Euktēmōn, the Crown rises; it is stormy.

On the 8th, according to Eudoxos, the Pleiades rise <at nightfall>.[k]

On the 10th, according to Eudoxos, <the Crown>[l] rises in the morning.

On the 12th day, according to Eudoxos, Scorpio begins to set at nightfall; a storm follows; and a great wind blows.

On the 17th, according to Eudoxos, the whole of Scorpio sets at nightfall.[m] According to Kallippos, the Claws[20] begin rising; it signifies.

On the 19th, according to Eudoxos, north and south winds blow.

---

[14] the Harbinger of the Vintage appears (*phainetai*). A reference to the morning rising.

[15] the Bird. Our Cygnus.

[16] Arcturus is visible: beginning of autumn. The morning rising of Arcturus was understood as the real beginning of autumn. See comment 21.

[17] the Goat. Our Capella, α Aur.

[18] The Wheat-Ear. Our Spica, α Vir.

[19] the Kids. The two small stars ζ and η Aur. They belong to the she-goat, Capella.

[20] the Claws. "Claws" (of the Scorpion) was an alternative, early name for the constellation Libra.

On the 22nd, according to Eudoxos, the Hyades rise at nightfall.

On the 28th, according to Kallippos, the tail of Taurus sets; it signifies.

On the 29th, according to Eudoxos, the north and south winds blow.

On the 30th, according to Euktēmōn, a strong storm at sea.

The Sun passes through Scorpio in 30 days.

On the 3rd, according to Dositheus, it is stormy.

On the 4th day, according to Dēmokritos, the Pleiades set with the dawn[n]; wintry winds[21] for the most part, cold weather and frost already. It is likely to blow. The trees begin greatly to shed their leaves. According to Kallippos, the forehead of Scorpio rises; windy.

On the 5th, according to Euktēmōn, Arcturus sets in the evening, and great winds blow.

On the 8th, according to Eudoxos, Arcturus sets at nightfall[o]; and it signifies; and a wind blows.

On the 9th, according to Kallippos, the head of Taurus sets; rains.

On the 10th, according to Euktēmōn, the Lyre rises in the morning[p]; it storms with rain.

On the 12th, according to Eudoxos, Orion begins to rise at nightfall.

On the 13th, according to Dēmokritos, the Lyre rises with the rising Sun; and the air is stormy for the most part.

On the 14th, according to Eudoxos, rainy weather.

On the 15th, according to Euktēmōn, the Pleiades set; and it signifies; and Orion begins <to set: whether beginning>,[q] or in the middle, or finishing, it is stormy.

On the 16th, according to Kallippos, the bright star in Scorpio[22] rises; it signifies; the Pleiades set visibly.[23]

On the 18th, according to Eudoxos, Scorpio begins to rise in the morning.

On the 19th, according to Eudoxos, the Pleiades set in the morning and Orion begins to set[r]; it is stormy.

On the 21st, according to Eudoxos, the Lyre rises in the morning.

On the 27th, according to Euktēmōn, the Hyades set; and it rains.

On the 28th, according to Kallippos, the horns of Taurus set; rainy weather.

On the 29th, according to Eudoxos, the Hyades set[s]; it storms violently.

---

[21] wintry winds. The morning setting of the Pleiades was the traditional sign of the beginning of the season of bad weather, as many ancient writers testify. For example: "Winter lasts from the setting of the Pleiades to the spring equinox, spring from the equinox to the rising of the Pleiades, summer from the Pleiades to the rising of Arcturus, autumn from Arcturus to the setting of the Pleiades" (Hippokratēs, *Regimen III* lxviii. W.H.S. Jones, trans.).

[22] the bright star in Scorpio. Antares, α Sco.

[23] the Pleiades set visibly. That is, the Pleiades make their morning setting. The modifier "visibly" is of no particular significance here, as all the risings and settings in a *parapēgma* are visible ones.

The Sun passes through Sagittarius in 29 days.

On the 7th, according to Euktēmōn, the Dog sets; and it is stormy. According to Kallippos, Sagittarius begins to rise and Orion sets visibly; it is stormy.[24]

On the 8th, according to Eudoxos, <the whole of> Orion sets in the morning.[t]

On the 10th, According to Euktēmōn, the stinger of Scorpio rises.

On the 12th, according to Eudoxos, the Dog sets in the morning; it is stormy.

On the 14th, according to Eudoxos, rain.

On the 15th, according to Euktēmōn, the Eagle rises; the south wind blows.

On the 16th, according to Dēmokritos, the Eagle rises with the Sun; and it is likely to signify thunder, lightning, and rain,[25] or else wind, or most often both. According to Eudoxos, the Dog rises at nightfall; southerly[u] [winds]. According to Kallippos, Gemini leaves off[v] setting; southerly [winds].

On the 19th, according to Euktēmōn, the Goat[w] sets.

On the 21st, according to Eudoxos, the <whole of > Scorpio rises in the morning[x]; and it is stormy.

On the 23rd, according to Eudoxos, the Goat sets in the morning.

On the 26th, according to Eudoxos, the Eagle rises in the morning.

The Sun passes through Capricorn in 29 days.

On the 1st day, according to Euktēmōn, winter solstice; it signifies. According to Kallippos, Sagittarius finishes rising; winter solstice; it is stormy.

On the 2nd, according to Euktēmōn, the Dolphin rises; it is stormy.

On the 4th, according to Eudoxos, winter solstice[26]; it is stormy.

---

[24] Sagittarius begins to rise and Orion sets visibly; it is stormy. The morning setting of Orion, around the end of November, was a traditional sign of winter and stormy weather (Hesiod, *Works and Days* 619–23). Aratos (*Phenomena* 665–78) points out that Orion sets while Sagittarius is rising. Like Hesiod, Aratos (*Phenomena* 300–304) identifies the morning rising of Sagittarius as a season too dangerous for sailing.

[25] the Eagle rises with the Sun; it is likely to signify thunder, lightning, and rain. Aratos (*Phenomena* 313–15) describes the Eagle as small in size but ill-tempered in its rising from the sea while night is departing. And he attempts to derive *aetos* ("eagle") from aētos ("stormy" or "furious").

[26] On the 4th, according to Eudoxos, winter solstice. This remark tells us something about how the ancient compiler combined the various *parapēgmata* into one. He began with the summer solstice and collated his three sources (the *parapēgmata* of Kallippos, Euktēmōn, and Eudoxos) day by day. Since Eudoxos did not recognize a zodiacal inequality in the movement of the Sun, his winter solstice fell 3 days later than the winter solstice of Kallippos. See appendix 2 for more detail.

On the 7th, according to Euktēmōn, the Eagle sets in the evening; and it is stormy.

On the 9th, according to Eudoxos, the Crown sets at nightfall.

On the 12th, according to Dēmokritos, the south wind blows for <the most part. According to Eudoxos, the Dolphin> rises <in the morning>.[y]

On the 14th, according to Euktēmōn, midwinter; a strong south wind blows stormy on the sea.[27]

On the 15th, according to Kallippos, Capricorn begins to rise; south wind.

On the 16th, according to Euktēmōn, a stormy south wind at sea.

On the 18th, <according to Eudoxos, the Eagle>[z] sets[A] at nightfall; and the south wind blows.

On the 27th, according to Euktēmōn, the Dolphin sets in the evening. According to Kallippos, Cancer finishes setting; it is stormy.

The Sun passes through Aquarius in 30 days.

On the 2nd, according to Kallippos, Leo begins to set; rainy weather.

On the 3rd, according to Euktēmōn, the Lyre sets in the evening; rainy weather. According to Dēmokritos, a storm.

On the 4th day, according to Eudoxos, the Dolphin sets at nightfall.

On the 11th, according to Eudoxos, the Lyre sets at nightfall; rain.

On the 14th, according to Eudoxos, fair weather; sometimes the west wind blows.[28]

On the 16th, according to Dēmokritos, the west wind begins to blow 43 days after the solstice and persists.[B]

On the 17th, according to Euktēmōn, it is time for the west wind to blow. According to Kallippos, Aquarius rises to its middle; the west wind blows.

---

[27] a strong south wind blows stormy on the sea. In the eastern Mediterranean, the prevailing winter wind is from the north, but there was a period just after the winter solstice when the south wind often blew—hence the many references to the south wind under the sign of Capricorn (days 12, 14, 15, 16, and 18). Theophrastos (*On Winds* 10) confirms that the south wind blows in the winter. According to Pliny (*Natural History* ii 48), the south wind whips up larger waves at sea than even the northeast wind, and is especially violent at night. Aristotle (*Meteorology* 363a16–18) mentions that the south wind is stronger than the north.

[28] fair weather; sometimes the west wind blows. *Zephyros*, the west wind, was classified by Aristotle as a warm wind (*Meteorology* 364b23). Its rise, about a month and a half after the winter solstice, was a sign of the end of winter, as in Pliny: "the spring opens the seas to voyagers; at its beginning the west winds soften the wintry heaven, when the Sun occupies the 25th degree of Aquarius; the date of this is February 8" (*Natural History* ii 47. H. Rackham, trans.). On the Tower of the Winds in Athens, Zephyros is represented as a youth with benign countenance, and is the only wind represented without a tunic or a vest. His loose mantle is gathered up to form a basket and is filled with flowers.

On the 25th, according to Euktēmōn, the Bird[C] sets in the evening; and it storms violently.

The Sun passes through Pisces in 30 days.

On the 2nd[D] it is time for the swallow to appear; and the bird winds blow.[29] According to Kallippos, Leo finishes setting; and the swallow appears; it signifies.

On the 4th, according to Dēmokritos, begin the changeable days, called halcyon days.[30] According to Eudoxos, Arcturus rises at nightfall; there is rain; the swallow appears; and for the following 30 days, north winds blow, especially those called "the harbingers of the birds."

On the 12th, according to Euktēmōn, Arcturus rises in the evening and the Harbinger of the Vintage is visible[31]; a cold north wind blows.

On the 14th, according to Dēmokritos, cold winds blow—those called the bird winds—for about 9 days. According to Euktēmōn, the Horse rises[E]; a cold north wind blows.

On the 17th, according to Eudoxos, it is stormy; and the kite appears. According to Kallippos, the southern one of the Fishes rises; the north wind ceases.

On the 21st, according to Eudoxos, the Crown rises at nightfall; the bird winds begin blowing.

On the 22nd, according to Euktēmōn, the kite appears; the bird winds blow until equinox.

On the 29th, according to Euktēmōn, the first stars of Scorpio set; a cold north wind blows.

---

[29] it is time for the swallow to appear; and the bird winds blow. The return of the swallow, two months after winter solstice, was a sign of the end of winter (Hesiod, *Works and Days* 564–69). The bird winds were so named because they were associated with the return of the swallow and other birds in the early spring. Aristotle (*Meteorology* 362a23–26) says that they begin to blow 70 days after the winter solstice, and that they correspond to the etesian winds, which blow after the summer solstice. In the same passage, Aristotle also discusses a southerly fair-weather wind (*Leukonotos*, the weather-clearing south wind). Some writers therefore identify the bird winds with this one (see Lee 1952, p. 178, note a). However, ancient authorities make it clear that the bird winds are northerly (Columella, *De re rustica* xi 2.21. Pseudo-Aristotle, *On the Cosmos* 395a4). A feeble, intermittent south wind (*Leukonotos*) blows shortly after the winter solstice. According to Aristotle (*Meteorology* 362a16), it may escape notice. Then, 60 or 70 days after the solstice, the northerly bird winds blow.

[30] halcyon days. The halcyon, or kingfisher, was fabled to brood in a nest floating on the ocean, and to charm the waves and wind into calmness. See, for example, Theokritos, Seventh Idyll.

[31] Arcturus rises in the evening and the Harbinger of the Vintage is visible. This is apparently a reference to the evening rising of the Harbinger of the Vintage. The compiler of the *parapēgma* has failed to be clear.

On the 30th, according to Kallippos, the northern one of the Fishes[F] finishes rising; the kite appears; the north wind blows.

The Sun passes through Aries in 31 days.

On the 1st, according to Kallippos, the Knot of Pisces[32] rises; spring equinox. According to Euktēmōn, equinox; and a fine drizzle; it storms violently[G]; it signifies.

On the 3rd, according to Kallippos, Aries begins rising; rain or snowstorm.

On the 6th, according to Eudoxos, equinox[33]; there is rain.

On the 10th, according to Euktēmōn, the Pleiades hide themselves in the evening.[H]

On the 13th, according to Eudoxos, the Pleiades set at nightfall and Orion begins to set at nightfall[I]; there is rain. According to Dēmokritos, the Pleiades hide themselves while the Sun is setting[J] and become invisible for 40 nights.[34]

On the 21st, according to Eudoxos, the Hyades set at nightfall.

On the 23rd, according to Euktēmōn, the Hyades hide themselves; and hail follows; and the west wind blows. According to Kallippos, the Claws begin setting; and in many places hail.[K]

On the 27th, according to Eudoxos, the Lyre rises at nightfall.

The Sun passes through Taurus in 32 days.

On the 1st [L] day, according to Eudoxos, Orion sets at nightfall[M]; rainy weather. According to Kallippos, Aries finishes rising; rainy weather, and in many places hail.

On the 2nd,[N] according to Euktēmōn, the Dog hides itself; and there is hail. On the same [day],[O] the Lyre rises <in the evening>.[P] According to Eudoxos, the Dog sets at nightfall; and there is rain. According to Kallippos, the tail of Taurus rises; southerly [winds].[Q]

On the 7th, according to Eudoxos, there is rain.

On the 8th, according to Euktēmōn, the Goat rises in the morning[R]; fair weather <or>[S] it rains with a southerly rain.

On the 9th, according to Eudoxos, the Goat rises in the morning.

---

[32] the Knot of Pisces. α Psc, the star at the knot where the two fishing lines join.

[33] On the 6th, according to Eudoxos, equinox. Eudoxos's equinox falls late because he took the four seasons to be all of roughly equal length. (See comment 26.) According to the *Celestial Teaching* of Leptinēs, however, it seems that Eudoxos's spring equinox should fall only 4 days later than that of the Geminos *parapēgma*. This would put Eudoxos's equinox on the 5th day of Aries.

[34] invisible for 40 nights. This is the classical length of the Pleiades' period of invisibility, as already in Hesiod: "For forty days and nights they hide themselves" (*Works and Days* 385).

On the 11th, according to Eudoxos, Scorpio begins to set in the morning; and there is rain.

On the 13th, according to Euktēmōn, the Pleiades rise; beginning of summer[35]; and it signifies. According to Kallippos, the head of Taurus rises; it signifies.

On the 21st, according to Eudoxos, the whole of Scorpio sets in the morning.

On the 22nd, according to Eudoxos, the Pleiades rise[T]; and it signifies.

On the 25th[U], according to Euktēmōn, the Goat[V] sets in the evening.

On the 30th, according to Euktēmōn, <the Bird>[W] rises in the evening.

On the 31st, according to Euktēmōn, the Eagle rises in the evening.

On the 32nd, according to Euktēmōn, Arcturus sets in the morning[X]; it signifies. According to Kallippos, Taurus finishes rising. According to Euktēmōn, the Hyades rise in the morning[Y]; it signifies.

The Sun passes through Gemini in 32 days.

On the 2nd, according to Kallippos, Gemini begins to rise; southerly [winds].

On the 5th, according to Eudoxos, the Hyades rise in the morning.

On the 7th, according to Eudoxos, the Eagle rises at nightfall.

On the 10th, according to Dēmokritos, there is rain.

On the 13th, according to Eudoxos, Arcturus sets in the morning.

On the 18th, according to Eudoxos, the Dolphin rises at nightfall.

On the 24th, according to Euktēmōn, the shoulder of Orion rises. According to Eudoxos, Orion begins to rise.[Z]

On the 29th, according to Dēmokritos, Orion begins to rise; it is likely to signify upon this.

---

[35] the Pleiades rise; beginning of summer. The morning rising of the Pleiades was a traditional sign of the beginning of summer. See comment 21.

Fragments 1 and 2, from Geminos's Other Works

Fragment 1,
From Geminos's *Philokalia*:
Geminos on the Classification of the
Mathematical Sciences

INTRODUCTION

Geminos was the author of a mathematical work of considerable length which discussed, among other things, the philosophical foundations of geometry. This book took up a great many mathematical subjects that had been the subject of dispute, for example the classification of lines, the relation among axioms and postulates, and the status of Euclid's parallel postulate. A large number of passages from this work are preserved by Proklos (fifth century A.D.) in his *Commentary on the First Book of Euclid's Elements*. Indeed, Proklos cites Geminos some two dozen times. The exact title of Geminos's work is uncertain, but at the end of a discussion of the classification of lines that do not meet, motivated by Euclid's definition of parallel lines, Proklos concludes, "So much I have selected from Geminos's *Philokalia* to elucidate the subject before us."[1] *Philokalia* means "love of the beautiful."

Whether *Philokalia* was the title of Geminos's entire mathematical treatise, an alternative title, or perhaps the title of one of its books, has been a subject of debate. Pappos (third century A.D.), in his *Mathematical Collection*, cites a work of Geminos "on the classification of mathematics."[a, 2] But Eutokios (sixth century), at the beginning of his commentary on the *Conics* of Apollōnios, cites Geminos from "the sixth book of the doctrine of mathematics."[b] Tannery points out that Pappos's title suits the long extract (our fragment 1) that Proklos gives in part 1 of his Prologue, but not most of the rest of Proklos's citations of Geminos. Both Tannery and Heath therefore prefer the title given by Eutokios, *The Doctrine* (or *Theory*) *of Mathematics*.[3] But the title mentioned by Prok-

---

[1] Friedlein 1873, 177; Morrow 1970, 139.

[2] Hultsch 1876–78, vol. 3, 1026; Ver Eecke 1933, vol. 2, 813. This comes in the course of Pappos's own discussion of the branches of mechanics (Ver Eecke 1933, vol. 2, 809–14).

[3] On the title issue, see Tannery 1887, 18–19; Heath 1926, vol. 1, 39; Aujac 1975, xi–xii.

los, *Philokalia,* is clearly a real title, and it seems broad enough for, and even especially appropriate to, the fragments and citations of the work that we possess. Moreover, Proklos seems to have known Geminos's work intimately. For brevity and definiteness, we shall refer to Geminos's lost mathematical book as the *Philokalia,* without meaning to imply that the question of the title can definitely be decided.

In 1864, Hultsch[4] included in his edition of the geometrical works of Hero of Alexandria several pages of anonymous remarks (found in some Byzantine manuscripts of Hero's *Definitions*) that he took to be fragments of the same work by Geminos that Proklos cites so often (our *Philokalia*). These anonymous fragments carry such titles as "Definition of geometry," "What is the goal of geometry?" "On logistic," "What is the subject matter of logistic?" (Historians of Greek mathematics sometimes call these anonymous fragments the *Variae Collectiones.*) Hultsch attributed some of them to Hero himself, some to Euclid, Anatolios of Alexandria, or Proklos, as well as some to Geminos. In 1912, Heiberg, at the end of his edition of Hero's *Definitions,* attributed a substantially larger array of these fragments to Geminos.[5] For example, some material that Hultsch had attributed to Hero's *Catoptrics* Heiberg now ascribed to Geminos. One of the sections newly attributed to Geminos carried the title "What is scenography?" The oldest manuscripts of Hero's *Definitions* are of the fourteenth century and do not all contain the complete selection of the *Variae Collectiones.* And, of course, there is no way to know just when in the manuscript history of the *Definitions* the text acquired this additional material, which some copyist must have seen as relevant to understanding Hero. The attribution to Geminos remains controversial, and, as we have seen, authorities disagree about which fragments should be attributed to which authors.[6] But whether by Geminos or by another writer interested in the classification of the mathematical sciences, and perhaps writing under Geminos's influence, these fragments do shed light on Geminos's remarks in fragment 1.[7]

[4] Hultsch 1864, 246–49.

[5] Heiberg 1912, 96–108, 165 (Greek text and German translation). In making the attribution to Geminos, Heiberg was influenced by the arguments of Martin 1844, 113.

[6] See Tannery 1887, 43–46. Tannery gives several arguments against Hultsch's attribution of much of this material to Geminos, preferring to ascribe a good deal of it instead to Anatolios. On the other hand, Tannery sees the influence of Geminos on a fragment on optics that Hultsch had ascribed to the *Catoptrics* of Hero.

[7] Some of the same anonymous passages are also found at the end of fourteenth-century manuscripts of a short work on optics by Damianos, who was perhaps of the fourth to sixth centuries A.D. Schöne 1897, 22–31 (Greek text and German translation). On Damianos, see Todd 2003a.

Finally, scholia to the first book of Euclid's *Elements* preserve considerable portions of Geminos's remarks on Euclid's definitions, and
frequently mention Geminos by name.[8] The scholiast apparently used
the same work of Geminos that Proklos relied upon, but according to
Heath the scholia drawn from Geminos "are valuable in that they give
Geminos pure and simple, whereas Proklos includes extracts from
other authors."[9] The situation is not so simple, however, since the
scholiast relied on Proklos himself (rather than on Geminos directly)
for a considerable number of passages. This is clear, for example, in
the passage "So much I have selected from Geminos's *Philokalia* to
elucidate the subject before us," which occurs in the scholia as a ver-
batim duplicate of the remark in Proklos.[10] In any case, Geminos's
Philokalia seems to have attracted a fair amount of attention and comment in Antiquity. This stands in contrast to Geminos's only extant
work, the *Introduction to the Phenomena*, which is not cited by any
ancient writer.

Let us now sketch the context in which Proklos quotes Geminos on
the classification of the mathematical sciences. In the course of the Prologue to his *Commentary on the First Book of Euclid's Elements*, Proklos has just completed a discussion of the Pythagorean division of
mathematics. According to Proklos, the Pythagoreans divided mathematical science into four parts. First of all, one-half of mathematics is
concerned with quantity (*poson* = "how many?") and the other half
with magnitude (*pēlikon* = "how large?"). And then each of these divisions is itself twofold. Thus, a quantity can be considered either in its
own regard or in relation to some other quantity; and a magnitude can
be regarded either as stationary or as in motion. Arithmetic studies
quantity as such, while music studies the relations between quantities.
Geometry is devoted to magnitude at rest, and spherics (i.e., astronomy) to magnitude in motion. As we learn from Plato's *Timaeus*,[11]
arithmetic is prior to music, since number comes into being before
ratio. Similarly, geometry is prior to spherics because rest precedes motion. It is after this discussion of the Pythagorean doctrine that Proklos
takes up Geminos's position on the division of mathematics. The
reader may find it helpful to refer to fig. I.14, which presents Geminos's scheme in diagram form. Sec. 9 of the Introduction provides a
detailed discussion.

---

[8] See Heiberg 1888, 81 line 4; 82 line 29; 107 line 20; 108 line 17; 134 line 12.
[9] Heath 1921, vol. 2, 224.
[10] Heiberg 1888, 108, lines 16–18. Compare with Friedlein 1873, 177, lines 24–25.
[11] Plato, *Timaeus* 35a–41a.

Translation of Fragment 1[12]

This, then, is the position of the Pythagoreans and their division of the four sciences.[13]

But others, such as Geminos, think it proper to divide mathematics according to another scheme. They make one branch concerned with mental things only, and one concerned with perceptible things or touching on them. They call mental, of course, those objects of contemplation that the soul itself calls up, separating itself from material forms. In the branch engaged with mental things, they place arithmetic[14] and geometry as the two first and most important parts. In the branch that operates with perceptible things they place six parts: mechanics, astronomy, optics, geodesy, canonics,[15] and logistic.[16]

On the other hand, they do not think it right to call tactics one of the parts of mathematics, as others do. Rather, they hold that it uses sometimes logistic, as in the tallying up of the companies,[c] and sometimes geodesy, as in the division and measurement of camps. Still less do they consider history or medicine to be part of mathematics, even if those who write histories often use mathematical theorems in writing on the positions of the *klimata* or in inferring the sizes, diameters, and perimeters of cities,[d] and the physicians often clarify their own discussions by such methods: for Hippokratēs[17] and all those who have said anything on seasons and places make clear the utility of astronomy to medicine. In the same way, then, the tactician will use the theorems of mathematics, although he is not a mathematician, if, when he wishes to show the smallest size he should form the camp in a circle, or to show the largest, in a square, pentagon, or some other polygon.[18]

[12] The translation is based on the text of Friedlein 1873, 38–42. All significant departures from Friedlein's text are indicated in the textual notes. In making our own translation of this passage we have benefited by consulting those of Tannery (1888, 38–42), Ver Eecke (1948, 31–36), Aujac (1975, 114–117) and Morrow (1970, 31–35). This passage was also discussed by Heath (1921, vol. 1, 10–18).

[13] the four sciences. See Plato (*Republic* vii 520a–32c) for a famous justification of the Pythagorean quadrivium of arithmetic, geometry, music, and astronomy for the education of the guardians of the state.

[14] arithmetic. *arithmētikē* (*technē* understood) is the "arithmetical (art)," the science of pure number, which is not to be confused with the elementary computation taught to school children. We would characterize *arithmētikē* as number theory, of the sort that originated with the Pythagoreans. See sec. 9 of the Introduction.

[15] canonics. *kanonikē* (*technē* understood) is the mathematical theory of musical scales.

[16] logistic. *Logistikē* is the art of practical computation.

[17] Hippokratēs. In Greek medical writing, the heliacal risings and settings of the fixed stars were sometimes used to indicate the time of year. For examples, see Hippokratēs, *Regimen* III, 68; Hippokratēs, *Epidemics* i, 13–14.

[18] pentagon or some other polygon. This paragraph recalls Plato, *Republic* vii 522c–26d, in which Socrates' interlocutors propose military applications of arithmetic and

These, then, are the species of mathematics as a whole. Geometry is in turn divided into the theory of the plane and stereometry.[19] There is no branch of study specially devoted to points and lines, inasmuch as no figure would arise from them without planes or solid bodies: and, indeed, it is always the task of geometry, both plane and solid, to construct [figures] or to compare or to divide those that have already been constructed.[20]

Similarly, there is a division of arithmetic into the study of linear numbers, plane numbers, and solid numbers[21]: for it examines the classes of number in themselves, as they proceed from the unit, and the generation of the plane numbers, both similar and dissimilar,[22] and the progression to the third dimension.

Geodesy and logistic are analogous to these [i.e., to geometry and arithmetic], since they do not discourse about mental numbers or figures, but rather about perceptible ones. For the task of geodesy is to measure not a cylinder or a cone, but rather heaps as cones and pits as cylinders; and to measure not by means of mental straight lines, but by means of perceptible ones, sometimes more precise ones, such as rays of sunlight, and sometimes coarser ones, such as ropes or a carpenter's rule. Nor, again, does the calculator (*logistikos*) consider the very properties of numbers in themselves, but rather as present in perceptible things, which is why he takes the names for them from the things being counted, calling some apples[e] and others bowls. And he does not concede that

---

geometry before Socrates convinces them that true arithmetic and geometry lead the soul to higher things.

[19] stereometry. *Stereometria* is the "measurement of solids," which Geminos uses here to mean "solid geometry." But Hero of Alexandria wrote a work under the title *Stereometrica* that was concerned more with practical applications, i.e., the mensuration of solids. See sec. 9 of the Introduction.

[20] to divide those that have already been constructed. This is an evident reference to an ancient tradition of division of figures, which goes back to Babylon and is witnessed also by Euclid's *On the Division of Figures*. See Heath 1921, vol. 1, 425–30.

[21] linear, plane, and solid numbers. A linear number is a number regarded as a single factor. Linear numbers are analogous to lengths and are the measures of lengths. A plane number, analogous to an area, is a number regarded as a product of two factors. Thus, 6 regarded as 2×3 is a plane number. A solid number, analogous to a volume, is the product of three factors. Thus, 6 regarded as $1 \times 2 \times 3$ is a solid number. See Euclid, *Elements* vii, def. 16. See also Theōn of Smyrna, *Mathematical Knowledge Useful for Reading Plato* i 7; i 22.

[22] similar and dissimilar plane numbers. Two plane numbers are similar (*homoioi*) if the factors composing them stand in the same ratio. Thus $6 = 2 \times 3$ and $24 = 4 \times 6$ are similar plane numbers, because 3:2 :: 6:4. If 6 and 24 are regarded as the areas of rectangles, one measuring 2 by 3 and the other 4 by 6, the two rectangles are similar in the geometer's sense. See Euclid, *Elements* vii, def. 21. See also Theōn of Smyrna, *Mathematical Knowledge Useful for Reading Plato* i 22.

something is least, as does the arithmetician, since he assumes his least in relation to a particular class: one man, as unit, is his measure for a crowd.

Again, optics and canonics are offspring of geometry and arithmetic. The former uses visual lines and the angles formed from them. It is divided into: a part called optics proper, which explains the cause of false appearances due to the distances of the things observed, such as the meeting of parallels or the perception of squares as circles; a part called general catoptrics, which is concerned with all the varieties of reflections and involves the study of images; and the part called scene-painting,[23] which explains how one may represent appearances in pictures without disproportion or distortion when the things drawn are at a distance or raised to a height. Canonics, in turn, treats the apparent ratios of musical notes by discovering the divisions of the *kanōn*,[24] relying always on sense-perception and, as Plato says,[25] putting the ears before the mind.

Besides these sciences there is the one called mechanics, which is a part of the study of perceptible and material objects. Under this comes the construction of instruments useful in war, such as the engines for defense that Archimēdēs[26] is said to have built during the siege of Syracuse, and also the science of wonder-working,[27] which works its contrivances by means of air, as both Ktēsibios[28] and Hero[29] describe, or by means of

[23] scene-painting. *Skēnographikē*, or scenography, is the art of applied perspective, useful for the realistic representation of buildings in theatrical sets.

[24] *kanōn* (see comment 15, above) is here used as a synonym for *monochordos*, the monochord, the ancient instrument with a single taut string and a movable bridge.

[25] Plato (*Republic* vii 531a–c) disapproves of those who rely upon experiment to determine the rules governing musical harmony, rather than inquiring into number itself to determine which numbers are inherently concordant and why. Geminos himself does not seem to object to putting the ears before the mind.

[26] Archimēdēs is said to have contrived amazing engines for the defense of the city when the Romans under Marcellus were besieging Syracuse (212 B.C.). According to Plutarch (*Marcellus* xvii) these were so effective that the appearance of any stick or rope over the city walls was sufficient to terrify the Roman soldiers. Nevertheless, the city fell and Archimēdēs was among those killed during its sack.

[27] wonder-working. *Thaumatopoiïkē* was the art of devising automata and other gadgets, often operated by means of fluid or air pressure.

[28] Ktēsibios. Ktēsibios of Alexandria (second or third century B.C.) was the inventor of a force-pump, a water organ (a musical instrument in which a constant air pressure is supplied by means of a water pump), and several elaborate water clocks. One of the most difficult problems solved by Ktēsibios in his design of water clocks was the regulation of the length of the seasonal hour, which varies in the course of the year. Ktēsibios's book on his inventions is lost. The fullest surviving account of them is given by Vitruvius, *On Architecture* ix 7; x 7–8, with excellent commentaries in the Budé volumes of Vitruvius. Brief descriptions of the pump and the water clock are given in Landels 1978, 75–76, 193–94. See Drachmann 1948 for more detail.

[29] Hero. Hero of Alexandria (probably mid-first century A.D.). If a date in the first century B.C. for Geminos's work is correct, this mention of Hero is an interpolation by Proklos or by a later copyist. See sec. 6 of the Introduction, in which Geminos's date is discussed. Hero is perhaps best known for his invention of a toy steam engine, described in

weights whose disequilibrium is the cause of motion and whose equilibrium is the cause of rest, as the *Timaeus* has established,[30] or, finally, by means of cords and ropes mimicking the tugs and movements of living creatures. Also under mechanics come the general science of things in equilibrium and the determination of what are called centers of gravity, as well as sphere-making[31] in imitation of the celestial motions, such as Archimēdēs practiced, and, generally, all that is concerned with matter in motion.

There remains astronomy, which treats the cosmic motions, the sizes and shapes of the heavenly bodies, their illuminations and their distances from the Earth, and all such questions. It benefits greatly from sense perception but also has a good deal in common with physical theory. Its parts are: gnomonics, which is engaged with the measurement of the hours through the placement of gnomons; meteoroscopy,[32] which discovers the different altitudes [of the pole][f] and the distances of the stars and teaches many other complex matters from astronomical theory; and dioptrics, which examines the positions[g] of the Sun, Moon, and the other stars by means of just such instruments [i.e., dioptras].

Such are the writings on the parts of mathematics that we have received from the ancients.

---

his *Pneumatics*. He wrote on many of the branches of applied science listed by Geminos: mechanics (including both military engineering and wonder-working), optics, dioptrics, and geodesy. For a summary of Hero's works, see Landels 1978, 199–208.

[30] as the *Timaeus* has established. According to Plato (*Timaeus* 57d–58a), the cause of motion is always to be found in nonuniformity or unevenness (*anōmalotēs*), by which he means a nonuniformity in the arrangement of bodies. If all was uniform, it would be impossible for one body to move and for another to be moved. Rest is therefore associated with uniformity. These remarks come in the midst of a discourse on the nature of the elements and of their changes and transformations. Plato had nothing to say on mechanics proper.

[31] sphere-making. *Sphairopoiïa* is the science of constructing models of the heavens, including celestial globes, armillary spheres, and orreries.

[32] meteoroscopy. *Meteōroskopikē* is most likely concerned with the armillary sphere as a specialized instrument of observation. See sec. 9 of the Introduction.

Fragment 2,
From Geminos's *Concise Exposition of the Meteorology of Poseidōnios*:
Geminos on the Relation of Astronomy to Physics

INTRODUCTION

In a meteorological work that has not come down to us, Geminos gave a discussion of the relationship between astronomy and physics that is important for our understanding of the Greeks' attitudes toward these sciences. His discussion is preserved, in a more or less direct quotation, by Simplikios in his *Commentary on Aristotle's Physics*. The supposed title of Geminos's lost work is given by Simplikios, as he quotes Geminos "from the *Concise Exposition of the Meteorology of Poseidōnios*."[1, a]

Simplikios, who flourished about A.D. 540, produced commentaries on a number of Aristotle's works, including the *Physics* and *On the Heavens*. Because Simplikios still had access to a large philosophical literature of commentary and criticism, much of which is now lost, he is a valuable source for the history of interpretations of and reactions to Aristotle. But Simplikios had no direct knowledge of Geminos's lost meteorological treatise. Rather, as he makes clear, he quotes this extract from an earlier commentator, Alexander of Aphrodisias, who had copied it out of Geminos.

Alexander of Aphrodisias, a philosopher and commentator on Aristotle, flourished about A.D. 200. His works are the earliest complete commentaries on Aristotle that have come down to us, and are therefore of great interest for the history of philosophy.[2] Among the works we have from Alexander are a treatise *On Fate*, as well as commentaries on the *Metaphysics*, the *Meteorology*, and the logical works of Aristotle. But the passage in which Alexander quoted this extract from Geminos is not to be found in Alexander's extant work. Since Simplikios was writing a commentary on Aristotle's *Physics*, it is likely that he took this passage from Alexander's own commentary on the *Physics*, which has not survived.

---

[1] *Concise Exposition of the Meteorology of Poseidōnios*. We have followed Jones 1999a, 255, for this rendering of the title.

[2] On Alexander of Aphrodisias and his place in the Aristotelian tradition, see Todd 1976, 1–20.

Alexander does allude, however, to Geminos in one of his extant works, in a brief passage in his *Commentary on Aristotle's Meteorology*:

Those who follow Geminos and Ailios, in proof of the rainbow being a reflection, use the [fact that] when [one is] approaching it, it seems to approach, and when going away, to go away, just as one sees [things] making [their] appearance in mirrors.[3]

Ailios cannot be identified with certainty, but Aristotle had already attempted to explain the rainbow as a reflection.[4] Geminos's explanation of the rainbow presumably came from the same meteorological work as fragment 2. According to Simplikios, Geminos's book was an epitome of Poseidōnios's *Meteorology*. But in the only fragments of Geminos's book that we possess (that is, fragment 2 and the remark on the rainbow), Geminos follows Aristotle rather closely. This is not too surprising, since Poseidōnios must have taken Aristotle's *Meteorology* for his own point of departure.

There is a third, very brief, ancient citation of Geminos's work, made by Priscianus Lydus (sixth century A.D.), who says in his *Explanation of Problems for King Chosroes* that he has taken some things "from Geminos's Commentary on the *Peri Meteōrōn* of Poseidōnios."[b, 5]

It may seem odd that Geminos and Poseidōnios should have discussed the relation between physics and astronomy in works called *Meteorology*. For the Greeks, however, meteorology is the "study of things raised on high." It has potentially as much to do with astronomy as with things in the upper air, such as rainbows. Ancient authors preserve several different titles for Poseidōnios's lost meteorological work and, indeed, there may have been more than one of them. The preserved fragments suggest that while Poseidōnios's work included a treatment of atmospheric phenomena, such as the rainbow, it probably opened with a general discussion of the cosmos, and also discussed the fiery nature of the Sun.[6] Thus, the beginning portions of Poseidōnios's *Meteorology* may have somewhat resembled the corresponding parts of Kleomēdēs' *Meteōra*.[7] Fragment 2 could have fit quite logically into such a work.

Let us turn now to the intellectual context of fragment 2. Just before giving his quotation from Geminos, Simplikios is in the course of discussing a

---

[3] For the Greek text, see Hayduck 1899, or Manitius 1898, 285 (which quotes Ideler's older text).

[4] Aristotle, *Meteorology* 373a32–74a3.

[5] The passage is quoted in Edelstein and Kidd 1989, 22, and Kidd 1999, 52, from Bywater 1886, *Solutiones ad Chosroem, Prooemium*, 42.8–11. Priscianus wrote in Greek, but this work survives only in a Latin translation of the ninth century.

[6] See Kidd 1999, p. 72, and fragments 14, 15, 16, 17.

[7] But this is only one of several possible titles for Kleomēdēs' book. See note 22 in our Introduction.

passage from Aristotle's *Physics*,[8] in which Aristotle distinguishes physics from mathematics on the one hand and from metaphysics on the other. In the course of this investigation, Aristotle raises the question of whether astronomy is a separate science or a part of physics. For, as Aristotle points out, one who seeks to know what the Sun and the Moon are can hardly avoid inquiry into their essential natures. Moreover, he continues, physical thinkers have discussed the shape of the Sun and the Moon and have asked whether the Earth and the cosmos are spherical. In contrasting physics with mathematics, Aristotle remarks that the mathematician deals with lines and surfaces, not as the boundaries of natural bodies, but rather in abstraction. He also mentions optics, harmonics, and astronomy as sciences that partake of both the physical and the mathematical, though they are more physical than mathematical in nature. Geometry and optics stand, as it were, in opposite relationships to the mathematical and the physical. Thus the geometer may deal with physical lines, but only in their mathematical, not their truly physical, nature. In the opposite way, optics deals with mathematical lines, but as physical lines, not truly mathematical ones.

Some elements of Aristotle's discussion find echoes in Geminos, not only in fragment 2, under discussion here, but also in Geminos's classification of the mathematical sciences (fragment 1). But Geminos states the relationship of astronomy to physics in a much clearer and simpler way. Indeed, his remarks constitute the clearest statement of this relationship we find in any of the Greek astronomical writers. Though he may have begun with Aristotle (or with Poseidōnios, who began with Aristotle), it is clear that Geminos had done some hard thinking on this subject himself. For a detailed discussion of this passage, see sec. 10 of the Introduction.

TRANSLATION OF FRAGMENT 2[9]

Alexander, in his diligent way, provides, from the *Concise Account of the Meteorology of Poseidōnios*, a passage by Geminos that takes its starting point from Aristotle.[10] Here it is:

It is [the task] of physical theory to inquire into the essence of the heaven and the stars, their power and quality, their origin and destruction;

---

[8] Aristotle, *Physics* 193b22.

[9] The translation of fragment 2 is based on the text of Diels 1882, 291–92. English translations of this passage have been published by Heath 1932, 123–25, and by Bowen and Todd 2004, 199–204, which we have consulted with profit.

[10] takes its starting point from Aristotle. See Aristotle's discussion beginning at *Physics* 193b22.

and, by Zeus, it can even make demonstrations concerning their size, form, and arrangement. Astronomy, on the other hand, does not attempt to speak about any such thing, but demonstrates the arrangement of the heaven, presenting the heaven as an orderly whole, and speaks about the shapes, sizes, and distances of the Earth, Sun, and Moon, about eclipses and conjunctions of the stars, and about the quality and quantity of their motions. Therefore, since it deals with the investigation into quantity, magnitude, and quality in relation to form, it naturally needed arithmetic and geometry for this. And concerning these things, the only ones of which it undertook to give an account, astronomy has the capacity to reach conclusions by means of arithmetic and geometry.

Now in many cases both the astronomer and the physicist[11] will propose to demonstrate the same point, such as that the Sun is large or that the Earth is spherical, but they will not proceed by the same paths. One [the physicist] will prove each point from considerations of essence or inherent power, or from its being better[12] to have things thus, or from origin and change; but the other [the astronomer] will prove them from the properties of figures or magnitudes, or from the amount of motion and the time appropriate to it. Again, the physicist will often reach the cause by looking to creative force; but the astronomer, when he makes demonstrations from extrinsic properties, is not competent to perceive the cause, as when, for example, he makes the Earth and the stars spherical.[13] Sometimes he does not even desire to take up the cause, as when he discourses about an eclipse;[14] but at other times he invents by way of hypothesis

[11] the astronomer and the physicist. "Astronomer" = *astrologos*. "physicist" = *physikos*, which could also be translated "natural [philosopher]," since, as the following passages from Geminos will make clear, the approach to nature of the *physikos* was that of a philosopher in the Aristotelian tradition, not that of a modern physicist.

[12] from its being better. This is a kind of argument recommended by Aristotle. See *Physics* ii 198b4.–9.

[13] the stars are spherical. Aristotle, *On the Heavens* (291b11–23), proves this, both by appeal to physical principles and by astronomical methods. Physically, the stars must be spherical. Nature does nothing without purpose. Therefore the stars, which do not move of themselves, should have the figure least suited to motion. The figure least suited to motion is the sphere, since it has no instrument to serve that purpose. (This recalls Plato's remark about the cosmos as a whole: the demiurge made the cosmos without hands and feet, for it had no need to grasp anything, nor any need for stepping. *Timaeus* 33d.) Astronomically, it is clear that the Moon is spherical, from the evidence of its monthly cycle of phases, and, Aristotle concludes, if one of the heavenly bodies is spherical, the others must be spherical also.

[14] eclipse. Geminos's remark here may seem slightly strange, since knowledge of the cause of eclipses goes back at least to Anaxagoras (fifth century B.C.). Any astronomer of Geminos's day was perfectly capable of setting forth the cause of an eclipse. But in the use of eclipses in some practical calculation, he might very well omit to discuss the cause. Eclipses of the Moon might be used, for example, in a discussion of lunisolar cycles. Or

and grants certain devices, by the assumption of which the phenomena will be saved.[15]

For example, why do the Sun, Moon, and planets appear to move irregularly? [The astronomer would answer] that if we assume that their circles are eccentric, or that the stars go around on an epicycle, their apparent irregularity will be saved. And it will be necessary to fully examine in how many ways it is possible for these phenomena to be brought about,[16] so that the treatment of the planets befits a causal explanation[17] according to the accepted way. And thus a certain person, Hērakleidēs of Pontos,[18, c] coming forward, says that even if the Earth moves in a certain way and the Sun is in a certain way at rest, the apparent irregularity with regard to the Sun can be saved.

For it is certainly not for the astronomer to know what is by nature at rest and what sort [of bodies] are given to movement. Rather, introducing hypotheses that certain [bodies] are at rest and others are moving, he considers from which hypotheses the phenomena in the heaven will [actually] follow. But he must take from the physicist the first principles,[19] that the motions of the stars are simple, uniform, and

---

again, the time of onset of a lunar eclipse, observed at two different localities, might be used to deduce the difference in geographical longitude of the two places of observation. In either of these applications there would be no need to discuss the cause of eclipses.

[15] the phenomena will be saved. On the slogan and program of "saving the phenomena," see sec. 10 of the Introduction.

[16] examine in how many ways it is possible for these phenomena to be brought about. This echoes a remark of Hipparchos. According to Theōn of Smyrna (*Mathematical Knowledge* iii 26.3), Hipparchos said that research into the explanation of the same phenomena by hypotheses that are quite different is a task worthy of the attention of the mathematician. Hipparchos, Theōn, and Geminos were all alluding to the fact that different mathematical models are capable of accounting for the same phenomena.

[17] causal explanation, *aitiologia*. For the meaning of this term, see the discussion in sec. 10 of the Introduction.

[18] Hērakleidēs of Pontos. Thus Diels and the mss. Most scholars agree, however, that this name does not belong here. Probably, the text originally said that a "certain person (*tis*), coming forward, says . . . ," and then the name Hērakleidēs of Pontos was interpolated by some later copyist. Aristarchos of Samos, not Hērakleidēs, was responsible for the view that the Earth moves in a circle about a stationary Sun. We have clear testimony from Aëtios (*Opinions of the Philosophers* iii 13.3), as well as from Simplikios's *Commentary on Aristotle's* On the Heavens (293b30 and 289b1; Heiberg 1894, 519, and 444–45), that Hērakleidēs espoused the daily rotation of the Earth on its axis, but not the motion of the Earth around the Sun. For translations of the relevant passages, see Heath 1932, 93–94. A partially heliocentric system has sometimes been ascribed to Hērakleidēs. In this system, Venus and Mercury orbit the Sun while the Sun and the other planets orbit the Earth. Although such a system is described by several ancient writers (e.g., Theōn of Smyrna, *Mathematical Knowledge Useful for Reading Plato* iii 33), the ascription of this view to Hērakleidēs is mistaken. For a discussion of all the evidence, see Eastwood 1992.

[19] first principles. See *Introduction to the Phenomena* i 18–21 for Geminos's discussion of circular motion as the first principle of astronomy.

orderly, from which he will demonstrate that the dance[20] of all [the stars] is circular, with some turning round on parallel circles, some on oblique ones.[21]

In this manner, then, does Geminos, or rather Poseidōnios in Geminos, give the distinction between physics and astronomy, taking his starting point from Aristotle.

[20] dance, *choreian*. Plato (*Timaeus* 40c3) spoke of the choric dances (*choreias*) of the planets, and Geminos's readers might well have made this association. But the word could also be used, by analogy, of any circling motion.

[21] parallel and oblique circles. On the parallel circles see v 1–9. The fixed stars move from east to west on parallel circles, i.e., circles parallel to the equator. In the course of any one day or night, the Sun, Moon, and each of the planets may be considered to move on a parallel circle. But the Sun, Moon, and planets each have an additional and slower motion from west to east along the zodiac, a circle that is oblique to the equator.

Textual Notes to Geminos's *Introduction to the Phenomena*

TITLE

^a Geminos's *Introduction to the Phenomena*, Γεμίνου εἰσαγωγὴ εἰς τὰ φαινόμενα, B and M. Γεμίνου εἰσαγωγὴ εἰς τὰ φαινόμενα, V, but with εἰσαγω- γὴ inserted above the line in the same hand. The title and author's name are want- ing in A. Γεμίνου εἰσαγωγὴ εἰς τὰ μετέωρα, C. Γεμίνου τὰ φαινόμενα, Am- brosianus C 263 inf. (Milan).

CHAPTER I

^a The circle of the signs. Geminos refers to the zodiac indifferently as ὁ τῶν ζῳδίων κύκλος or ὁ ζῳδιακὸς κύκλος. The first will be translated as "the circle of the signs," and the second as "the zodiac circle," but no significance should be attached to the distinction.

^b Leo, ὁ Λέων, following the mss. ἡ Παρθένος ("Virgo"), Manitius, followed by Aujac. Virgo does extend over a greater range in longitude (some 46°) than any other zodiacal constellation. But because Leo also takes up more than the canonical 30°, there is nothing wrong with the reading of the mss.

^c Autumnal equinox. Here begins a lacuna of some thirty lines in the Greek text, which Manitius (followed by Aujac) has filled from the medieval Latin translation. The text supplied from the Latin is enclosed between the marks < >.

^d The Latin text of this sentence appears to be corrupt: *Et conversio estiva fit, quando sol pervenit ad propinquiora loca sont capitum in habitationibus nostris et elevatur ab orizonte nostro ultimiore suis elevationibus*. . . . The Latin word *sont* is glossed a little further on by *cenit*, i.e., zenith (Aujac 1975, 3n2). We con- jecture that this originated as an imperfect transcription of the Arabic *samt*, which means "direction," and is the normal word for "zenith" in astronomical texts.

^e <Stilbōn,>. Added by Manitius (followed by Aujac) in analogy to Kleomēdēs (*Meteōra* i 2.35), supported by the medieval Latin version of Geminos.

CHAPTER II

^a antiskian, ἀντίσκια, our emendation. ἀντισυζυγίαν, the mss., Manitius, Aujac. Manitius interprets *antisyzygy* as if it were an alternative name, used by

some, for *syzygy*. Aujac treats *antisyzygy* as a fifth distinct aspect, added by some. But to our knowledge, ἀντισυζυγία occurs in no other astronomical or astrological writer, while ἀντίσκια is a common term that is indeed used synonymously with "in syzygy."

ᵇ occur, συντελοῦνται, following Aujac and the mss. ἄρχονται, Manitius.

ᶜ of spring, of summer, of fall, of winter. Thus the mss. and Aujac. Manitius brackets these words for deletion.

ᵈ <the . . . Earth,>. The passage in brackets < > is missing in the Greek mss. but has been supplied by Manitius (followed by Aujac) from the medieval Latin translation.

ᵉ the first degree of Libra culminates below the Earth. This whole statement (ii 21) is corrupt in all the manuscripts. For the statement to be correct, the first degree of Capricorn must be at the meridian, and not on the horizon. Thus Geminos must have written either: (1) "when the first degree of Aries is setting, the first degree of Cancer culminates, the first degree of Libra rises, and the first degree of Capricorn culminates below the Earth"; or (2) "when the first degree of Libra is setting, the first degree of Capricorn culminates, the first degree of Aries rises, and the first degree of Cancer culminates below the Earth." As may be confirmed with an armillary sphere or a celestial globe, it is only when the equinoctial points are on the horizon that the ecliptic is divided into four equal parts by the horizon and the meridian. In general, the ecliptic is *not* divided into four equal parts by the horizon and the meridian, as Geminos himself points out (ii 22).

ᶠ contained, ἐμπεριλαμβανόμενα, accepting Manitius's emendation. Geminos uses ἐμπεριλαμβάνεται in just this way at ii 36, 39, and 44. παραλαμβανόμενα, Aujac and the mss.

ᵍ perceivable through reason, λόγῳ θεωρητόν.

CHAPTER III

ᵃ whole, ὅλῳ, accepting Manitius's emendation after the Latin version and in analogy to the preceding statement about Lyra. Not in the Greek mss. or Aujac.

ᵇ hand. Before this word, Manitius adds ἀριστερᾷ, "left," without ms. authority.

ᶜ the Wild Animal that Centaurus is holding, and the Thyrsus-lance that Centaurus is holding according to Hipparchos. Θηρίον, ὃ κρατεῖ ὁ Κένταυρος, καὶ Θυρσόλογχος, ὃν κρατεῖ ὁ Κένταυρος καθ᾽ Ἵππαρχον, following Manitius and A, B, C. Aujac, following I, P, T, M and the medieval Latin version, deletes καὶ Θυρσόλογχος, ὃν κρατεῖ ὁ Κένταυρος, thus making the passage read "the Wild Animal that Centaurus is holding according to Hipparchos." Aujac reads the deleted phrase as a spurious addition that originated as a gloss drawn from Ptolemy's or Hipparchos's mention of the thyrsus as a part of Centaurus. The best argument for this view is that the phrase is missing in the Latin version of Geminos, which was based (through an Arabic intermediary) on a Greek manuscript older than any we now possess. Deleting the phrase, however, produces a reading that attributes the Wild Animal to Hipparchos, even though the Wild Animal is well attested before Hipparchos (see the discussion following Table 3.3).

ᵈ the Caduceus according to Hipparchos, Κηρύκειον (Κηρύκιον, Manitius) καθ᾿ Ἵππαρχον, following Manitius and the mss. *et Abrachis nominavit eam purpuream*, Latin. Aujac deletes this phrase as too problematical.

CHAPTER V

ᵃ all these circles are inscribed . . . in astronomy, Συγκαταγράφονται δὲ οὗτοι πάντες εἰς τὴν σφαῖραν διὰ τὸ πρὸς μὲν ἄλλας πραγματείας τῶν ἐν τῇ ἀστρολογίᾳ πολλὰ συμβάλλεσθαι, following Aujac and the mss. Manitius changes συγκαταγράφονται ("are inscribed together") to οὐ καταγράφονται ("are not inscribed") on the authority of the Latin, *non signantur*. It is unlikely that 182 parallel circles were drawn on many ancient celestial globes. But Manitius's reading seems to require an adversative or concessive construction of some kind in place of διὰ τό . . . πολλὰ συμβάλλεσθαι. In either case, Geminos's meaning is clear: the astronomer may need many parallels to specify the declinations of stars, or to work out the lengths of all the days of the year; but for the beginner the five principal parallels will suffice, and these are the only ones customarily shown on globes and armillary spheres.

ᵇ For the horizon in Greece. A, B, and C have a lacuna here. The words in brackets < > follow Aujac's guess, supported by the Latin version, *orizon autem regionis grecorum qui nominantur elenes dividit orbem tropici estivi*. Manitius prefers "For the horizon at the Hellespont." The Hellespont was traditionally associated with the *klima* of 15 hours, as in Ptolemy (*Almagest* ii 6) and Kleomēdēs (*Meteōra* ii 1.442), and the 15-hour figure is more accurate for the Hellespont than for Greece. But Aratos did not specify the latitude to which his poem applies, and Neugebauer (1975, 711) points out the error of assuming the Hellespont whenever the 15-hour *klima* is mentioned. (In vi 8, Geminos puts Rome in the 15-hour *klima*.) Although Geminos interpreted Aratos as having intended a specific *klima*, we have no way of knowing whether Geminos meant to identify this *klima* narrowly with the Hellespont, or more generally with Greece.

ᶜ our *oikoumenē*, τὴν ἡμετέραν οἰκουμένην, following Aujac and A, B, C. Manitius prints οἴκησιν ("dwelling place"), which occurs in a ms. of the *Sphere of Proklos* (L). Even with a broad definition of the *oikoumenē*, the text is sensible. Moreover, Geminos uses *oikoumenē* in a similar way at v 3.

ᵈ one makes the division in declination in the following way, διαιρεῖται οὕτως κατὰ πλάτος, thus Aujac, following A, B, C. Manitius prints <ὁ μεσημβρινὸς> διαιρεῖται οὕτως, "<the meridian> is divided in the following way."

ᵉ the one perceptible [to our sight] and the other perceivable by reason, εἷς μὲν ὁ αἰσθητός, ἕτερος δὲ ὁ λόγῳ θεωρητός.

ᶠ inclination, ἔγκλιμα.

ᵍ whole, ὅλη, following Aujac and the mss. Manitius marks this word for deletion.

ʰ slants <through> the tropic circles, λελόξωται <διὰ> τῶν τροπικῶν κύκλων, following Manitius and A ; <διὰ> added by Manitius. λελόξωται τῷ τροπικῷ κύκλῳ, Aujac, B, and C.

Chapter VI

^a the longest day turns out to be two months long. διμηνιαίαν τῶν ἡμερῶν τὴν μεγίστην ἡμέραν συμβαίνει γίνεσθαι, Aujac. We follow Manitius in deleting τῶν ἡμερῶν.

^b <by the horizon>. Inserted by Manitius.

^c This is why. Manitius brackets this whole statement (vi 31) for deletion, branding it an inept interpolation. But the same cause for the intensification of the heat around summer solstice is invoked at xvii 28, and we see no need to doubt that the argument is Geminos's.

^d it (the subject of the verb simply understood), following Manitius. ὁ ἥλιος, Aujac after the Latin version. Not in A or B. αὐτός, C.

^e The cause of the inequality in the lengthening of the days is the obliquity of the zodiac circle. The mss. read: Αἰτία δέ ἐστι τῆς ἀνισότητος καὶ τῆς τῶν ἡμερῶν παραυξήσεως ("The cause of the inequality and of the lengthening of the days is . . ."). Following Aujac, we suppress καί. Manitius suppresses τῆς ἀνισότητος καί, on the grounds that the cause of the inequality of the days has already been given at vi 24. Thus Manitius would read, "The cause of the lengthening of the days is the obliquity of the zodiac circle." At vi 24, however, Geminos is explaining only why some days are longer than others. By contrast, in the present paragraph he is addressing the fact that the changes in the length of the day (from one day to the next) are of different sizes at different times of the year.

^f and the nights. Manitius brackets these words for deletion.

^g moreover, the increases have nearly a constant difference, καὶ σχεδὸν τὴν αὐτὴν παραλλαγὴν ἔχουσιν αἱ παραυξήσεις, more literally, "the increases have nearly the same difference." Manitius brackets these words as a gloss. But this hypothesis of constant differences is connected intimately with the remark that follows, i.e., that the daily increases around the equinox are ninety times as large as the daily increases around the solstices. See sec. 12 of the Introduction for Geminos's use of arithmetic progressions. (Note the typographical error in παραλλαγὴν in Aujac's text.)

^h Pisces. After the sentence, Manitius adds <νότιόν ἐστι>, "they are southern," in analogy to the remark at the end of vi 40.

Chapter VII

^a For this reason . . . in the same place. Manitius deletes this sentence as a gloss.

^b arc, περιφέρεια, following Manitius and Aujac. περιφορά, the mss.

^c degrees, μοιρῶν, following Manitius. μερῶν, Aujac and the mss.

^d rising. In the mss. this is followed by καὶ τὴν δύσιν, "and the setting." Following Manitius, we delete these words as an imprecise interpolation, for the signs that rise in the greatest time actually set in the least time. What these words are meant to suggest is that signs that set when the zodiac is upright take the most time to set; i.e., that what goes for rising goes for setting, too.

^e <it results that>, <συμβαίνει>, following Manitius.

ᶠ the rising. Followed in the mss. by καὶ τῆς δύσεως, "and the setting," which we delete, following Manitius (see textual note d above).

ᵍ as one another, κατ᾽ ἄλληλα, following Aujac. κατ᾽ ἄλλα, the mss., deleted by Manitius.

ʰ Capricorn . . . Gemini. The mss., Manitius, and Aujac all have Κριοῦ . . . Παρθένου, "Aries . . . Virgo." The text states that the zodiac is lowest on the horizon when the first degree of Capricorn is culminating. This is correct for the northern midlatitudes. The problem is that the text goes on to claim that the semicircle of the zodiac from Aries 0° to Libra 0° therefore rises in a short time. In fact, this semicircle takes exactly 12 hours to rise at any latitude. This error is so elementary and so easily detected on a globe or armillary sphere that it is impossible to believe that Geminos could have committed it. A correct claim would be that the semicircle *centered around* Aries 0° (rather than *starting from* Aries 0°) would take a short time to rise. These six signs (from Capricorn 0° to Gemini 30°) are just those signs that rise in the course of the night on summer solstice. The version of the text is probably due to an early, and unskillful, emendation.

ⁱ Cancer . . . Sagittarius. The mss., Manitius and Aujac all have Ζυγοῦ . . . Ἰχθύων, "Libra . . . Pisces." This error is analogous to that committed in the preceding sentence. See textual note h.

ʲ or when the first degree of Libra is culminating. Manitius suppresses this parenthetical remark as an interpolation. This could well be right, for Geminos discusses the situation when Libra (the Claws) is culminating at vi 22, immediately below.

ᵏ <and at this time>. Inserted by Manitius after the model of the preceding statements (vii 19–20).

ˡ they asserted, ἀπεφήναντο, adopting Manitius's emendation in analogy to the preceding statements. ὥστε, Aujac and the mss.

ᵐ Virgo rises. Manitius inserts after these words, without manuscript authority, <but Pisces sets>. This emendation, like many others that Manitius makes to later parts of this passage, is not required for either meaning or clarity. Throughout the remainder of the chapter, Manitius makes many additions that we have ignored, and it seems unnecessary to mention them all here. Manitius's emendations do make the discussion perfectly balanced and complete, but whether Geminos really so belabored the point is another question.

ⁿ [signs] that are equally distant from the solstitial and equinoctial points rise and also set in equal time. The text is sloppy here: see the discussion in comment 13. Still, Manitius's emendation of this sentence does not result in correct astronomy, but only states the error in more emphatic terms: "From these things it is clear that [signs] that are equally distant from the solstitial <points rise in equal time, and also set,> and <signs that are equally distant from> the equinoctial points <both> rise and set in equal time."

CHAPTER VIII

ᵃ 29 + 1/2 + 1/33. Written $\overline{κθ}$ L' $λγ'$ in the mss., as is normal.

ᵇ fall. Before this word, Manitius inserts <τὴν>, thus giving the passage the sense, "and the summer festival to be also a winter one, and <the> fall [festival to be] also a spring one."

^c spring [festival]. Here the Latin version adds: *immo permutantur in omnibus temporibus 4 et eorum revolutiones sunt in omnibus diebus anni, in omnibus mille et quadringentis et sexaginta annis*. This appears to be a gloss based on viii 24.

^d most of the Greeks suppose the winter solstice according to Eudoxos to be at the same time as the feasts of Isis [reckoned] according to the Egyptians, Ὑπολαμβάνουσι γὰρ οἱ πλεῖστοι τῶν Ἑλλήνων ἅμα τοῖς Ἰσίοις κατ' Αἰγυπτίους καὶ κατ' Εὔδοξον εἶναι χειμερινὰς τροπάς. See the discussion in comment 10.

^e again. Accepting Manitius's transposition of πάλιν from near the beginning of the sentence to near the end.

^f they sought. Adopting Manitius's punctuation of this passage and his deletion of ὅθεν immediately before these words.

^g The time of the period, ὁ τῆς περιόδου χρόνος, following Aujac and the mss. Manitius deletes these words without comment.

^h <the feasts>, <αἱ ἑορταί>, accepting the insertion by Manitius and the earlier editors. Not in Aujac.

ⁱ The lunar year is reckoned at 354 days. Manitius deletes this entire paragraph (viii 34–35) as spurious, but its role in the development is clear: it provides a capsule summary of the lunar year used in the *octaetēris* immediately before Geminos takes up the defects of the *octaetēris*.

^j 30 days. These words are followed in the mss. and Aujac by πρὸς τὰς τοῦ ἡλίου ὥρας, "with respect to the seasons of the Sun." Following Manitius, we delete this phrase as a gloss.

^k For the monthly period has not been taken accurately. After these words, the mss. contain two additional sentences:

> For the monthly period, taken accurately, is 29;31,50,8,20 days. Because of this, it will sometimes be necessary to intercalate 4, instead of 3, embolismic days in the 16 years.

We delete this passage as an interpolation. Manitius deletes all of viii 43–45. Aujac deletes only the third sentence of viii 43.

The value for the length of the synodic month, 29;31,50,8,20 days, appears in the Latin translation and in some of the Greek manuscripts—the others showing variants. This is, in fact, a more accurate value for the length of the month than the value usually used by Geminos, $29 + 1/2 + 1/33$ days. The value 29;31, 50,8,20 days is given by Ptolemy (*Almagest* iv 2), who says that it was obtained by Hipparchos, "by calculations from observations made by the Chaldeans and in his time." This parameter is actually of Babylonian origin and plays a role in system B of the Babylonian lunar theory (see Neugebauer 1975, 483). In view of his familiarity with other details of Babylonian astronomy, there is no historical reason why Geminos could not have known the value 29;31,50,8,20. Neugebauer (1975, 585) accepts Geminos's mention of this parameter as genuine.

The difficulty is that the Hipparchian-Babylonian value is mentioned but once, while Geminos uses on several occasions the value $29 + 1/2 + 1/33$ days ($= 29;31,49,5,27$), which he introduced at the beginning of chapter viii. When Geminos returns to lunisolar cycles in chapter xviii, he uses the value $29 + 1/2 + 1/33$ days again, and does not mention the better Hipparchian-

Babylonian value. Moreover, in the mss. the Hipparchian-Babylonian value for the synodic month is given in base-60 (as by Ptolemy), while throughout chapter viii Geminos operates with unit fractions. And when Geminos does introduce base-60 notation in chapter xviii, he is careful to explain his notation to his reader. It would be odd if Geminos used base-60 fractions earlier in the book without explanation.

Finally, the Hipparchian-Babylonian month length has no function in the argument. Geminos states (viii 42) that the *octaetēris* fails in the length of the month, as well as in the number of embolismic months. Then (at viii 43–45) he points out that because the month is longer than 29½ days, the *octaetēris*'s basic pattern of alternating full and hollow months makes the month too short. The already-mentioned length of the month, 29 + 1/2 + 1/33 days, is sufficient for this argument, and there is no need to introduce the Hipparchian-Babylonian value. Then (at viii 46–47) Geminos, fulfilling his promise in viii 42, shows how the *octaetēris* fails to give the correct number of embolismic months. Here he introduces a new parameter, the lunar year of 354⅓ days, which implies a slight modification of his adopted length of the month. Again, there is no role in the argument for the Hipparchian-Babylonian value. Thus it appears that the Hipparchian, base-60 parameter in chapter viii was interpolated as a comment drawn from the *Almagest*, or some source dependent on it.

^{l} The year according to them is therefore 365 5/19 days. This sentence is rejected by Manitius as an inept addition. But it follows immediately from the length of the nineteen year period, 6,940 days, which Geminos has stated in the preceding sentence.

^{m} and the days total 7,050. These words are followed in the mss. by the sentence: "But it was necessary to count 110 hollow, for which reason there are 6,940 days according to the Moon for the nineteen-year period." Following Manitius, we delete this sentence as an interpolation and, moreover, one that breaks the flow of the argument and anticipates the conclusion reached in viii 54.

^{n} therefore reckon, οὖν ἄγουσι, following Manitius. συνάγουσι, Aujac and the mss.

CHAPTER IX

^{a} faces toward the east. After these words, Manitius adds, without manuscript authority, <and when it rises after sunrise, the bright [side] of it faces toward the west.> We have no reason to think that Geminos so belabored the issue.

^{b} <the distance of the Moon>. Added by both Manitius and Aujac, following the earlier editors.

^{c} at the latest, βραδύτατον, following Aujac. Deleted by Manitius.

^{d} becomes, γίνεται, following Manitius's conjecture. ἀνατέλλει, Aujac and the mss.

^{e} The entire monthly period. Manitius rejects this entire paragraph (ix 16) as an inept repetition of the opening lines of chapter viii.

CHAPTER XI

ᵃ for all the eclipses of the Moon occur in this space. Manitius rejects these words as a gloss; but they are an essential clarification of the meaning of τὸ ἐκλειπτικόν.

ᵇ for the magnitudes of the eclipses are in due relation to the Moon's motion in latitude, πρὸς λόγον γὰρ τῆς κατὰ πλάτος κινήσεως τῆς σελήνης τὰ μεγέθη τῶν ἐκλείψεων σύμφωνα γίνεται. More literally, ". . . are harmonious in relation to. . . ." Manitius deletes σύμφωνα because it normally governs the dative case. In the mss. and Aujac, κινήσεσως is followed immediately by τῆς ἡμερησίου: thus, "the Moon's 'daily' motion in latitude." But as Manitius points out, this is surely a later and irrelevant addition, for the argument has nothing to do with 1 day's worth of the Moon's motion in latitude.

CHAPTER XII

ᵃ For the sign following the Sun is always invisible because of the rays of the Sun, while the one preceding it is visible. Manitius, without explanation, brackets this sentence for deletion, and the sentence could be read as an awkward gloss. It does make sense, however, if we regard it (like many others in this passage) as applying to observations made at sunrise. The situation is reversed for observations made at sunset: then part of the sign following the Sun is visible, while the one preceding it is invisible, for it sets before the Sun does.

ᵇ to have stood away a distance of two signs, δύο ζῳδίων διάστημα ἀφεστηκός. Manitius, without explanation, brackets these words for deletion.

ᶜ as the night advances it departs toward the east from the carefully noted star; and <the star, toward the> west <from the Moon>. Accepting Manitius's emendation: <ὁ ἀστὴρ ἀπὸ τῆς σελήνης πρὸς>

ᵈ while moving toward the same parts as the cosmos. Adopting, with Manitius and the earlier editors, κινούμενοι τῷ κόσμῳ. κινουμένου τοῦ κόσμου, Aujac.

ᵉ the motion proper to it were from west to east. Adopting Manitius's text, ἰδίας ὑπαρχούσης αὐτῷ τῆς κινήσεως <τῆς> [μὲν ἀπ' ἀνατολῆς ἐπὶ δύσιν, τῆς δὲ] ἀπὸ δύσεως ἐπ' ἀνατολήν ("the motion proper to it were [from east to west and] from west to east.") As Manitius conjectures, the phrase "from east to west" is most easily explained as a marginal note that worked its way into the text. Aujac prints the mss. text and translates "at the same time from east to west and from west to east."

ᶠ but nothing, οὐδὲν δέ, accepting Manitius's emendation. οὐδέ, Aujac and the mss.

CHAPTER XIII

ᵃ for this reason. Manitius brackets this entire statement (xiii 10) for deletion, because it is not found in the Latin version.

ᵇ again, πάλιν. Deleted by Manitius.

ᶜ for the last time, τὸ ἔσχατον (our emendation). The mss. read πρῶτος, "for the first time." However, "having escaped for the first time from the rays of the Sun," is inappropriate. The visible evening rising is the *last*, not the first, of the star's risings to be visible during the annual cycle. (Autolykos, *On Risings and Settings* i 2, as well as the initial definitions, Aujac 1979, 69. This is also apparent from fig. 13.1.) Statement 13 (on the last rising to escape the rays of the Sun) can be set off against statement 9 (on the first rising to escape the rays of the Sun). The concluding portion of statement 13 is correct. On successive evenings following the VER, once the sky becomes dark enough for stars to be seen, star *S* will be seen higher and higher above the eastern horizon. Note that, in this situation, "escaping the rays of the Sun" can only be applied to the *rising* of the star and not to the star itself: the star itself is visible during some part of the night both before and after the visible evening rising.

ᵈ whenever both the Sun and the star are <exactly> on the horizon. Following Manitius, who inserts <ἀπαραλλάκτως> immediately after ὅταν, by analogy to xiii 17. We point also to xiii 6, and 11, where κατὰ ἀλήθειαν is used for a similar purpose.

ᵉ for the first time, τὸ πρῶτον. The mss. have τὸ ἔσχατον, "for the last time." Manitius suggested (p. 274) but did not actually adopt this emendation, lest it excuse the author (Geminos or his excerptor) of repeated error. Aujac tries to rescue the text by translating τὸ ἔσχατον as "à la dernière limite," meaning that the star is seen setting at the last moment before sunrise. It seems better to concede that the text is corrupt.

ᶠ For the <stars>. From here to the end of xiii 27, the text is badly corrupted. Every proposition in xiii 21–23 and 27 is either wrong or problematical as stated.

The Difficulties

The text does not state whether the propositions in xiii 21–27 apply to true or to visible phases. One might suppose that visible phases are meant, since in the *parapēgmata* and in literary references to heliacal risings and settings, unqualified phases are, almost invariably, visible ones. But the propositions in xiii 29 also are unqualified; and these propositions are correct as stated only for the *true* phases.

A second difficulty with assuming that visible phases are meant is that the text never states a quantitative visibility rule. Autolykos, for example, assumes that the visible risings and settings occur when the star is on the horizon and the Sun is half a zodiac sign below the horizon, measured along the ecliptic. Thus, the visible phases precede or follow the true ones by about 15 days. This simple rule, a crude but reasonable approximation, allows quantitative predictions about the dates of the visible phases. That Geminos never mentions such a rule might seem to imply that his propositions apply to true rather than to visible phases. It should be pointed out, however, that a good deal can be done without a quantitative visibility rule, simply by using Geminos's proposition xiii 19 and some notions of symmetry.

Let us now examine what the propositions assert.

- For stars on the zodiac circle:

xiii 21. The time from evening rising to morning rising is 6 months, as is the time from evening setting to morning setting. In short-hand notation:

$$ER \to MR = 6 \text{ months}$$
$$ES \to MS = 6 \text{ months}$$

This proposition is false for visible phases. It is correct for true phases; but it is correct for *all stars*, and thus there would be no reason to single out zodiacal stars.

- For stars north of the zodiac:

xiii 22. $ER \to MR > 6$ months
xiii 27. $MS \to ER < 6$ months

The first proposition is false for true phases (the time should be = 6 months). It is correct for visible phases, but is correct for *all* stars, not just northern ones. The second proposition is false for both true and visible phases.

- For stars south of the zodiac:

xiii 23. $ER \to MR < 6$ months
xiii 27. $MS \to ER > 6$ months

The first proposition is false for true phases (the time should be = 6 months). It is also false for the visible phases (the time should be > 6 months). The second proposition is false for true phases (the time should be << 6 months). For the visible phases, the proposition is indeterminate: it is correct for some southern stars but false for others, depending upon just how far south the particular star is.

Emendations

All difficulties with these propositions would be solved by the following simple emendations.

- For stars on the zodiac circle:

xiii 21. $MR \to MS = 6$ months
            $ES \to ER = 6$ months

- For stars north of the zodiac:

xiii 22. $MR \to MS > 6$ months
xiii 27. $ES \to ER < 6$ months.

- For stars south of the zodiac:

xiii 23. $MR \to MS < 6$ months
xiii 27. $ES \to ER > 6$ months.

All of the emended propositions are correct *for both true and visible phases*. Moreover, their applicability to the visible phases does not depend on adopting any particular numerical value for the visibility rule. Indeed, the emended propositions may be proved for the visible phases simply by invoking xiii 19 and a simple symmetry requirement. We need only assume that the visible morning rising follows the true one by just as much time as the visible morning setting follows the true one. Similarly, we assume that the VES precedes the TES by just as much time as the VER precedes the TER. Finally, the emended propositions are equivalent to those in Autolykos, *On Risings and Settings* i 4–5.

ᵍ extend above the Earth. Immediately after these words the mss. have τοῖς (τὸ B) ἀεὶ μᾶλλον πρὸς ἄρκτον κειμένοις, which, following Manitius, we delete as an accidental duplication of the beginning of the sentence.

ʰ because smaller segments [of the star's diurnal circles] extend above the Earth, διὰ τὸ ἐλάττονα τμήματα ὑπὲρ γῆν φέρεσθαι, following Manitius and C. ἐλάττονα γὰρ τὰ τμήματα ἢ ὅμοια φέρονται οἱ πρὸς μεσημβρίαν ἀστέρες κείμενοι, Aujac, A, and B. Absent in the Latin version. In C, immediately after φέρεσθαι is the following, rejected by Manitius: τοῖς ἀεὶ μᾶλλον πρὸς μεσημ–βρίαν κειμένοις· ἀπὸ δὲ ἑῴας δύσεως μεχρὶς ἑσπερίας ἐπιτολῆς μείζων ὁ χρόνος γίνεται διὰ τὸ μείζονα τμήματα ὑπὸ γῆν φέρεσθαι. Although the exact form of the text is uncertain here, its basic meaning is clear. The expression ἐλάττονα . . . ἢ ὅμοια ("smaller than similar") in A and B is meant to compare one arc to another in terms of angular extent, rather than absolute length. The same expression is used by Autolykos (*On Risings and Settings* i 11) and by Euclid (*Phenomena*, Berggren and Thomas 1996, 74n47.) Arcs of circles of different radii are *similar* if they subtend the same angle from their respective centers. One arc is *smaller than similar*, compared to another, if it subtends a smaller angle, regardless of the absolute sizes of the circles to which the arcs belong.

ⁱ morning setting. Thus the manuscripts and Manitius. Aujac emends to "morning rising." The proposition stated by the manuscripts is certainly false, but Aujac's emendation makes the proposition correct for the visible phases of *all* stars, not just northern ones, for which reason we have not adopted it. (Aujac's emendation would not be correct for the true phases.) See textual note f for a proposed solution.

ʲ morning setting. Thus A, B, and Manitius. Aujac emends to "morning rising," but this still does not make the astronomy correct. See textual note f for a suggested emendation of this passage.

ᵏ while for those toward the south the time from morning setting till evening rising is more than 6 months. Thus A and B. Missing in C and the Latin version.

CHAPTER XIV

ᵃ the fixed stars. After these words, Manitius inserts <οἱ πρὸς ἄρκτον κείμενοι> by analogy to xiv 3. But see textual note b.

ᵇ those lying in the south, οἱ ἐπὶ (πρὸς, Manitius) μεσημβρίαν κείμενοι. We have punctuated the sentence differently than Manitius or Aujac by placing a comma before this phrase rather than after it.

ᶜ some rise at the same time, and some later, ἅμα μὲν ἀνατέλλει [καὶ δύνει, τὰ μὲν πρότερον], τὰ δὲ ὕστερον, following Manitius. ἅμα μὲν ἀνατέλλει, τὰ μὲν πρότερον, τὰ δ' ὕστερον, Aujac.

ᵈ These. This word is followed in the Greek mss. by πάντα. Missing in the Latin version and deleted by Manitius, whom we follow.

CHAPTER XV

[a] which is approximately 100,000 stades in length and about half that in width, ἐπὶ μὲν τὸ μῆκος οὖσα ὡς ἔγγιστα περὶ ῑ μυριάδας σταδίων, ἐπὶ δὲ τὸ πλάτος ὡς ἔγγιστα τὸ ἥμισυ. Aujac rejects this phrase as an early gloss that worked its way into the text. This may well be right, but it must have happened early, since the phrase also occurs in the Latin version. See Aujac 1975, 49n3. See also comment 3 to chapter xv, p. 209.

CHAPTER XVI

[a] world maps. γεωγραφίας.

[b] in proportion. κατὰ λόγον.

[c] <that lies>. Accepting Manitius's emendation <, ὃς κεῖται>. The same construction occurs at xvi 8.

[d] antarctic. Manitius's correction. Here and in the next line, the mss. have "arctic." But see xvi 11, where "antarctic" is correctly used.

[e] one rising occurs before the other, and one setting before the other. The Greek is very compact: literally, "there are forerisings and foresettings."

[f] circle, κύκλον, following Manitius. κύκλως, Aujac.

[g] the spherical construction that accords with reality. τὴν κατ᾽ ἀλήθειαν σφαιροποιίαν.

[h] alien to the spherical construction in accord with nature. τῆς δὲ κατὰ φύσιν σφαιροποιίας ἀλλότρια.

[i] on, ἐπί, following Aujac and the mss. περὶ ("around"), Manitius.

[j] about, περί, following Aujac and the mss. ὑπὸ ("under"), Manitius; *sub*, the Latin version.

[k] For this reason . . . 40 days. Manitius brackets this sentence for deletion, without explanation.

[l] which is why the lengths of the days receive large increases around the equinoxes. Manitius brackets these words for deletion, without explanation.

[m] <no> delay, ἐπιμονῆς μὲν <οὐ>, Manitius, followed by Aujac. ἐπιτολῆς μὲν, the mss. ("rising").

[n] Sun, ἡλίου, Manitius and Aujac, following the earlier editors. κύκλου, "circle," the mss.

CHAPTER XVII

[a] not . . . ten stades. Accepting Manitius's text (followed by Aujac), μὴ δέκα στάδια. μηδ᾽ ἕκαστα διὰ, the mss.

[b] Atabyrion (Ἀταβύριον). Thus Aujac, following Petavius. Σαταβύριον, the mss. and Manitius.

[c] 10 stades. A,B,C (with some questions of reading) have ͵δ, that is, 4. The Latin version, *10 stadia*. Manitius suggests δέκα, 10. Aujac prints ῆ, 8, without manuscript authority.

ᵈ <in connection with>. <ἐπὶ>, Manitius.

ᵉ which define the seasons, τοὺς καιροὺς ἀφωρισμέναις. Manitius and Aujac, following the earlier editors, insert <κατὰ> before this phrase, thus giving the sense "which are fixed in the course of the seasons."

ᶠ <twice>. Accepting Manitius's addition. Cf. vi 31 and xvii 28.

ᵍ at their risings. These words are followed in A, B, and C (but not the Latin version) by "and settings," which is deleted by both Manitius and Aujac.

ʰ true, ἀληθινὴν. Thus T, Manitius, Aujac. Missing from A, B, and C.

ⁱ for each [locality]. ἑκάστῳ, Aujac, following the consensus of the mss. Manitius proposes ἐκ δὲ τούτων (see, e.g., xvi 25, xvii 23); thus: "it is clear from these things that . . ."

ʲ is some other one of the stars in [another] constellation. More literally, "some other one of the *constellated* (κατηστερισμένων) stars." Manitius rejects this word, "because the Dog is not a constellation but a star." But both the star and the constellation go by the name Dog. (See iii 14.)

ᵏ foreknowledge. πρόγνωσιν, following Manitius. ἐπίγνωσιν, Aujac and the mss.

ˡ he omitted as mistaken. ὡς διεψευσμένας παρέλιπε, after Aujac, following one ms. (T). διεψευσμένων, consensus of the principal mss. <ᾠήθη εἶναι> διεψευσμένας, Manitius, following a conjecture of the earlier editors: "he deemed mistaken. . . ."

ᵐ the, τὰς, Aujac and the mss. Manitius conjectures τινας by analogy to the wording in the first sentence of xvii 48.

CHAPTER XVIII

ᵃ the Moon. After Aujac, following A, B, and C. Deleted by Manitius.

ᵇ <smallest>. Accepting Manitius's insertion <ἐλαχίστην>.

ᶜ and always a greater one on the succeeding days. Accepting Manitius's emendation καὶ ἀεὶ μείζονα <ἐν> ταῖς ἑξῆς ἡμέραις. Aujac prints καὶ μείζονα ἔτι ταῖς ἑξῆς ἡμέραις (ἔτι H. εἰ A, B, and C).

ᵈ <then, in turn,> always, <εἶτ' αὖθις> ἀεί. Thus Manitius, following the earlier editors. The insertion of εἶτ' αὖθις occurs in V, in a second hand. ἀεί is Manitius's conjecture. Aujac prints εἶτα, following H. εἰς τὰ A, B, and C.

ᵉ <they asked>, <ἐζήτησαν>, added by Manitius, followed by Aujac. *donec scitur*, the Latin version.

ᶠ <Then, dividing the> number <of degrees>. Bracketed words added by Manitius in analogy to the Latin version. Followed by Aujac.

ᵍ The sixtieth of one degree. This entire statement (xviii 8) is deleted by both Manitius and Aujac as a later interpolation, but it is present in all the best Greek mss., as well as the Latin version. Although the remark seems elementary, it is possible that Geminos did not expect his readers to be familiar with base-60 fractions. It is noteworthy that in the other chapter (viii) devoted to computations with lunar periods, Geminos works entirely with unit fractions, rather than sexagesimal fractions. Perhaps he felt that a word of explanation was required here.

ʰ exactly, ἐπ᾽ ἀκριβές. Present in the best mss., but deleted by Manitius. The expression is contrasted with κατὰ τὸ ὁλοσχερές ("in a rough way") in the following sentence.

ⁱ <added together>. Inserted by Manitius in analogy to the Latin version. Not in Aujac.

Parapēgma

ᵃ finishes, λήγει. A conjecture of Wachsmuth, followed by Manitius and Aujac. δύνει, the mss.

ᵇ the Eagle sets in the morning. Manitius, without explanation, deletes "in the morning." The meaning is not changed.

ᶜ the Crown sets. Manitius adds <in the morning>, which does not change the sense.

ᵈ the Horse rises <in the evening>. Accepting Manitius's addition, for agreement with the sphere. The absence of a word indicating the *evening* rising is probably due to an early copying error. In the absence of such a modifier, a morning phase is generally to be understood.

ᵉ the Harbinger of the Vintage sets at nightfall, Προτρυγητὴρ ἀκρόνυχος δύνει, Manitius, following Wachsmuth (who prints ἐπιδύνει). The mss. and Aujac have ἐπιτέλλει, "rises," instead of δύνει, "sets." Aujac (1975, 159n7) points out that Manitius's correction of the mss. is inconsistent with Hofmann's theoretically calculated date for the visible evening setting of the Harbinger of the Vintage (Vindemiatrix, ε Vir) in 45 B.C. (For Hofmann's tables, see *Paulys Realencyclopädie der classsischen Altertumswissenschaft. Neue Barbeitung*, Band 6 (1909), s.v. Fixsterne, cols. 2427–30.) However, Hofmann's date for the evening setting of Vindemiatrix (22 November) is clearly impossible, for it differs by only a few days from the date printed for the evening setting of Gemma, α CrB, a star that is quite far away. Consultation of a celestial globe adjustable for precession shows that the evening setting is the only possible phase for Vindemiatrix at this time of year; indeed, the fit is good.

ᶠ the Bird, Ὄρνις, following a conjecture of Diels and Rehm (1904, 105). The mss. and Aujac have Ὀϊστός, Arrow (modern Sagitta). This cannot be right, as Sagitta made its morning setting a month and a half earlier. (A similar error involving the Arrow in Euktēmōn's *parapēgma* occurs at the 25th day of Aquarius.) Manitius conjectures Ἵππος, Horse, our Pegasus, which is astronomically possible.

ᵍ at dawn, ὄρθρου. Manitius suppresses this word as a gloss.

ʰ great star in the Charioteer. Manitius marks these words for deletion.

ⁱ <in the evening>. Accepting Manitius's addition. The absence of this modifier is surely due to an early copying error.

ʲ the Pleiades appear in the evening. In the mss., these words are followed by ἐκ τοῦ πρὸς ἔω, which Manitius rejects as a gloss on φαίνονται. Aujac translates the whole sentence thus: "les Pléiades sont visibles le soir, après l'avoir été à l'aube." She explains in a note that the Pleiades were visible both morning and evening during the time between the evening rising and the morning setting.

However, the events registered in a *parapēgma* are risings and settings, not mere visibility in any part of the sky. Moreover, the Pleiades are in the category of the dock-pathed, so there is no day of the year on which their rising and setting may both be seen. It seems best to interpret this phrase as an unfortunate gloss. The glossarist was perhaps led astray by the misuse of φαίνονται here: properly, a star "appears" at the time of its morning rising, when it emerges from a period of invisibility. The word ought not be used of an evening rising, as it evidently is here.

[k] <at nightfall>. Accepting Manitius's addition.

[l] <the Crown>. Conjecture of Manitius. The constellation name is missing in the Greek mss. The conjecture seems sound, not only because it agrees with the sphere, but because the *parapēgma* contains dates due to Eudoxos for the three other phases of the Crown. The Latin translation mentions *azula.* Aujac (1975, 159n6) interprets this as a reference to a blue star and identifies it with α CrB, the Pearl of the Crown. This may instead be a corruption of *Azulafe*, a name used for Lyra in some versions of the star lists associated with the Alphonsine Tables (Allen 1899, 284). The morning rising of Lyra would, however, be a month premature. It is likely that the lacuna in the Greek text is older than the Latin version.

[m] the whole Scorpion sets at nightfall. Σκορπίος ἀκρόνυχος αἴξ ὅλος δύνει, the mss. Manitius and Aujac, whom we follow, suppress αἴξ (Goat), which perhaps originated as a marginal note. According to Eudoxos, the Goat made its evening rising on the 4th day of Libra.

[n] set with the dawn, δύνουσιν ἅμα ἠοῖ, following the mss. and Aujac. Manitius suggests δύνουσιν ἅμα ἡλίῳ <ἀνίσχοντι>, in analogy to the notice by Dēmokritos for the 13th day of Scorpio.

[o] at nightfall. In the mss., this word is followed by πρωΐας, rejected by both Manitius and Aujac.

[p] in the morning. Manitius deletes this modifier, which does not change the sense.

[q] <to set: whether beginning>, <δύνειν· καὶ ἀρχομένῳ>, added by Wachsmuth, followed by Aujac and Manitius, who, however, prefers another form of the infinitive (δύεσθαι).

[r] Orion begins to set. Manitius adds "in the morning," which does not change the meaning.

[s] the Hyades set. Manitius adds "in the morning."

[t] <the whole of> Orion sets in the morning. <ὅλος>, accepting Manitius's addition. Eudoxos has Orion beginning to set on the 19th day of Scorpio. So this notice no doubt refers to the completion of the process, when Orion completely disappears. The astronomy works well.

[u] southerly, νότια, following Aujac. ὅτι, A, B, C. ὑετία, Manitius. νότια occurs in the following line. This adjective also tends to imply dampness (see LSJ, s.v. νότιος). Manitius translates it by *Scirocco* in the line below.

[v] leaves off, μετίασι, following Aujac and the mss. Manitius prints μεσοῦσι, "are in the middle of."

[w] the Goat, Αἴξ. Manitius, followed by Aujac. In the mss., the name of Eudoxos occurs in the place of a star name. The Goat is our Capella.

^{x} the <whole of> Scorpio rises in the morning. Accepting Manitius's addition, <ὅλος>. Eudoxos has the Scorpion *beginning* to rise on the 18th day of Scorpio.

^{y} according to Dēmokritos, the south wind blows for <the most part. According to Eudoxos, the Dolphin> rises <in the morning>. The lacuna in the Greek mss. has been filled by Wachsmuth (followed by Manitius and Aujac) from the Latin version: *secundum sententiam democriti flat auster plus illo. et secundum considerationem audikios oritur delfinus ortu matutino.* The astronomer's name is strange in the Latin version; but the ascription of this notice to Eudoxos completes his set of four phases for the Dolphin.

^{z} <according to Eudoxos, the Eagle>. This lacuna is filled by Manitius's conjecture (followed by Aujac). This would complete the set of four phases for the Eagle attributed to Eudoxos, but other possibilities (listed by Manitius) cannot be excluded.

^{A} sets. Followed in the mss. by ὁ Περσεύς, Perseus. This constellation name is deleted as spurious by Manitius (followed by Aujac). The astronomy is wrong for Perseus; moreover, this constellation is not mentioned anywhere else in the *parapēgma*.

^{B} and persists. These words are deleted by Manitius. We follow Aujac, who transposes them to the end of the sentence. The mss. read: ζέφυρος πνεῖν ἄρχεται καὶ παραμένει ἡμέρας (ἡμέραις, Manitius and Aujac) ῑ καὶ μ̄ ἀπὸ τροπῶν.

^{C} the Bird, Ὄρνις, following a conjecture of Diels and Rehm. (Ornis is our Cygnus.) The Greek mss. have a lacuna. The Latin version, *Sagitta*, followed by Aujac, who prints <'Ὀϊστός>. The astronomy is impossible, however, for the Arrow made its evening setting a month earlier. Manitius proposes Ἵππος, Horse (our Pegasus), which is a viable alternative. It is remarkable that the first Miletus *parapēgma* has the Arrow setting on the 24th day of Aquarius (Diels and Rehm 1904, 105). This stone *parapēgma*, dated to about 110 B.C., does not explicitly mention its authorities, but is clearly based in part upon Euktēmōn. From the agreement of all the sources, it is likely that we are dealing here with a very ancient corruption of Euktēmōn's *parapēgma*. (A similar error involving the Arrow in Euktēmōn's *parapēgma* occurs at the 10th day of Virgo.)

^{D} On the 2nd. Manitius conjectures <according to Euktēmōn>. Wachsmuth, <according to Metōn>.

^{E} rises. Thus Manitius, followed by Aujac, for fidelity to the globe. A, B, C have "sets in the morning," which is impossible.

^{F} the northern one of the Fishes. Thus the mss. Manitius would emend to "the southern one" (cf. day 17). The southern Fish does begin its morning rising before the northern Fish, but there appears to be no reason to depart from the mss.

^{G} According to Euktēmōn, equinox; and a fine drizzle; it storms violently. The passage is corrupt. We have thought it best to translate Manitius's text, which renounces any attempt at restoration: Εὐκτήμονι ἰσημερία, καὶ ψεκὰς λεπτή· χειμαίνει σφόδρα. A, B, C have ἴση ἰσημερινὸς immediately after λεπτή. Aujac, following one ms. (G), corrects ἴση to ὕσει ("it will rain"). Wachsmuth transposes Εὐκτήμονι ἰσημερία to immediately follow λεπτή. Böckh (1863, 403) conjectures that a name has dropped out of the text. The name of another authority could originally have been placed after λεπτή.

ᴴ the Pleiades hide themselves in the evening. Stars that have an annual period of invisibility "hide themselves" (κρύπτονται) at the time of their evening setting. The use of ἑσπέριοι ("in the evening") with κρύπτονται is not necessary, and is even somewhat irregular. Manitius therefore deletes this modifier. But as we have seen on many occasions, the compiler of our *parapēgma* is not very strict in his use of technical vocabulary.

ᴵ to set at nightfall, δύνειν ἀπὸ ἀκρονύχου.

ᴶ setting, δύνοντι. A correction by Manitius, followed by Aujac, of ἀνίσχοντι ("rising"), the mss. The reference is again to the evening setting, when the Pleiades disappear for a month and a half.

ᴷ and, in many places, hail. Manitius adds before these words, without manuscript authority, <ὑετία,> "rainy weather."

ᴸ On the 1st. Thus Manitius, followed by Aujac. "3rd," A, B, C. *et in primo eorum*, Latin.

ᴹ Orion sets at nightfall. Manitius adds, without manuscript authority, <ὅλος>. Thus, "the whole of Orion sets at nightfall."

ᴺ On the 2nd. Thus the mss. and Aujac. Manitius adopts Wachsmuth's emendation, "on the 4th."

ᴼ on the same [day], τῇ δὲ αὐτῇ, following Aujac and the mss. τῷ δ᾽ αὐτῷ, Manitius, following an emendation of Wachsmuth ("according to the same," i.e., Euktēmōn). Lehoux (forthcoming), following Böckh, suggests this is a corruption of τῇ δ᾽, Εὐκτήμονι: thus "[on] the 4th, according to Euktēmōn." As Aujac points out, the whole passage from "On the 2nd" to "Eudoxos" is missing in the Latin translation, which suggests an early corruption of the text.

ᴾ the Lyre rises <in the evening>. Manitius's emendation. The globe shows that the evening rising is intended. The modifier apparently dropped out at an early stage in the manuscript tradition.

ᑫ southerly [winds], νότια. As for note u, above.

ᴿ in the morning. Manitius brackets this modifier for deletion, which does not change the sense.

ˢ <or>. Accepting Manitius's addition.

ᵀ the Pleiades rise. Manitius unnecessarily adds <in the morning>.

ᵁ On the 25th. . . . On the 30th. . . . These two statements occur out of sequence in the mss., where they are found immediately after the entry for day 13 of Taurus. Aujac prefers to eliminate these lines, both of which are visibly corrupt.

ⱽ the Goat, Αἴξ (our Capella). Thus Manitius, for fidelity to the globe. Ἀετός, "Eagle," the mss. This appears to be a copying error occasioned by the proximity of the entry for day 31.

ᵂ <the Bird>, <Ὄρνις>, our Cygnus. A constellation name is missing in the mss. This seems the most plausible way to fill the lacuna. Manitius, following a conjecture of Wachsmuth, prefers <Ὀϊστὸς>, the Arrow.

ˣ in the morning. Manitius deletes this modifier. The meaning is not affected.

ʸ in the morning. As for Note X.

ᶻ Orion begins to rise. Manitius adds, without manuscript authority, <in the morning>. The meaning is not affected.

Fragment 1

[a] ἐν τῷ περὶ τῆς τῶν μαθημάτων τάξεως. Hultsch 1876–78, vol. 3, 1026.

[b] ἐν τῷ ἕκτῳ τῆς τῶν μαθημάτων θεωρίας. Heiberg 1891–93, vol. 2, 170.

[c] the companies, λόχων, following Tannery (1912, vol. 9, 126) and Morrow. λόγων, Friedlein.

[d] sizes, diameters, and perimeters of cities. Omitting the repetition καὶ διαμέτρους ἢ περιμέτρους, which follows immediately after in Friedlein's text. See Tannery 1912, vol. 9, 126 (followed by Morrow).

[e] apples. There is some question whether apples or sheep are meant. Following the scholiast to Plato's *Charmides*, LSJ defines a μηλίτης (ἀρίθμος) as an arithmetical question about a number of sheep, thus deriving μηλίτης from μῆλον = sheep. Morrow's translation of Proklos (1970, 33) is in accord. But Heath and Tannery derive the word from μῆλον = apple, in which Aujac concurs. (Heath 1921, vol. 1, 14–15. Tannery 1887, 40, 48n. Aujac 1975, 165n1.) Plato (*Laws* vii 819b) recommends the use of problems involving apples, garlands, and bowls to make lessons in ciphering more attractive to children. Moreover, as Aujac points out, in the arithmetical epigrams of the *Greek Anthology*, problems involving apples are numerous (xiv 3, 48, 117, 118, 119).

[f] altitudes [of the pole]. Friedlein's text has simply τῶν τε ἐξαρμάτων. ἔξαρμα (from ἐξαίρω) is a "raising up" or an "elevation" of a body. In astronomical writers it frequently appears in the expression τὸ ἔξαρμα τοῦ πόλου, "the altitude of the pole." Geminos, *Introduction to the Phenomena* vi 24; xiv 2. Hipparchos, *Commentary* i 3.6–7; i 3.12.

[g] positions, τὰς ἐποχάς, following Tannery (1912, vol. 9, 126). ε ἀποχάς, Friedlein.

Fragment 2

[a] ἐκ τῆς ἐπιτομῆς τῶν Ποσειδωνίου Μετεωρολογικῶν ἐξηγήσεως. Diels 1882, 291.

[b] ex commento Gemini Posidonii de Μετεώρων. Edelstein and Kidd 1989, 22.

[c] Hērakleidēs of Pontos, Ἡρακλείδης ὁ Ποντικός, Diels and the manuscripts. Bracketed for deletion by Aujac, following Tannery. We concur that the name should be expunged.

The Geminos *Parapēgma*

SOURCES OF THE GEMINOS *PARAPĒGMA*

The Geminos *parapēgma* is a compilation based upon six authorities:

Metōn	fl. 430 B.C.	1 citation
Euktēmōn	430	46 citations
Dēmokritos	400	11 citations
Eudoxos	360	61 citations
Kallippos	330	34 citations
Dositheus	230	4 citations

The number of citations is the number of times that an authority's name occurs in the text of the *parapēgma*. A single citation may involve several items of information: a star phase, a weather prediction, a notice of an equinox or solstice, etc.

Some scholars have maintained that this compilation was not made by Geminos himself, but by some earlier writer.[1] These scholars point to the fact that Dositheus, the latest source, is a century and a half earlier than Geminos. Moreover, although Geminos mentions Hipparchos in chapter iii of the *Introduction to the Phenomena*, and we know that Hipparchos compiled a *parapēgma*, the Geminos *parapēgma* makes no use of it. Another argument against Geminos's authorship is the fact that the season lengths in the *parapēgma* do not agree with the Hipparchian lengths given in chapter i. Finally, Geminos's attack in chapter xvii on the belief that the stars cause changes in the weather can be seen as inconsistent with the authoring of a *parapēgma*. According to this view, then, the compilation was made after Dositheus but before Hipparchos, and thus shortly after 200 B.C. Perhaps the *parapēgma* was attached to the *Introduction to the Phenomena* some time after Geminos's day. Or perhaps, as Lehoux has suggested, Geminos himself attached an already existing compilation to his own book.[2]

Against these arguments we may note that Geminos seems to have known only one work by Hipparchos—a work on the constellations. He

<hr>

[1] Böckh 1863, 22; Manitius 1898, 281; Neugebauer 1975, 580, 587.
[2] Lehoux 2000, 34.

does not mention Hipparchos, for example, in connection with the season lengths in chapter i. Thus it is possible that Geminos simply did not have access to Hipparchos's *parapēgma*. Moreover, at viii 50, Geminos mentions Philippos (Philip of Opus), but the Geminos *parapēgma* makes no reference to Philippos's *parapēgma*. This certainly does not imply that Geminos *parapēgma* was compiled before the time of Philippos (mid-fourth century B.C.). And, if Geminos were compiling a grand *parapēgma* based on the opinions of ancient authorities, he might have taken the season lengths simply as he found them in one of the *parapēgmata* (probably that of Kallippos). It was bad enough that his three principal authorities (Euktēmōn, Eudoxos, and Kallippos) did not agree about the lengths of the seasons. Moreover, there is no incompatibility between Geminos's inclusion of a *parapēgma* and his firm belief that the stars do not cause the weather. For in chapter xvii he remarks more than once that, although the stars are not *causes* of the weather, they can serve as *signs*. Star phases and weather signs held an important place in the Greek astronomical tradition, and therefore fit naturally into an *Introduction to the Phenomena*. After all, Leptinēs had earlier included a short *parapēgma* in his *Celestial Teaching*.[3] Finally, as Aujac points out, the manuscript tradition unanimously places this *parapēgma*, without any interruption of any sort, at the end of Geminos's book.[4] Whether this *parapēgma* was compiled by Geminos or someone else is a question we must leave unanswered.

The Minor Authorities

Metōn, Dēmokritos, and Dositheus are mentioned infrequently. Indeed, Metōn is cited only once—his date for the morning rising of the Dog is given as the 25th day of Cancer. It might be thought that this notice was due to a confounding of his name with that of Euktēmōn, since these two are often mentioned together. But this appears not to be the case, for the text attributes a different date to Euktēmōn for the rising of the Dog (day 27 of Cancer. See also day 1 of Leo.) Most likely, the compiler possessed only scattered notices, drawn from some secondary source, for Metōn, Dēmokritos, and Dositheus. These few notices do not add much to the *parapēgma*. In every case but one (ES of Vindemiatrix), the dates of the star phases attributed to Metōn, Dēmokritos, or Dositheus merely provide slightly different alternatives for dates attributed to Euktēmōn, Eudoxos, or Kallippos. The handful of weather predictions due to Metōn, Dēmokritos, and Dositheus might have been of more interest to an ancient reader.

[3] Tannery 1893, 293.
[4] Aujac 1975, 157.

The Major Authorities

The Geminos *parapēgma* is substantially based on three *parapēgmata* that the compiler may have possessed in their entirety—those of Euktēmōn, Eudoxos, and Kallippos. The portion of Euktēmōn's *parapēgma* preserved in our text is thus the oldest substantial extract we have of this genre among the Greeks. Moreover, the treatment of the star phases shows definite differences in the three *parapēgmata*. These differences stand out with remarkable clarity in table A2.1. Thus we can recognize an evolution in the *parapēgma* over the century that separated Euktēmōn and Kallippos.

Euktēmōn gave the dates of the heliacal risings and settings of a number of traditionally important stars or small groups of stars:

Sirius	Arcturus	Kids	Hyades
Vindemiatrix	Capella	Pleiades	

To these he added dates for a few constellations:

Aquila	Delphinus	Corona Borealis	Scorpius
Lyra	Orion	Pegasus	Cygnus (? uncertain)

These stars and constellations are distributed all over the sky (four stars or groups from among the zodiacal constellations, nine from the northern constellations, and two from the southern constellations). Euktēmōn made no effort at systematic coverage of the globe: most of the stars and constellations he selected had a traditional significance. Some of them were used to mark the seasons, notably the Pleiades, Arcturus, Sirius, and Vindemiatrix. Others served as weather signs, e.g., Capella, the Kids, the Hyades, Aquila, Orion, and Scorpius—constellations that are all given special mention as weather signs in Aratos's *Phenomena*.

For five of the stars or groups (Arcturus, Capella, the Pleiades, Aquila, Lyra) the Geminos *parapēgma* preserves Euktēmōn's dates for all four phases. For several others, it preserves three of the four. Thus, it seems likely that Euktēmōn provided a complete treatment for the stars he had selected.

Eudoxos's *parapēgma* shows a great similarity to that of Euktēmōn. The stars and constellations of Eudoxos preserved in the Geminos *parapēgma* are these:

Sirius	Capella	Hyades
Arcturus	Pleiades	

Aquila	Delphinus	Corona Borealis
Lyra	Orion	Scorpius

This is quite similar to Euktēmōn's list. So Eudoxos, too, concentrated on stars of traditional significance. The compiler seems, if anything, to have paid greater heed to Eudoxos than to Euktēmōn, no doubt because of his greater reputation. For nine stars or groups, all four of Eudoxos's phases are preserved in the Geminos *parapēgma*.

Kallippos's *parapēgma* shows a number of departures from those of Euktēmōn and Eudoxos. First, as we can see in table A2.1, Kallippos introduced a systematic treatment of the zodiacal constellations. Some of Kallippos's phases are preserved for all twelve. (Libra, the Balance, is present under its ancient guise as the Claws of the Scorpion.)

Second, Kallippos shows much greater selectivity in his use of traditional nonzodiacal stars. Kallippos includes the following:

Twelve zodiacal constellations, plus

Sirius Pleiades Arcturus Orion

This list differs strikingly from those for Euktēmōn and Eudoxos. The purpose of noting star phases was twofold: to tell the time of the year and to forecast the weather. In Kallippos, we see a shift away from weather prediction toward systematic time reckoning. This is evident, of course, in his systematic use of zodiacal constellations. But it is equally clear in his selection of other stars. Sirius, Arcturus, and the Pleiades were all important markers of the agricultural (not astronomical) seasons. The morning setting of the Pleiades marked the beginning of winter; their morning rising, the beginning of summer. The morning rising of Arcturus indicated the beginning of fall. The morning rising of Sirius marked the height of the summer heat. Of Kallippos's four groups listed above, only Orion lacks such a special role. Interestingly, Kallippos has omitted such traditional weather signs as the Hyades, the Kids, and Aquila.

Third, Kallippos introduces a systematic treatment of extended constellations in their parts. Thus, he tells us when Virgo starts to make her morning rising, when she has risen as far as her shoulders, when she has risen to her waist, when the wheat ear (Spica) has risen, and when Virgo has finished rising. To be sure, Euktēmōn and Eudoxos had already done a bit of this, but Kallippos carries it much further. This is another manifestation of Kallippos's effort to make the *parapēgma* a more precise tool. In this he may be seen as a precursor of, and influence on, Ptolemy, who abandoned constellations altogether, in favor of individual stars.

Finally, assuming that the compiler has not given us an unrepresentative sample, Kallippos does not bother to record all four phases. In every case, Kallippos confines himself to morning risings and morning settings. The morning phases are psychologically the most important. The morning rising of a star is the first rising to be visible in the course of the year. Similarly, the morning setting is the first setting to be visible. The evening risings and settings are the last to be visible. Thus, the morning phases are more certain: you know when you've seen Sirius rise for the first time; but it may take a few days to be sure whether you've seen it rise for the last time. Kallippos's use of only morning phases may be another manifestation of his desire to achieve better precision in telling the time of year. If so, this particular innovation did not last, for Ptolemy returned to the traditional practice of citing all four phases.

Synopsis of the Star Phases in the Geminos Parapēgma

Table A2.1 provides a synopsis of the star phases in the Geminos *Parapēgma*, arranged chronologically for the three different star classes.

Examples (compare with the entries in Table A2.1)

Cancer: According to Kallippos, Cancer begins its morning rising on day 1 of Cancer, and finishes on day 27 of Cancer. Also according to Kallippos, Cancer finishes its morning setting on day 27 of Capricorn.

Scorpio: According to Euktēmōn, the stinger of Scorpio makes its morning rising on day 10 of Sagittarius. According to Kallippos, the head of Scorpio makes its morning rising on day 4 of Scorpio.

Arrangement of the Table

The table is divided into three sections, according to the location of the stars on the sphere. Such a division makes it easier to list the phases of each star in the order of their occurrence through the year. Stars near the ecliptic (called "dock-pathed" stars by Ptolemy) have their phases in the order MR, ER, MS, ES. Stars far enough south of the ecliptic ("night-pathed" stars) have their phases in the order MR, MS, ER, ES. Stars that are far enough north of the ecliptic ("doubly -visible" stars) follow the order MR, ES, ER, MS. (See sec. 11 of the Introduction.)

Note on Bird (Ornis), Horse (Hippos), and Arrow (Oïstos)

A good deal of confusion surrounds these constellations in the Geminos *parapēgma*. The *parapēgma* cites Euktēmōn for two phases of the Arrow (MS, 10♍; and ES, 25♒). Unfortunately, both of the dates given in the *parapēgma* are astronomically impossible. A similar error involving the (evening) setting of the Arrow occurs already in one of the stone *parapēgmata* found at Miletus. It appears that we are here confronted with a very early error in the tradition of Euktēmōn's parapēgma. Diels

TABLE A2.1
Star Phases in the Geminos *Parapēgma*. (Key to symbols follows table.)

Dock-Pathed Stars (Stars Near the Ecliptic)		Euktēmōn	Eudoxos	Kallippos
Cancer	MR			1♋ b 27♋ f
	ER			
	MS			27♑ f
	ES			
Leo	MR			30♋ b 12♌ m
	ER			
	MS			2♒ b 2♓ f
	ES			
Virgo	MR			29♌ b* 5♍ ^
				17♍ m 5♎ f
	ER			
	MS			
	ES			
Spica (αVir)	MR			24♍
	ER			
	MS			
	ES			
Harbinger of Vintage (Vindemiatrix, ε Vir)	MR	10♍		
	ER	12♓*		
	MS			
	ES			
Claws (Libra)	MR			17♎ b
	ER			
	MS			23♈ b
	ES			
Scorpio	MR	10♐ +	18♏ b 21♐ f*	4♏ o
	ER			
	MS	29♓b	11♉ b 21♉ f	
	ES		12♎ b 17♎ f	
Antares (α Sco)	MR			16♏
	ER			
	MS			
	ES			
Sagittarius	MR			7♐ b 1♑ f
	ER			
	MS			
	ES			
Capricorn	MR			15♑ b
	ER			
	MS			
	ES			

TABLE A2.1 *(continued)*

Dock-Pathed Stars (Stars Near the Ecliptic)

		Euktēmōn	Eudoxos	Kallippos
Aquarius	MR			17♒ m
	ER			
	MS			
	ES			
Pisces	MR			17♓ s 30♓ n(f)
	ER			
	MS			
	ES			
Knot	MR			1♈
(α Psc)	ER			
	MS			
	ES			
Aries	MR			3♈ b 1♉ f
	ER			
	MS			1♎ b
	ES			
Taurus	MR			2♉+ 13♉o 32♉ f
	ER			
	MS			28♎+ 9♏o 28♏<
	ES			
Gemini	MR			2♊ b
	ER			
	MS			16♐ f
	ES			
Pleiades	MR	13♉	22♉	
	ER	5♎	8♎	
	MS	15♏	19♏	16♏
	ES	10♈	13♈	
Hyades	MR	32♉	5♊	
	ER		22♎	
	MS	27♏	29♏	
	ES	23♈	21♈	
Orion	MR	24♊ ^ 13♋ f	24♊ b 11♋ f	
	ER		12♏ b	
	MS	15♏ b*	19♏ b 8♐ f*	7♐
	ES		13♈ b 1♉ f*	
Horse	MR	14♓*		
	ER	17♌*		
	MS			
	ES			

TABLE A2.1 (continued)

Doubly Visible Stars (Northern Stars)

		Euktēmōn	Eudoxos	Kallippos
Crown	MR	7♎	10♎*	
	ES		9♑	
	ER		21♓	
	MS		10♌	
Eagle	MR	15♐	26♐	
	ES	7♑	18♑*	
	ER	31♉	7♊	
	MS	28♋	5♌	
Lyre	MR	10♏	21♏	
	ES	3♒	11♒	
	ER	2♉*	27♈	
	MS	17♌	22♌	
Dolphin	MR	2♑	12♑*	
	ES	27♑	4♒	
	ER		18♊	
	MS		18♌	
Arcturus	MR	10♍ 20♍	19♍	17♍
	ES	5♏	8♏	
	ER	12♓	4♓	
	MS	32♉	13♊	
Goat	MR	8♉	9♉	
(Capella,	ES	25♉*		
α Aur)	ER	20♍*	4♎	
	MS	19♐*	23♐	
The Kids	MR			
(ζ, η Aur)	ES			
	ER	3♎		
	MS			
Bird	MR			
	ES	25♒*		
	ER	30♉*		
	MS	10♍*		

Night-Pathed Stars (Southern Stars)

		Euktēmōn	Eudoxos	Kallippos
Dog Star	MR	27♋ 1♌	27♋	30♋
	MS	7♐	12♐	
	ER		16♐	
	ES	2♉	2♉	

TABLE A2.1 (*continued*)

Star Phases Attributed to the Minor Authorities

Metōn:

Dog	MR	25♋

Dēmokritos:

Pleiades	MS	4♏
	ES	13♈
Orion	MR	29♊ b
Eagle	MR	16♐
Lyre	MR	13♏

Dositheus:

Crown	MS	16♋ b
Dog	MR	23♋
Harbinger of Vintage	ES	18♌ *

KEY TO THE SYMBOLS USED IN THE TABLE

Names of the phases:

MR morning rising   ER evening rising
MS morning setting   ES evening setting

Zodiac signs

♈ Aries	♋ Cancer	♎ Libra	♑ Capricorn
♉ Taurus	♌ Leo	♏ Scorpio	♒ Aquarius
♊ Gemini	♍ Virgo	♐ Sagittarius	♓ Pisces

Stages in the heliacal rising or setting:

b begins   m in middle   f finishes

Parts of a constellation:

o head or forehead	^ shoulder(s)	n northern part
< horns	+ stinger or tail	s southern part

An asterisk * means that some aspect of the notice is more or less uncertain.

and Rehm[5] suggested that Euktēmōn originally wrote "Bird," which seems the most plausible solution, in terms both of astronomy and of language. Manitius[6] preferred to give these two phases to the Horse, which is astronomically possible; in this case, the four phases of the Horse according to Euktēmōn are filled out. The *parapēgma* also contains the notice of an evening rising according to Euktēmōn at 30♉, with no constellation name. We have assigned this phase to the Bird. Manitius[7] assigns it to the Arrow: for Manitius, this then becomes the only mention of the Arrow in his version of the *parapēgma*—a solution with little plausibility. In any case, no great certainty can be attributed to the three phases of the Bird given in table A2.1.

RELATION OF THE GEMINOS PARAPĒGMA TO OTHER GREEK PARAPĒGMATA

By comparing the Geminos *parapēgma* to Ptolemy's and to the second Miletus *parapēgma*, we can gain an impression of how accurately the Geminos *parapēgma* preserves the star phases and weather predictions of its authorities.

Second Miletus Parapēgma

Fig. I.18 is a photograph of a portion of the first Miletus *parapēgma*, discussed in sec. 11 of the Introduction. The second Miletus *parapēgma*[8] differs in two major ways from the first. (1) The first Miletus *parapēgma* does not make weather predictions, but lists only star phases and signs of the season. In contrast, the second *parapēgma* includes day-by-day weather predictions. (2) The first Miletus *parapēgma* lists no authorities for its star phases. But the second (surprisingly for a work in stone) includes multiple notices of the star phases, attributed to various authorities. Because the second Miletus *parapēgma* cites its authorities, its notices of star phases can be directly compared with those in the Geminos *parapēgma*, as in table A2.2.

In Table A2.2 we have omitted notices from each *parapēgma* that have no counterparts in the other. For example, we have omitted the references to Kallaneus the Indian in the Miletus *parapēgma*. Conjectural restorations are enclosed by pointed brackets < >. The intervals between

[5] Diels and Rehm 1904, 105.
[6] Manitius 1898, 214, 226.
[7] Manitius 1898, 232.
[8] For the text of the second Miletus *parapēgma*, see Diels and Rhem 1904. A complete English translation will soon be available in Lehoux (forthcoming).

Table A2.2
Comparison of the Second Miletus *Parapēgma* and the Geminos *Parapēgma*

Second Miletus Parapēgma	Geminos Parapēgma
(456D Left Column)	
o\<Eudoxos> & Egyptians: \<Scorpio> ES	17♎ Eudoxos: Scorpio ES
o	18
o Eudoxos & Egyptians: \<North an>d south winds	19 Eudoxos: North and south winds
o	20
o	21
o Eudoxos & Egyptians: \<Hyad>es ER	22 Eudoxos: Hyades ER
o	23
(456A Right Column)	
o Euktēmōn: \<Goat> E\<S>	25♋ Euktēmōn: \<Goat> ES
o	26
o	27
o	28
o	29
o Euktēmōn: Eagle ER	30 Euktēmōn: \<Bird> ER
o Euktēmōn: Arcturus MS, signifies	31 Euktēmōn: Eagle ER
o	32 Euktēmōn: Arcturus MS, signifies

events in the Miletus *parapēgma* are more certain than the absolute day count: because of missing day holes, we cannot be sure just where these phases were meant to fall in the zodiac sign. We have matched the beginnings of the two extracts against the corresponding entries in the Geminos *parapēgma*.

The close correspondence for the Eudoxos entries in the first extract suggest that, at least over short stretches of time, the Geminos *parapēgma* and the second Miletus *parapēgma* are in fair agreement. This suggests that both reproduce reasonably faithfully the sources on which they were based. Since each includes notices omitted by the other, neither can have been derived from the other.

The Euktēmōn entries in the second extract show an imperfect correspondence, with a slip of 1 day occurring between the first entry and the Eagle entry 5 days later. This could be a simple copying error by one of the writers. An alternative explanation of the discrepancy is that the compiler of the Geminos *parapēgma* has stretched out Euktēmōn's phases in order to cover Kallippos's long spring season. But we shall see below that this is unlikely.

Ptolemy's Parapēgma

Ptolemy's *parapēgma*, which forms a part of his *Phaseis*,[9] introduced a number of innovations. Notably, Ptolemy carried to its logical conclusion the improvement in precision that had been begun by Kallippos: Ptolemy does not give the dates of the heliacal risings and settings of constellations or parts of constellations, but only of individual stars. He includes fifteen stars of the first magnitude and fifteen of the second. In this way, he eliminates the uncertainty in the first or last appearances of extended constellations, such as Orion or Cygnus.

Thus, Ptolemy was unable to use the traditional dates of star phases due to Euktēmōn, Eudoxos, Kallippos, etc. Rather, he began with the heliacal risings and settings for the *klima* of Alexandria. He then computed the dates on which the stars ought to make their heliacal risings and settings in other *klimata*. The "calculations" may well have been performed with the aid of a celestial globe. Thus, although Ptolemy gives a complete set of heliacal risings and settings for five different *klimata* (from 13½ to 15½ hours, by ½-hour steps), he does not report any star phases for the older authorities. He does, however, give an ample selection of weather predictions attributed to specific authorities. Thus, in comparing the Geminos *parapēgma* with Ptolemy's *parapēgma*, we cannot use the star phases, but must restrict ourselves to weather predictions.

Unfortunately, in the case of weather signs and "signifying," it is often hard to identify corresponding lines in the two *parapēgmata*. This is especially so for the predictions of Eudoxos. It is clear that neither Ptolemy nor the compiler of the Geminos *parapēgma* included all the weather predictions that Eudoxos had put in his original *parapēgma*. Thus it is often hard to know which Eudoxan "signifying" in Ptolemy corresponds to which Eudoxan "signifying" in Geminos. So, in attempting to match Ptolemy's *parapēgma* against the Geminos *parapēgma*, we must restrict ourselves to singular or otherwise especially clear events.

This turns out to be especially easy to do for the predictions of Kallippos. Table A2.3 shows that the weather signs of Kallippos reported by Ptolemy bear a simple and steady relationship to those reported in the Geminos *parapēgma*. The left column ("Day") is the day number in the year, counting from the summer solstice. The Geminos *parapēgma* begins with the summer solstice, but Ptolemy's does not. But counting in this way allows us to at least roughly match the corresponding days of the year. The second column gives the date as identified by Ptolemy, in terms of the Alexandrian calendar, together with the predictions of Kallippos that

[9] For the Greek text of Ptolemy's *Phaseis*, see Heiberg 1898–1903, vol. 2. An English translation of the *parapēgma* will be available in Lehoux (forthcoming).

TABLE A2.3
Weather Signs of Kallippos in Ptolemy and in the Geminos *Parapēgma*

Day	Ptolemy's Parapēgma	Geminos Parapēgma
1	1 Epiphi Summer solstice	1♋ Summer solstice
...	...	...
67	2 Thoth	5♏ Kallippos: etesian winds cease
68	3	6
69	4 Kallippos: stormy, etesian winds cease	7
70	5	8
...	...	...
86	21	24 Kallippos: rain
87	22	25
88	23 Kallippos: rain	26
...	...	...
131	6 Athyr	9♏ Kallippos: rain
132	7	10
133	8 Kallippos: rain	11
...	...	...
150	25	28 Kallippos: rain
151	26	29
152	27 Kallippos: rain	30
...	...	...
168	13 Choiak	16♐ Kallippos: south wind
169	14	17
170	15 Kallippos: south wind, signifies	18
...	...	...
196	11 Tybi	15♑ Kallippos: south wind
197	12	16
198	13 Kallippos: south wind	17
...	...	...
242	27 Mecheir	2♓ Kallippos: swallow appears, signifies
243	28	3
244	29 Kallippos: swallows appear, windy	4
...	...	...
273	28 Phamenoth	3♈ Kallippos: rain or snowstorm
274	29	4
275	30 Kallippos: rain or snowstorm	5

Ptolemy selected for this time of the year. The third column gives the corresponding information from the Geminos *parapēgma*: the date in terms of the day within the zodiac sign, together with the predictions attributed to Kallippos. Table A2.3 includes all the weather predictions of Kallippos for which it is possible to identify the corresponding entries in the two *parapēgmata*. As is readily seen, the predictions of Kallippos reported by Ptolemy run a steady 2 days behind those reported by Geminos. The return of the swallow, the cessation of the etesian winds, etc., provide clear correspondences. Moreover, the number of Kallippos's weather signs reported by Ptolemy and by Geminos is not great. Thus, even rainstorms often clearly line up with the usual 2-day offset. Ptolemy and Geminos each report weather signs due to Kallippos that the other leaves unmentioned, which proves that Ptolemy drew upon some source other than the Geminos *parapēgma*. In view of this fact, the correspondence between Ptolemy and Geminos for the case of Kallippos is very striking. This shows that the Geminos *parapēgma* reliably preserves the sequence of events in Kallippos's *parapēgma*, and it shows that Ptolemy does, too.[10]

The import of this 2-day offset is not obvious, but it seems that one of the compilers (and perhaps both) has adjusted the date of Kallippos's summer solstice. If the predictions of Euktēmōn in the Geminos *parapēgma* are compared with those in Ptolemy's *parapēgma*, a similar offset can be discerned—but in this case it is only of a single day. That is, the predictions attributed to Euktēmōn (counted from the summer solstice) tend to run 1 day later in Ptolemy than in Geminos. There is some scatter, and the correspondence is not as perfect as for Kallippos. When the predictions of Eudoxos in the Geminos *parapēgma* are compared with those in Ptolemy's *parapēgma*, there is no offset. That is, the predictions (counted from the summer solstice) tend to occur on the very same day in the two *parapēgmata*, although there is some scatter. But again, the pattern is much less clear than is the case with the predictions of Kallippos.

STRUCTURE OF THE GEMINOS *PARAPĒGMA*

Now we are in a position to say something about how the compiler of the Geminos *parapēgma* most likely proceeded. It seems clear that the

[10] One additional near-correspondence with a 2-day offset occurs on days 257 and 259 of the year. Geminos has, according to Kallippos, "north wind stops" (*lēgei boreas*, day 17 of Pisces). But Ptolemy has, according to Kallippos, "cold north wind blows" (*boreas psuchros pnei*, day 14 of Phamenoth). This seems to be an error in one of the two texts, perhaps occasioned by *epipnei boreas psuchros* in the preceding entry in Geminos (day 14 of Pisces).

compiler relied upon the *parapēgma* of Kallippos for the basic structure. This is suggested by the four explicit mentions of the beginnings of the seasons "according to Kallippos" (at day 1 of Cancer, Libra, Capricorn, and Aries). No other authority is mentioned for the beginnings of all four seasons. Moreover, the lengths of the seasons are in fair (though not perfect) accord with the season lengths attributed to Kallippos in the *Celestial Teaching* of Leptinēs. In the Geminos *parapēgma*, the lengths of the seasons are (in days, and beginning with summer): 92, 89, 89, 95. The season lengths attributed to Kallippos by Leptinēs are: 92, 89, 90, and 94.[11] Finally, the close agreement between the *parapēgmata* of Geminos and Ptolemy regarding the predictions of Kallippos suggests that the Geminos *parapēgma* preserves the sequence of events in Kallippos's original calendar, though we cannot know which (if either) of these texts preserves the exact starting day with respect to the summer solstice.

Exactly how the events due to Eudoxos and Euktēmōn were added is not certain. The compiler might have proceeded in two different ways. (1) He might have simply slipped each prediction of Eudoxos or Euktēmōn in at its appropriate day number in the course of the year, ignoring the fact that Eudoxos and Euktēmōn had different lengths for the seasons than did Kallippos. This is the same as preserving the time intervals between successive events. Or (2) he might have preserved the place of an event within a particular season, or (perhaps) within a particular zodiac sign, e.g., making sure that an event scheduled for the 25th day of Taurus remains on the 25th day of Taurus.[12] We might already suspect that the compiler followed the first course for Eudoxos, in view of the notice of Eudoxos's winter solstice (day 4 of Capricorn). But now there is good reason for believing that the compiler also followed this course for Euktēmōn. It is probable that Ptolemy's and Geminos's *parapēgmata* are the results of independent collations, in view of the fact that each includes many notices that the other omits. The reasonably consistent 1-day offset for the predictions of Euktēmōn, which is maintained over the course of the year, seems then to imply that each compiler simply wrote the predictions out in order from the first to the last day of the year. In any case, it is clear from the scatter in the comparison between Geminos and Ptolemy that the texts of the *parapēgmata* of Euktēmōn and Eudoxos had already become less reliable by their time than the text of Kallippos.

[11] Tannery 1893, 294.
[12] This was the position that van der Waerden (following Rehm) took in his attempted reconstruction of the *parapēgma* of Euktēmōn (van der Waerden 1983, 104).

Glossary of Technical Terms
in Geminos's *Introduction to the Phenomena*

The glossary is arranged topically. Thus, under **Sun** will be found the terms for the tropical year, the four seasons, the equinoxes and solstices, etc. The glossary not only enables a reader to determine which Greek word corresponds to a particular English word used in the translation, but also provides capsule definitions for most technical terms. For each technical term, a reference is provided to a passage in which the term is used in a defining or characteristic way. (If several consecutive terms listed here occur in the same passage in the text, a reference to that passage is given only after the last term, and only a chapter number is given if a term appears throughout a chapter.) No attempt has been made to list every occurrence of every term, or every lexical form of a given word. (For these, see the "Index graecitatis" at the back of Manitius 1898. Aujac 1975 includes an alphabetically arranged *Lexique des Termes Techniques*, which includes philosophical vocabulary.)

THE COSMOS AND THE OBJECTS WITHIN IT

The astronomical scene for Geminos's book is the **cosmos**, *kosmos*, which has the **Earth**, *gē*, like a **point**, *sēmeion*, in its **middle**, *mesē* (xvi, 29). The cosmos is bounded by **the sphere of the fixed stars**, *hē tōn aplanōn asterōn sphaira* (i 23). Its one motion is a daily **revolution**, *peristrophē* (v 4), in which the following heavenly bodies participate:

The Sun

The **Sun**, *Hēlios* (i 7), marks the beginning of the day, *hēmera* (vi 1), by its **rising**, *anatolē* (xiii, also applied to the daily rising of stars), and the beginning of **night**, *nux* (vi), by its **setting**, *dusis* (vi 1). "Day" (vi 1) can also refer to the whole period from one sunrise to the next. And *dusis* is used for both the diurnal setting of stars (xiii 2) and their heliacal setting (xiii 4).

The **year**, *eniautos* (i 7), and **the annual period**, *eniausios chronos* (i 7), both refer to the tropical year of the Sun's **circuit**, *peridromē* (viii 5), around the **zodiacal circle**, *zōidiakos kuklos* (i 3). Another synonym for this kind of year is **the year by the Sun** or **the solar year,** *kath' hēlion eniautos* (viii 5), which Geminos uses when he wishes to distinguish the tropical year from a lunar year of 12 synodic months. When it is a question of counting years, however, the word used throughout the work is *etos* (vii 24, 33), **year.**

Geminos defines the seasons, **spring,** *ear*; **summer,** *theros*; **autumn,** *phthinopōron*; and **winter,** *cheimōna* (i 9), by the **equinoxes,** *isēmeriai*, and **solstices,** *tropai*. Thus, the **spring** or **vernal equinox** (i 9) marks the beginning of spring, the **autumnal equinox** (v 6) that of autumn, while the **summer solstice** (i 9; v 4) and **winter solstice** (v 7) mark the beginnings of those seasons. On the days surrounding a solstice there is an apparent **tarrying,** *epimonē* (vi 30), of the Sun at the corresponding tropic.

The **solar circle,** *hēliakos kuklos* (i 33), refers to the actual circle on which the annual, **proper displacement,** *parodos* (v 16), of the Sun takes place. It lies closer to the Earth than, and is **eccentric,** *ekkentros* (i 34), to the zodiac. Much looser in meaning is the **course of the Sun,** *hēliakos dromos* (i 34, referring to the Sun's eccentric circle; viii 32, or referring simply to the progress of the Sun around the zodiac). The word *dromos* is also used for the Sun's (ii 23), or a star's (xiv 1) **course** or **path** across the sky from rising to setting.

The Moon

Geminos describes the **phases,** *schēmatismoi* (ix 11), of the **Moon,** *selēnē* (xi), as follows: **crescent Moon,** *mēnoides* (lit., "Moon-shaped"); **first** or **third quarter,** *dichotomos* (lit., "cut in half"); **gibbous,** *amphikurtos* (lit., "curved on both sides"); and **full Moon,** *panselēnos* (lit., "whole Moon") (ix 11).

The Moon defines two periods of time: the **synodic month** (from new Moon to new Moon), *mēn* (viii 1), and the **lunar year,** *kata selēnēn eniautos* (viii 46), of 12 lunar months. Synonymous with *mēn* is *mēniaios chronos* (viii 2), the **monthy period.** Some days of the month are named for the phases of the moon, so the **1st day of the month** is the *noumēnia* (viii 11), and the **middle** of the month the *dichomēnia* (viii 1). The **30th day** of the month is called the *triakas* (viii 12), after the word "thirty." On the last day of the month the Moon **runs under,** *hupotrochazei* (ix 10), the Sun.

The **civil month,** *kata polin mēn*, is the month as reckoned in a city's calendar, and it may be **full,** *plērēs*; or **hollow,** *koilos*, i.e., it may contain either 30 or 29 days, respectively. A full and a hollow month together constitute a **double month,** *dimēnon*, of 59 days (viii 3).

The Stars

Star, *astēr* (iii 4), or *astron* (xvii 7). All but five stars are **fixed** or **non-wandering**, *aplaneis* (vii 2), as distinguished from the five planets (below). Like other ancient civilizations, the Greeks grouped the stars into **constellations**, *katastērigmena zōidia* (iii 1). (Manitius prefers the spelling *katēsterismena*.) The term is also used (i 4) specifically for the *zodiacal constellations*, as distinguished from the zodiacal signs. For the names of individual stars and constellations, see the tables in chapter iii. There is a rich vocabulary associated with the heliacal risings and settings of the stars or, as they are sometimes called, **phases**, *phaseis* (xvii 26, where *phasis* is translated as "appearance"). The same word is also applied to the phases of the Moon (viii 11). The terms for fixed star phases are discussed in sec. 11 of the Introduction.

The Planets

The **wandering stars**, *planētes asteres* (xii 22; xvii 38), are sometimes simply called the **planets**, *planētēs* (i 19). Each of the five planets known to the ancients has a proper name, but is also called the star of a certain god. For example, **Saturn** is *Phainōn* ("Shiner") but also "the star of Kronos" (i 24). For further details on the planet names, see comment 20 to chapter i. The designation "wanderer" originated in the fact that, although the planets share in the westward diurnal motion of the cosmos, they also have a proper eastward motion along the zodiac, which is said to be **opposite that of the cosmos**, *hupenantiōs tōi kosmōi* (xii 6). Moreover, the planets pass by the fixed stars while moving eastward, and occasionally make a retrograde motion to the west, with respect to the fixed stars. Between these two motions they will stand still with respect to the fixed stars at what is called a **station**, *stērigmos* (xii 22). Despite this apparent irregularity, each planet's motion is somehow **circular**, *egkuklios*, and **uniform** (or at constant speed), *homalos* (i 19).

To explain the apparent **anomaly**, *anōmalia* (i 20), Geminos refers vaguely to a **spherical construction** or **spherical system**, *sphairopoiïa* (xii 22, 26), for each planet, which, through a combination of spheres (or perhaps only circles), produces the apparently irregular movement of a planet. Geminos uses *sphairopoiïa* in a more general way, for the **spherical arrangement** of the cosmos, through which some stars rise and set each day while others do not (xiv 9; with related uses at xvi 19, 27, 29). *Sphairopoiïa* can also indicate a branch of applied mechanics, devoted to the construction of models of the heavens (fragment 2). For a full discussion of this term, see sec. 10 of the Introduction.

Each of the planets lies **lower**, *tapeinoteros* (i 23), than the sphere of stars, in the sense that the planets are closer to the Earth. Saturn is **higher**, *meteōroteros* (i 23), than Jupiter.

ECLIPSES

An **eclipse**, *ekleipsis*, may be solar (x) or lunar (xi). Not every **conjunction**, *sunodos* (viii 1), of the Sun and Moon gives rise to an eclipse of the Sun, because the Moon's path varies above and below the zodiacal circle. The **covering**, *epiprosthesis*, of the Sun by the Moon causes a solar eclipse (x 1) while the covering of the Moon by the Earth (xi 3) causes a lunar eclipse. During a lunar eclipse, the Moon **falls into the shadow of the Earth**, *empiptei eis to skiasma tēs gēs* (xi 2.) All the eclipses of the Moon fall into an **eclipse zone**, *ekleiptikon* (xi 6, 7), a belt that extends 2° above and below the ecliptic.

MATHEMATICS

The implicit definition of a **sphere**, *sphaira* (i 23), is a three-dimensional figure whose boundary points are all equidistant from its **center**, *kentron* (i 32). Any straight line passing through the center and joining two points on the surface of the sphere is a **diameter**, *diametros* (ii 1). *Periphereia* can mean either an **arc** (i 18) or the **whole circumference** (v 49) of a circle, *kuklos* (i 1; i 11). Geminos sometimes divides a circle on the sphere into 60 **parts**, *merai*, which are called **sixtieths**, *hexēkosta* (v 46). But he also sometimes divides the circle into 360 **degrees**, *moipai* (i 8), each of which is divided into **sixtieths**, *hexēkosta* (xviii 7). A **minute**, or **60th of a degree**, is also called *prōton lepton* ("first division"), and a **second**, *deuteron lepton* ("second division") (xviii 8). But units of time, e.g., days, can also be divided into sixtieths in the same way (xviii 10). Any **great circle**, *megistos kuklos* (v 70), divides the sphere into two equal **hemispheres**, *hēmisphairia* (v 54).

Other geometrical terms are: **to cut**, *temnein* (i 33); **section** (of a sphere), *ektmēma* (xvi 5); **perimeter**, *perimetron* (xvi 9); **side**, *pleura* (ii 7); **oblong**, *paramēkēs* (xvi 4); **round**, *stroggulos* (xvi 4, 5); **square**, *tetragōnon* (ii 1); **triangle**, *trigōnon* (ii 1); and **boundary**, *peras* (i 40).

Among the arithmetical terms are: **to exceed**, *pleonazein* (viii 40); **excess**, *huperochē* (v 20); **difference**, *parallagē* (vi 33); **a quarter**, *tetartēmorion* (i 18); **proportion**, *logos* (xvi 4) but also *summetria* (xvi 5); **double**, *diplasios* (xvi 3); **to multiply**, *polu / polla-plasiazein* (viii 4); **to multiply by eight**, *oktaplasiazein* (viii 38); **approximate**, *oloscherēs* (ii 20); **to add,** *suntithenai* (i 17); an **increase,** *parauxēsis* (vi 29); a **diminution,** *meiōsis* (xviii 5).

THE CELESTIAL SPHERE AND ITS CIRCLES

The **sphere of the fixed stars,** *hē tōn aplanōn asterōn sphaira* (i 23), is centered on the Earth and has all the fixed stars on its surface. It rotates each day around a diameter known as the **axis,** *axōn* (iv 1), which meets the surface of the sphere at two points known as the **poles of the cosmos,** *poloi tou kosmou* (iv 1). The one visible to us is the **north pole,** *boreios polos,* and the one invisible to us is the **south pole,** *notios polos* (iv 2).

The plane tangent to the surface of the Earth at a given locality defines the **horizon,** *horizōn* (v 54–63), which separates the part of the cosmos that is **visible,** *phaneron* (v 54), from the part that is **invisible,** *aphanes* (v 54), at that location. Because of the minute size of the Earth relative to the cosmos, the horizon may be regarded as **bisecting,** *dichotomōn* (v 54), the cosmos. Perpendicular to the horizon, passing through its **south point,** *mesēmbria* (v 8), and **zenith,** *kata koruphēn sēmeion* (v 64), is the **local meridian** circle, *mesēmbrinos kuklos* (ii 25, 26).

Parallel Circles

Parallel circles, *parallēloi kukloi* (v 1), are those circles on the sphere that are parallel to the paths traced by the fixed stars in the course of the diurnal rotation of the cosmos. The greatest of these is **the equator circle,** *isēmerinos kuklos* (v 6). An important group of these circles is the **always-visible circles,** *aei theōroumenoi kukloi* (v 2), described by the motions of the stars that never rise or set. The largest of these circles is the **arctic circle,** *arktikos kuklos* (v 2), which separates stars that are always above the horizon from the stars that rise and set. Equal in size and parallel to the arctic circle, but lying always below the horizon, is the **antarctic circle,** *antarktikos kuklos* (v 9).

Of the parallel circles, two mark the northern and southern limits of the Sun during its annual course, namely the **summer tropic circle,** *therinos tropikos kuklos* (v 4), and the **winter tropic circle,** *cheimerinos tropikos kuklos* (v 7).

These parallel circles are generally to be understood as circles on the celestial sphere. The homonymous circles on the Earth with which most modern readers are familiar may be regarded as central projections of their celestial counterparts, or—as Geminos puts it—they lie under them. Thus he says that the **equator circle on the Earth,** *en tēi gēi isēmerinos kuklos,* lies **under,** *hupo,* the **equator circle in the cosmos,** *en* *tōi kosmōi isēmerinos kuklos* (xv 3), i.e., under the celestial equator. The word *dunamis* (v 41, 42), **power**, in the context of parallel circles refers to their properties.

Oblique Circles

Among the **oblique**, *loxos* (v 51), circles is the zodiac. (The aforementioned **horizon** is another). The **circle of the signs**, *kuklos tōn zōidion* (i 1), and the **zodiac circle**, *zōidiakos kuklos* (i 3), are synonymous expressions. The individual signs are most often **slanted**, *plagios* (vii 11), to the horizon when they rise or set. A sign is said to be **following**, *epomenon* (xii 6), another if it rises after the other. Thus, Taurus follows Aries, Gemini follows Taurus, etc.

Any one of the twelve zodiacal signs may be referred to either as a **sign**, *zōidiōn* (lit. "small figure") (i 2), or as a **twelfth part**, *dōdekatēmorion* (i 1), of the zodiac. The latter term is used to make a clear distinction between a sign of the zodiac, which is a 30°-long geometrical **segment**, *tmēma* (i 1), and a zodiacal constellation, which may have the same name but is of irregular size and shape.

The zodiac is a band of 12° **width**, *platos* (v 53), and is represented by three parallel circles (v 51). The upper and lower circles define the width of the band, and the third (middle) circle is the **circle through the middles of the signs**, *dia tōn mesōn tōn zōidiōn kuklos* (ii 21). This is our "ecliptic," the apparent path of the Sun. Its points of **section**, *tomē* (vi 36), with the equator are the equinoctial points. The angle at which the ecliptic is slanted with respect to the equator is called its **obliquity**, *loxotēta* (ii 24).

Another pair of great circles are the **colure circles**, *kolouroi kukloi* (v 49), one of which passes through the poles and the solstitial points, and the other of which passes through the poles and the equinoctial points. Both are, therefore, **perpendicular**, *orthos* (vi 23; vii 10), to the equator.

The **Milky Way**, *ho tou galaktos kuklos* (v 68), is also an oblique circle.

INSTRUMENTS

For a discussion of astronomical instruments, see sec. 8 of the Introduction. Geminos knows two kinds of celestial globes, **solid globes**, *sphairai stereai* (xvi 12), and **armillary spheres**, *krikōtai sphairai* (xvi 12). The **sundial**, *hōrologion* (ii 38) / *hōroskopeion* (ii 35) / *skiothēron* (vi 32), marks the passage of the hours. It was common enough that Geminos repeatedly refers to phenomena regarding the shadow cast by its **gnomon**, *gnōmōn* (ii 35), to illustrate points about solar motion and geography (xvi 13). The *dioptra* (i 4) was an instrument that could be used for sighting objects or measuring angles. Several different kinds of instrument were called by the same name, some with applications to astronomy (including its teaching) and some applied to surveying.

ZODIACAL ASPECTS AND OTHER ASTROLOGICAL TERMS

Said to be **in opposition**, *kata diametron* (ii 1), are two diametrically opposite zodiacal signs. Such a pair, e.g., Aries and Libra, are separated by five signs. A pair of signs is **in trine**, *kata trigōnon* (11 7), when they are separated by three signs, as are Aries and Leo. By extension, the term may be used of two zodiacal points that are 120° apart, just as the previous term (in opposition) may be used for zodiacal points 180° apart. A pair of signs is **in quartile**, *kata tetragōnon* (ii 13), when they are separated by two others, as are Aries and Cancer. Again, the term may be used of zodiacal points 90° apart. Said to be **in syzygy**, *kata suzugian* (ii 27), are two signs that rise (and therefore set) on the same arcs of the horizon. For that reason they are contained by the same two parallel circles.

Geminos refers to the **significance**, *episēmasia*, of certain celestial events, such as the heliacal risings and settings of the stars, as weather signs. **Power**, *dunamis* (ii, xvii), refers to the ability of stars to exert influences on terrestrial life when their **positions**, *epochai* (ii 6), fall in certain signs. One example of such power is the ability of stars to create a **sympathy**, *sumpatheia* (ii 5, 12), i.e., a resonance or common destiny, between two individuals born under signs in a certain relation to one another.

GEOGRAPHY

A **spherical**, *sphairoeidēs* (vi 21), Earth was a central tenet of Greek astronomy from Eudoxos onward. Its north and south poles lie directly below their cosmic counterparts. The meridians passing through these poles and a given locality define local **north**, *boreios* (ii 8), and **south**, *notios* (ii 10). (These two adjectival forms are derived from the names of the respective winds, *boreas* (ii 8) and *notos* (ii 10)). The phrase "toward the bear," *pros arkton* (v 16, 21), and related forms also indicate the north, *arktos* being the Greek for "bear," i.e., the Great Bear, Ursa Major.

Four circles on the surface of the Earth, parallel to the equator, divide the Earth into belts or **zones**, *zōnai* (xv 1). The northernmost zone is **frigid**, *katapsugmenē* (xv 1), and uninhabited because of its extreme cold. Next is the northern **temperate**, *eukratos* (xv 2), zone, in which the Greeks and most other nations known to them lived. Then, between the northern and southern tropics, one finds the **torrid**, *diakekaumenē* (xv 3), zone, thought by some (but not by Geminos) to be uninhabitable because of its extreme heat. Proceeding southward, one then comes to the southern temperate zone and, finally, to the southern frigid zone. These two share the climatic characteristics of their northern counterparts.

A word used of places on the Earth is *oikēseis* (i 12), **regions**, or, more traditionally, **habitations**, from the verb *oikeō* ("I inhabit"). That the use of the word actually implies nothing about the habitability of the place is shown by v 38, where Geminos speaks, as Theodosios does in his *Peri oikēseōn*, of the *oikēsis* "beneath the pole," i.e. the north (or south) pole! An important, though somewhat flexible, geographical term is *oikoumenē* (xvi 3), the **inhabited** ("regions" understood). *Oikoumenē* can mean "the whole inhabited Earth," "the northern inhabited Earth," or "the Earth inhabited by us Greeks." For a detailed discussion, see comment 3 to chapter 5.

Geminos has four special terms for the inhabitants of the Earth: *synoikoi*, those dwelling within 90° longitude on either side of us in "our" zone (i.e., the northern temperate zone); *perioikoi*, those in the other hemisphere in our zone; *antoikoi*, those in the southern temperate zone in our hemisphere; and *antipodes* (xvi 1), those in the other hemisphere in the southern temperate zone.

To specify location north or south along a meridian, Geminos employs one of two notions. The first is the **elevation of the pole**, *exarma tou polou* (vi 24), or, the **inclination of the cosmos**, *egklima tou kosmou* (vi 24), both referring to the acute (or, at the poles, right) angle that the axis of the sphere makes with the horizon of a given locality, an angle numerically equal to the geographical latitude. The second notion Geminos uses is that of the **longest day or night**, *megistē hēmera / nuchta* (i 10/12) at a given locality. Knowledge of the length of the longest day is equivalent to knowledge of the latitude. A third, less technical, term for **elevation**, *meteōrismos*, is also applied to the pole (vi 25) and (in a slightly different form) to stars (xii 1).

The elevation of the pole gave rise to the division of the Earth according to *klimata* (i 10), belts parallel to the equator consisting of localities where the pole has roughly the same elevation. A common measure of distance between localities is the **stade**, *stadion* (v 56).

Geminos refers to **Ocean**, *Ōkeanos* (xvi 28), which Homer describes as encircling a flat Earth, but he denounces the error of those who take Homer at his word and draw disklike maps. (Nor did any **mathematician**, *mathēmatikos* (xvi 23), Geminos claims, believe that Ocean spread out over the torrid zone between the tropics.) The Earth should be represented, he says, by a **world map**, *geōgraphia* (xvi 4), drawn on an **oblong panel**, *pinax paramēkēs* (xvi 4), whose **length**, *mēkos*, is double its **width**, *platos*.

TIME AND CALENDRICAL MATTERS

Hours, *hōrai*, are of two kinds. The **equinoctial hour**, *isēmerinē hōra* (vi 6), is 1/24 of a night and day. The other, now called a "temporal" or "seasonal" hour, Geminos refers to by saying (vii 33) that "in every night six signs set in 12 hours." That is, they are twelfth-parts of a night or day. Finally, the word *hōra* may also denote one of the four seasons (ii 17).

A *parapēgma* (xvii 19) is a calendar that gave the dates, throughout the year, of the heliacal risings and settings of prominent stars or constellations and associated weather prognostications. Geminos's book closes with such a calendar.

Geminos discusses a number of **periods** or **cycles**, *periodoi* (viii 26), which were designed to comprise a whole number of solar years and a whole number of lunar months. The **8-year cycle**, *oktaetēris* (viii 27), consists of five years of 12 months and three years of 13 months. Thus it includes 3 intercalary, *embolimoi* (viii 26), months. But it slowly gets out of step with the Moon. Geminos describes a **16-year period**, *ekkaidekaetēris* (viii 39), that attempts to correct this defect. A better cycle is the longer, so-called Metonic cycle with its **19-year period**, *enneakaidekaetēris* (viii 48–58), incorporating twelve years of 12 months and seven years of 13 months. Longest of all is the **76-year period**, *hekkaiebdomēkontaetēris* (viii 59), of four Metonic cycles. Used in the computation of eclipses is a period called the *exeligmos* (xviii) with its whole number of days (19,756), synodic months (669), and revolutions in anomaly (717). These cycles are discussed in sec. 13 of the Introduction.

To return, *apokathistasthai*, is used of the Sun for its returning to a given point in the zodiac after a year (i 7), or of an Egyptian holiday returning to the same date in the calendar after 1,460 years (viii 24), or of the Moon coming back to the part of its monthly orbit where it has its greatest speed (xviii 2).

Epagomenal days, *epagomenai hēmerai* (viii 18), are days added according to some regular calendric scheme. In the Egyptian calendar of 365 days the last 5 days were epagomenal days.

Index of Persons Mentioned by Geminos

INTRODUCTION TO THE PHENOMENA (EXCLUDING THE *PARAPĒGMA*)

Aratos (fl. 270 B.C.)
v 24; vii 7, 13; viii 13; xiv 8; xvii 46, 48.
Aristotle (340 B.C.)
xvii 49.
Boēthos (second century B.C.)
xvii 48.
Chaldeans
ii 5; xviii 9.
Dikaiarchos (320 B.C.)
xvii 5.
Egyptians
viii 16, 20, 22, 23, 25.
Eratosthenēs (240 B.C.)
viii 24.
Euktēmōn (430 B.C.)
viii 50.
Hesiod (650 B.C.)
xvii 14.
Hipparchos (150 B.C.)
iii 8, 13.
Homer (750 B.C.)
vi 10, 16; xvi 27, 28; xvii 31.
Kallimachos (250 B.C.)
iii 8.
Kallippos (330 B.C.)
viii 50, 59.
Kleanthēs (270 B.C.)
xvi 21.
Kratēs [of Mallos] (160 B.C.)
vi 10, 16; xvi 22, 23, 27.
Philippos (330 B.C.)
viii 50.

Polybios (160 B.C.)
  xvi 32.
Pythagoreans
  i 19.
Pytheas (290 B.C.)
  vi 9.

PARAPĒGMA

Dēmokritos (400 B.C.)
Dositheus (230 B.C.)
Eudoxus (360 B.C.)
Euktēmōn (fl. 430 B.C.)
Kallippos (fl. 330 B.C.)
Metōn (430 B.C.)

FRAGMENT 1, FROM PROKLOS

*In the indexes for the two fragments, an asterisk * indicates a name that is used by the excerpter (either Proklos or Simplikios) and not by Geminos.*

Archimēdēs (fl. 250 B.C.)
Hero (first century A.D.)
Hippokratēs (420 B.C.)
Ktēsibios (250 B.C.)
Plato (390 B.C.)
Pythagoreans*

FRAGMENT 2, FROM SIMPLIKIOS

Alexander [of Aphrodisias]* (A.D. 200)
Aristotle* (340 B.C.)
Herakleidēs Pontikos (fl. 340 B.C.) (This name was probably interpolated by a copyist.)
Poseidōnios* (90 B.C.)

Bibliography

Achilleus (Achilles Tatius), *Introduction to the "Phenomena" of Aratos*. See Maass 1898, 25–85.

Aiton, E.J. 1981. "Celestial Spheres and Circles," *History of Science* 19, 75–114.

Alexander of Aphrodisias, *Commentary on Aristotle's Meteorology*. See Hayduck 1899.

Allen, Richard Hinkley 1899. *Star Names and Their Meanings* (New York: G.E. Stechert). Reprinted as *Star Names: Their Lore and Meaning* (New York: Dover, 1963).

Apollōnios of Pergē, *Conics*. See Heiberg 1893.

Aratos, *Phenomena*. See Mair and Mair 1955; Kidd 1997; J. Martin 1998.

Archimēdēs, *Works*. See Heath 1912; Dijksterhuis 1987.

Aristotle, *Metaphysics*. See Tredennick 1961–62.

————, *Meteorology*. See Lee 1952.

————, *On the Heavens*. See Guthrie 1939.

————, *Physics*. See Wicksteed and Cornford 1960–63.

Aristotle, Pseudo-, *On the Cosmos*. See Forster and Furley 1978.

Arnaud, P. 1984. "L'image du globe dans le monde romain: Science, iconographie, symbolique," *Mélanges de l'École Française de Rome* 96, 53–116.

Arnauldi, M., and K. Schaldach 1997. "A Roman cylinder dial: Witness to a forgotten tradition," *Journal for the History of Astronomy* 28, 107–17.

Ash, H.B. 1954–1960. ed. and trans. Columella, *On Agriculture*, 3 vols. (Cambridge: Harvard University Press; London: W. Heinemann).

Aujac, G. 1970. "La sphéropée, ou la mécanique au service de la découverte du monde," *Revue d'Histoire des Sciences* 23, 93–107.

———— 1975. ed. and trans. Géminos, *Introduction aux phénomènes* (Paris: Les Belles Lettres).

———— 1979. ed. and trans. Autolycos de Pitane, *La sphère en mouvement, Levers et couchers héliaques, Testimonia* (Paris: Les Belles Lettres).

Aujac, G., et al. 1987a. "The Foundations of Theoretical Cartography in Archaic and Classical Greece," in Harley and Woodward 1987, pp. 130–47.

———— 1987b. "The Growth of an Empirical Cartography in Hellenistic Greece," in Harley and Woodward 1987, pp. 148–60.

———— 1987c. "Greek Cartography in the Early Roman World," in Harley and Woodward 1987, pp. 161–76.

Autolykos of Pitanē. *On the Moving Sphere. Risings and Settings*. See Aujac 1970; Mogenet 1950; Bruin and Vondjidis 1971.

Babbitt, F.C. 1936. ed. and trans. Plutarch, *Moralia*, vol. 5. (Cambridge: Harvard University Press).

Baccani, D. 1989. "Appunti per oroscopi negli ostraca di Medinet Madi," *Analecta Papyrologica* 1, 67–77.

——— 1995. "Appunti per oroscopi negli ostraca di Medinet Madi (II)," *Analecta Papyrologica* 7, 63–72.

Bakhouche, B. 1996. *Les astres: Actes du colloque international de Montpellier 23–25 mars 1995*, 2 vols. (Montpellier: Université Paul Valéry).

Bara, J.-F. 1989. *Vettius Valens d'Antioche: Anthologies, Livre I* (Leiden: E.J. Brill).

Barker, A. 2000. *Scientific Method in Ptolemy's* Harmonics (Cambridge: Cambridge University Press).

Barton, T. 1994. *Ancient Astrology* (London: Routledge).

Berggren, J.L. 1976. "Spurious Theorems in Archimedes' Equilibrium of Planes: Book I," *Archive for History of Exact Sciences* 16, 87–103.

Berggren, J.L., and H. Eggert-Strand, forthcoming. "Theodosios on Habitations: A Translation and Commentary."

Berggren, J.L., and A. Jones 2000. *Ptolemy's Geography: An Annotated Translation of the Theoretical Chapters* (Princeton: Princeton University Press).

Berggren, J.L., and R.S.D. Thomas 1996. *Euclid's* Phaenomena: *A Translation and Study of a Hellenistic Treatise in Spherical Astronomy* (New York: Garland).

Bickerman, E.J. 1980. *Chronology of the Ancient World*, 2nd ed. (Ithaca: Cornell University Press).

Blass, F. 1883. De Gemino et Posidonio (Kiel: C.F. Mohr).

——— 1887. *Eudoxi ars astronomica* (Kiel: Schmidt and Klaunig). Reprinted in *Zeitschrift für Papyrologie und Epigraphik* 115 (1997) 3–25.

Böckh, A. 1863. *Über die vierjährigen Sonnenkreise der Alten, vorzüglich den Eudoxischen* (Berlin).

Boll, F. 1899. "Das Κηρύκιον als Sternbild," *Hermes* 34, 643–45.

——— 1901. "Die Sternkatalog des Hipparch und des Ptolemaios," *Bibliotheca Mathematica*, Folge 3, Band 2, 185–95.

Boll, F., and W. Gundel 1924. "Sternbilder, Sternglaube und Sternsymbolik bei Griechen und Römern," in W.H. Roscher, ed., *Ausführliches Lexikon der griechischen und römischen Mythologie*, vol. VI (Leipzig: Teubner), cols. 867–1071.

Bouché-Leclercq, A. 1899. *L'astrologie grecque* (Paris: Leroux).

Bowen, A.C. 2001. "La scienza del cielo nel periodo ptolemaico," in S. Petruccioli, ed., *Storia della scienza*, I. *La scienza greco-romana* (Rome: Instituto della Enciclopedia Italiana), pp. 806–39.

——— "Simplicius' Commentary on Aristotle, *De caelo* 2.10–12: An Annotated Translation (Part 1)," *SCIAMVS* 4 (2003) 23–58.

Bowen, A.C., and B.R. Goldstein 1989. "Meton of Athens and Astronomy in the Late Fifth Century B.C.," in E. Leichty et al., eds., *A Scientific Humanist: Studies in Memory of Abraham Sachs* (Philadelphia: Samuel Noah Kramer Fund), pp. 39–81.

——— 1996. "Geminus and the concept of mean motion in Greco-Latin astronomy," *Archive for History of Exact Sciences* 50, 157–85.

Bowen, A.C., and R.B. Todd 2004. trans. *Cleomedes' Lectures on Astronomy* (Berkeley: University of California Press).

Brown, D. 2000. *Mesopotamian Planetary Astronomy-Astrology* (Groningen: Styx Publications).

Bruin, F., and A. Vondjidis 1971. trans. *The Books of Autolykos* (Beirut: American University of Beirut).

Bury, R.G. 1987. ed. and trans. Sextus Empricus, *Against the Professors*, 4 vols. (London: William Heinemann; Cambridge: Harvard University Press).

Bywater, I. 1886. ed. *Prisciani Lydi quae extant. Metaphrasis in Theophrastum et Solutionum ad Chosroem liber*, in *Supplementum Aristotelicum*, 3 vols. (Berlin: Deutsche Akademie der Wissenschaften, 1885–1903), vol. 1, pt. 2.

Cajori, F. 1993. *A History of Mathematical Notations*, 2 vols. bound as one (New York: Dover. First pub. La Salle, Illinois: Open Court, 1928, 1929).

Carmado, D., and A. Ferrrara 1989. *Stabiae: Le Ville* (Castellamare di Stabia).

Catamo, M., et al. 2000. "Fifteen Further Greco-Roman Sundials from the Mediterranean Area and Sudan, *Journal for the History of Astronomy* 31, 203–21.

Catullus. See Goold 1983.

Censorinus, *On the Birth Day*. See Rocca-Serra 1980 or Sallmann 1988.

Charvet, P., and A. Zucker 1998. *Le ciel: Mythes et histoire des constellations: Les* Catastérismes *d'Érathosthène* (Paris: NiL Éditions).

Cherniss, H., and W.C. Helmbold 1957. eds. and trans. Plutarch, *Moralia*, vol. 12 (Cambridge: Harvard University Press).

Cicero, *De re publica*. See Keyes 1961.

———, *On the Nature of the Gods*. See Rackham 1961.

Columella, *On Agriculture*. See Ash 1954–60.

Cumont, F. 1935. "Les noms des planètes et l'astrolatrie chez les Grecs," *L'Antiquité Classique* 4, 5–43.

Cunliffe, B. 2002. *The Extraordinary Voyage of Pytheas the Greek: The Man Who Discovered Britain* (London: Penguin).

Cuvigny, H. 2004. "Une sphère céleste antique in argent ciselé," in P. Horak, ed. *Gedenkschrift Ulrike Horak* (Florence: Edizioni Gonnelli).

De Falco, V., M. Krause, and O. Neugebauer 1966. *Hypsikles. Die Aufgangszeiten der Gestirne*. Abhandlungen der Akademie der Wissenschaften in Göttingen. Philologisch-Historische Klasse, Dritte Folge, Nr. 62 (Göttingen: Vandenhoeck & Ruprecht).

Dicks, D.R. 1960. *The Geographical Fragments of Hipparchus* (London: Athlone Press).

——— 1972. "Geminus," in Gillispie 1970–80, vol. 5, pp. 344–47.

Diels, H. 1882. *Simplicii in Aristotelis physicorum libros quattuor priores commentaria* (Berlin: G. Reimer).

——— 1924. *Antike Technik* (Leipzig: Teubner).

Diels, H., and A. Rehm 1904. "Parapegmenfragmente aus Milet," *Sitzungsberichte der königlich Preussischen Akademie der Wissenschaften*, Jahrgang, 92–111.

Dijksterhuis, E.J. 1987. *Archimedes*. With a new bibliographic essay by Wilbur R. Knorr (Princeton: Princeton University Press. First pub. 1956).

Dilke, O.A.W. 1985. *Greek and Roman Maps* (Ithaca: Cornell University Press).

Diodoros of Sicily. See Oldfather 1947–67.

Diogenēs Laertios, *Lives and Opinions of Eminent Philsophers*. See Hicks 1958–59.

Dodge, B. 1970. ed. and trans. *The Fihrist of al-Nadim*, 2 vols. (New York and London: Columbia University Press).

Drake, S. 1989. "Hipparchus—Geminus—Galileo," *Studies in History and Philosophy of Science* 20, 47–56.

Dreyer, J.L.E. 1906. *History of the Planetary Systems from Thales to Kepler* (Cambridge: Cambridge University Press). Reprinted as *A History of Astronomy from Thales to Kepler* (New York: Dover, 1953).

Duhem, P. 1908. *ΣΩZEIN TA ΦAINOMENA. Essai sur la notion de théorie physique de Platon à Galilée* (Paris: A. Hermann et Fils).

——— 1969. *To Save the Phenomena: An Essay on the Idea of Physical Theory from Plato to Galileo*, trans. by Edmund Dolland and Chaninah Maschler (Chicago and London: University of Chicago Press).

Duke, D. 2006. "Analysis of the Farnese Globe," *Journal for the History of Astronomy* 37, 87–100.

Dupuis, J. 1892, *Théon de Smyrne, Philosophe platonicien: Exposition des connaissances mathématiques utiles pour la lecture de Platon* (Paris: Hachette. Reprinted, Bruxelles: Culture et Civilisation, 1966).

Duval, P.M., and G. Pinault 1986. "Les calendriers (Coligny, Villards d'Héria)," *Recueil des Inscriptions Gauloises*, vol. 3 (Paris: Editions du Centre national de la recherche scientifique).

Eastwood, B. 1992. "Heraclides and Heliocentrism: Texts, Diagrams, and Interpretations," *Journal for the History of Astronomy* 23, 233–60.

Edelstein, L., and I.G. Kidd 1989. *Posidonius*, Vol. I, *The Fragments*, 2nd ed. (Cambridge: Cambridge University Press).

Engels, D. 1985. "The Length of Eratosthenes' Stade," *American Journal of Philology* 106, 298–311.

Eratosthenēs, *Catasterisms*. See Olivieri 1897; Charvet and Zucker 1998.

Euclid, *Phenomena*. See Berggren and Thomas 1996.

Evans, J. 1984. "On the Function and the Probable Origin of Ptolemy's Equant," *American Journal of Physics* 52, 1080–89.

——— 1998. *The History and Practice of Ancient Astronomy* (New York: Oxford University Press).

——— 2003. "The Origins of Ptolemy's Cosmos," in S. Colafrancesco and G. Giobbi, eds., *Cosmology through Time: Ancient and Modern Cosmologies in the Mediterranean Area* (Milano: Mimesis), pp. 123–32.

Evans, J., and M. Marée (forthcoming). "A Minature Ivory Sundial and Equinox Indicator from Greek Egypt."

Field, J.V., and M.T. Wright 1984. "Gears from the Byzantines: A Portable Sundial with Calendrical Gearing," *Annals of Science* 42, 87–138.

Firmicus Maternus, *Mathesis*. See Monat 1992–97.

Fischer, I. 1975. "Another Look at Eratosthenes' and Posidonius' Determination of the Earth's Circumference," *Quarterly Journal of the Royal Astronomical Society* 16, 152–67.

Forster, E.S., and D.J. Furley 1978. ed. and trans. Aristotle, *On Sophistical Refutations and on Coming-to-Be and Passing Away; On the Cosmos* (Cambridge: Harvard University Press; London: William Heinemann).

Fotheringham, J.K. 1924. "The Metonic and Callipic Cycles," *Monthly Notices of the Royal Astronomical Society* 84, 383–92.

Frazer, J.G. 1931. ed. and trans. Ovid, *Fasti* (London: Heinemann; New York: G.P. Putnam's Sons).

Friedlein, G. 1873. *Procli Diadochi primum Euclidis elementorum librum commentarii* (Leipzig: Teubner).

Geminos, *Introduction to the Phenomena*. See Manitius 1898; Aujac 1975.

Gibbs, S. 1976. *Greek and Roman Sundials* (New Haven: Yale University Press).

Gillispie, C.C. 1970–80. ed. *Dictionary of Scientific Biography* (New York: Charles Scribner's Sons).

Goldstein, B. 1980. "The Status of Models in Ancient and Medieval Astronomy," *Centaurus* 24, 132–47.

—— 1997. "Saving the Phenomena: The Background to Ptolemy's Planetary Theory," *Journal for the History of Astronomy* 28, 1–12.

Goold, G.P. 1977. *Manilius. Astronomica* (Cambridge: Harvard University Press; London: William Heinemann).

—— 1983. *Catullus* (London: Duckworth).

Gössmann, F. 1950. *Planetarium Babylonicum oder Die Sumerisch-Babylonischen Stern-Namen*. A. Deimal, ed., *Sumerisches Lexikon*, Teil IV, Band 2 (Rome: Verlag des Päpstlichen Bibelinstituts).

Gould, J.B. 1970. *The Philosophy of Chrysippus* (Albany: State University of New York Press).

Grant, E. 1974. ed. *A Source Book in Medieval Science* (Cambridge: Harvard University Press).

Grenfell, B.P., and A.S. Hunt 1906. *The Hibeh Papyri, Part I* (London: Egypt Exploration Fund).

Gundel, H.G. 1992. *Zodiakos. Tierkeisbilder im Altertum* (Mainz am Rhein: Philip von Zabern).

Guthrie, W.K.C. 1939. ed. and trans. Aristotle, *On the Heavens* (London: William Heinemann; Cambridge: Harvard University Press).

Halma, [Nicolas] 1821. *Les Phénomènes d'Aratus de Soles, et de Germanicus César, avec les scholies de Théon, les Catastérismes d'Eratosthène, et la Sphère de Leontius*. Paris: Merlin.

Hannah, R. 2001. "From orality to literacy? The case of the parapegma," in J. Watson, ed., *Speaking Volumes: Oralitiy and Literacy in the Greek and Roman World* (Leiden: Brill), pp. 139–59.

—— 2002. "Euctemon's parapegma," in C.J. Tuplin and T.E. Rihll, eds., *Science and Mathematics in Ancient Greek Culture* (Oxford: Oxford University Press), pp. 112–32.

Harley, J.B., and David Woodward 1987. *The History of Cartography*, vol. 1, *Cartography in Prehistoric, Ancient, and Medieval Europe and the Mediterranean* (Chicago and London: University of Chicago Press).

Hayduck, M. 1899. *Alexandri in Aristotelis Meteorologicorum libros commentaria.* Commentaria in Aristotelem Graeca, iii, 2 (Berlin).

Heath, T.L. 1912. *The Works of Archimedes,* revised ed. (Cambridge: Cambridge University Press).

—— 1921. *A History of Greek Mathematics,* 2 vols. (Oxford: Clarendon Press).

Heiberg, J.L. 1885. *Euclidis Elementa,* vol. 5, *Continens Elementorum qui feruntur libros XIV–XV et Scholia in Elementa* (Leipzig: Teubner).

—— 1893. ed. *Apollonii Pergaei que graece exstant cum commentariis antiquis* (Leipzig: Teubner. Reprinted, Stuttgart: Teubner, 1974).

—— 1894. ed. *Simplicii In Aristotelis de Caelo commentaria.* Commentaria in Aristotelem Graeca, vii (Berlin).

—— 1898–1907. ed. *Claudii Ptolemaei Opera quae exstant omnia.* Vol. 1 (2 parts), *Syntaxis mathematica*; Vol. 2, *Opera astronomica minora* (Leipzig: Teubner).

—— 1912. *Heronis Alexandrini Opera quae supersunt omnia,* Vol. 4, *Heronis Definitiones cum variis collectionibus. Heronis quae feruntur geometrica* (Leipzig: Teubner).

Herodotos, *Histories.* See Sélincourt 1972.

Hicks, R.D. 1958–59. ed. and trans. Diogenes Laertius, *Lives of Eminent Philosophers,* 2 vols. (Cambridge: Harvard University Press; London: W. Heinemann).

Hildericus, E. 1590. *Gemini probatissimi philosophi, ac mathematici elementa astronomiae Graece, & Latine interprete Edone Hilderico.* (Altdorf, 1590; Leiden, 1603).

Hipparchos, *Commentary on the Phenomena of Aratos and Eudoxos.* See Manitius 1894.

Hippokratēs, *Regimen III.* See W.H.S. Jones 1923–1931, vol. 4.

Hölbl, G. 2001. *A History of the Ptolemaic Empire,* trans. T. Saavedra (London and New York: Routledge).

Hultsch, F. 1864. ed. *Heronis Alexandrini Geometricorum et stereometricorum reliquiae* (Berlin: Weidmann).

—— 1875–78. ed. *Pappi Alexandrini Collectionis quae supersunt,* 3 vols. (Berlin. Reprinted, Amsterdam: Verlag Adolf M. Hakkert, 1965).

Hunger, H., and D. Pingree 1989. *MUL.APIN. An Astronomical Compendium in Cuneiform.* Archiv für Orientforschung, Beiheft 24. (Horn, Austria: Ferdinand Berger & Söhne).

—— 1999. *Astral Sciences in Mesopotamia* (Leiden: Brill).

Hypsiklēs, *Anaphorikos.* See De Falco, Krause, and Neugebauer 1966.

Jones, A. 1983. "The Development and Transmission of 248-Day Schemes for Lunar Motion in Ancient Astronomy," *Archive for History of Exact Sciences* 29, 1–36.

—— 1994. "Peripatetic and Euclidean theories of the visual ray," *Physis* 31, n.s., fasc. 1, pp. 47–76.

—— 1999a. "Geminus and the Isia," *Harvard Studies in Classical Philology* 99, 255–67.

—— 1999b. *Astronomical Papyri from Oxyrhynchus.* Memoirs of the American Philosophical Society 233.

———— 2000. "Calendrica I: New Callippic Dates," *Zeitschrift für Papyrologie und Epigraphik* **129**, 141–58.

———— 2001a. "More Astronomical Tables from Tebtunis," *Zeitschrift für Papyrologie und Epigraphik* **136**, 211–20.

———— 2001b. "Pseudo-Ptolemy "*De Speculis*," SCIAMVS 2, 145–86.

Jones, H.L. 1959–1961. *The Geography of Strabo* (Cambridge: Harvard University Press; London: William Heinemann).

Jones, W.H.S. 1918–35. ed. and trans. *Pausanias' Description of Greece*, 5 vols. (London: W. Heinemann; New York: G.P. Putnam's Sons).

———— 1923–1931. *Hippocrates*, 8 vols. (Cambridge: Harvard University Press; London: W. Heinemann).

Kallimachos, *Aetia*. See Trypanis 1978.

Keyes, C.W. 1961. ed. and trans. Cicero, *De re publica, De legibus* (Cambridge: Harvard University Press).

Kidd, I.G. 1988. *Posidonius*, Vol. II, *The Commentary*, 2 parts (Cambridge: Cambridge University Press).

———— 1997. *Aratus. Phenomena*. (Cambridge: Cambridge University Press).

———— 1999. *Posidonius*, Vol. III, *The Translation of the Fragments* (Cambridge: Cambridge University Press).

Kleomēdēs, *Meteōra*. See Todd 1990; Bowen and Todd 2004.

Knorr, W. 1989. "Plato and Eudoxus on the Planetary Motions," *Journal for the History of Astronomy* **20**, 313–29.

Koch-Westenholz, U. 1995. *Mesopotamian Astrology: An Introduction to Babylonian and Assyrian Celestial Divination* (Copenhagen: Carsten Niebuhr Institute).

Kouremenos, T. 1994. "Posidonius and Geminus on the foundations of mathematics," *Hermes: Zeitschrift für Klassische Philologie* **122**, 437–50.

Kugel, A. 2002. *Sphères. L'art des mécaniques célestes* (Paris: J. Kugel).

Kunitzsch, P., and T. Smart 1986. *Short Guide to Modern Star Names and Their Derivations* (Wiesbaden: Otto Harrassowitz).

Künzl, E. 2000. *Ein römischer Himmelsglobus der mittlern Kaiserzeit* (Mainz: Römisch-Germanisches Zentralmuseum).

———— 2005. *Himmelsgloben und Sternkarten. Astronomie und Astrologie in Vorzeit und Altertum* (Stuttgart: Theiss).

Landels, J.G. 1978. *Engineering in the Ancient World* (Berkeley and Los Angeles: University of California Press).

Lasserre, F. 1966. ed. and trans. *Die Fragmente des Eudoxos von Knidos* (Berlin: Walter de Gruyter & Co).

Le Boeuffle, A. 1989. *Le ciel des Romans* (Paris: De Boccard).

Lee, H.D.P. 1952. ed. and trans. Aristotle, *Meteorologica* (Cambridge: Harvard University Press).

Lehoux, D.R. 2000. *Parapegmata, or, Astrology, Weather, and Calendars in the Ancient World* (Ph.D. dissertation, University of Toronto).

———— 2005. "The Parapegma Fragments from Miletus," *Zeitschrift für Papyrologie und Epigraphik* **152**, 125–40.

———— (forthcoming). *Astronomy, Weather, and Calendars in the Ancient World* (Cambridge: Cambridge University Press).

Leontios, *Construction of the Sphere of Aratos.* See Halma 1821; Maass 1898, 561–67.

Leptinēs, *Celestial Teaching (Ouranios Didascalea).* See Blass 1887; Tannery 1893, 283–94.

Lerner, M.-P. 1996. *Le monde des sphères,* 2 vols. (Paris: Les Belles Lettres).

Lewis, M.J. Taunton (2001). *Surveying Instruments of Greece and Rome* (Cambridge and New York: Cambridge University Press).

Liddell, H.G., and R. Scott, revised by H.S. Jones 1996. *A Greek-English Lexicon,* 9th ed. with New Supplement (Oxford: Clarendon Press).

Lloyd, G.E.R. 1973. *Greek Science after Aristotle* (New York: Norton).

——— 1978. "Saving the Appearances," *Classical Quarterly* 28, 202–22.

——— 1987. *The Revolutions of Wisdom: Studies in the Claims and Practices of Ancient Greek Science* (Berkeley: University of California Press).

——— 1991. *Methods and Problems in Greek Science* (Cambridge and New York: Cambridge University Press).

Lucretius, *On the Nature of Things.* See Rouse 1982.

Locher, K. 1989. "A further Hellenistic conical sundial from the theatre of Dionysius in Athens," *Journal for the History of Astronomy* 20, 60–62.

Maass, E. 1898. *Commentariorum in Aratum reliquiae* (Berlin: Weidmann; Reprinted, 1958).

Macrobius, *Commentary on the Dream of Scipio.* See Stahl 1952.

Mair, A.W., and G.R. Mair 1955. *Callimachus: Hymns and Epigrams. Lycophron. Aratus.* (London: William Heinemann; Cambridge: Harvard University Press).

Manilius, *Astronomica.* See Goold 1977.

Manitius, Carolus 1894. ed. and trans. *Hipparchi in Arati et Eudoxi Phaenomena commentariorum libri tres* (Leipzig: Teubner).

——— 1898. ed. and trans. *Gemini Elementa astronomiae* (Leipzig: Teubner; reprinted Stuttgart: Teubner, 1974).

——— 1909. ed. and trans. *Procli Diadochi Hypotyposis astronomicarum positionum* (Leipzig: Teubner; reprinted Stuttgart: Teubner, 1974).

Mansfield, J., and D.T. Runia 1997. *Aëtiana: The Method and Intellectual Context of a Doxographer, vol. 1, The Sources* (Leiden and New York: Brill).

Marsden, E.W. 1971. *Greek and Roman Artillery,* vol. 2, *Technical Treatises* (Oxford: Oxford University Press).

Martin, J. 1998. ed. and trans. Aratos, *Phénomènes,* 2 vols. (Paris: Les Belles Lettres).

Martin, T.-H. 1854. *Recherches sur la vie et les ouvrages d'Héron d'Alexandrie, disciple de Ctésibius, et sur tous les ouvrages mathématiques grecs, conservés ou perdus, publiés ou inédits, qui ont été attribués à un auteur nommé Héron.* Académie des Inscriptions et Belles-Lettres. Mémoires présentés pars divers savants. Première série, tome 4 (Paris: Imprimerie Impériale).

McClusky, S. 1998. *Astronomies and Cultures in Early Medieval Europe* (Cambridge: Cambridge University Press).

Mette, H.J. 1952. ed. *Pytheas von Massalia* (Berlin: W. de Gruyter).

Mittelstrass, J. 1962. *Die Rettung der Phänomene* (Berlin: Walter de Gruyter).

Mogenet, J. 1950. *Autolycus de Pitane, Histoire du Texte*. Publications Universitaires de Louvain, 3e série, fasc. 37.

Monat, P. 1992–97. ed. and trans. Firmicus Maternus, *Mathesis*, 3 vols. (Paris: Les Belles Lettres).

Morrow, G.R. 1970. trans. Proclus, *Commentary on the First Book of Euclid's Elements* (Princeton: Princeton University Press).

Mueller, I. 2004. trans. Simplicius, *On Aristotle's "On the Heavens 2.1-9"* (Ithaca: Cornell University Press).

——— 2005. trans. Simplicius, *On Aristotle's "On the Heavens 2.10-14"* (Ithaca: Cornell University Press).

Murray, A.T. 1936–39. ed. and trans. Demosthenes, *Private Orations*, 6 vols. (Cambridge: Harvard University Press; London: W. Heinemann).

Neugebauer, O. [1955]. *Astronomical Cuneiform Texts*, 3 vols. (London: Lund Humphries).

——— 1975. *A History of Ancient Mathematical Astronomy* (New York: Springer: Berlin).

Newcomb, S. 1898. "Tables of the Motion of the Earth on Its Axis and Around the Sun," *Astronomical Papers Prepared for the Use of the American Ephemeris and Nautical Almanac* 6.

Oates, J. 1986. *Babylon*, 2nd ed. (London: Thames and Hudson).

Oldfather, C.H. 1947–67. *Diodorus of Sicily*, 12 vols. (Cambridge: Harvard University Press; London: W. Heinemann).

Olivieri, A. 1897. *Mythographi Graeci* 16, 3, fasc. 1 (Leipzig).

Ovid, *Fasti*. See Frazer 1931.

Pappos of Alexandria, *Commentary on the Almagest*. See Rome 1931.

———, *Mathematical Collection*. See Hultsch 1875–78; Ver Eeeke 1933.

Pattenden, P. 1981. "A late sundial at Aphrodisias," *Journal of Hellenic Studies* 101, 101–12.

Pausanias, *Description of Greece*. See W.H.S. Jones 1918–35.

Pedersen, O. 1974. *A Survey of the Almagest* (Odense, Denmark: Odense University Press).

Petavius, Dionysius [Denis Petau] 1630. *Uranologion sive systema variorum authorum, qui de sphaera ac sideribus eorumque motibus Graece commentati sunt. Cura & studio Dionysii Petavii Aurelianensis e Societate Iesu* (Paris, 1630).

Pfeiffer, R. 1968. *History of Classical Scholarship: From the Beginnings to the End of the Hellenistic Age* (Oxford: Clarendon Press).

Pindar, *Odes*. See Sandys 1915.

Pingree, D. 1986. *Vettii Valentis Antiocheni Anthologiarum libri novem* (Leipzig: Teubner).

Pliny, *Natural History*. See Rackham 1947–63.

Plutarch, *Isis and Osiris*. See Babbitt 1936.

———, *On the Face in the Orb of the Moon*. See Cherniss and Helmbold 1957.

Porter, B., and R.L.B. Moss 1927–. *Topographical Bibliography of Ancient Egyptian Hieroglyphic Texts, Reliefs, and Paintings*, 7 vols. (Oxford: Clarendon Press).

Poseidōnios, *Fragments*. See Edelstein and Kidd 1989; Kidd 1988; Kidd 1999.

Price, D.J. 1969. "Portable Sundials in Antiquity," *Centaurus* 14, 242–66.

——— 1974. *Gears from the Greeks. The Antikythera Mechanism—A Calendar Computer from ca. 80 B.C.* Transactions of the American Philosophical Society, vol. 64, pt. 7.

Pricianus Lydus, *Explanation of Problems for King Chosroes.* See Bywater 1886.

Pritchett, W.K., and O. Neugebauer 1947. *The Calendars of Athens* (Cambridge: Harvard University Press).

Proklos, *Commentary on the First Book of Euclid's* Elements. See Friedlein 1878; Morrow 1970.

———, *Sketch of the Astronomical Hypotheses.* See Manitius 1909.

Pseudo-Aristotle, *On the Cosmos.* See Forster and Furley 1978.

Ptolemy, *Almagest.* See Heiberg 1898–1907, vol. 1; Toomer 1984.

———, *Geography.* See Berggren and Jones 2000.

———, *Optics.* See Smith 1996.

———, *Phaseis.* See Heiberg 1898–1907, vol. 2.

———, *Tetrabiblos.* See Robbins 1940.

Pytheas of Massalia, *On the Ocean.* See Mette 1952; Roseman 1994.

Rackham, H. 1947–63. ed. and trans. Pliny, *Natural History,* 10 vols. (Cambridge: Harvard University Press; London: W. Heinemann).

——— 1961. ed. and trans. Cicero, *De natura deorum; Academica* (Cambridge: Harvard University Press).

Rawlins, Dennis 1982. "The Eratosthenes-Strabo Nile Map," *Archive for History of Exact Sciences* 26, 211–19.

Reiner, Erica 1995. *Astral Magic in Babylon.* Transactions of the American Philosophical Society, vol. 85, pt. 4.

Reinhardt, K. 1921. *Poseidonios.* (Munich: C.H. Beck).

——— 1926. *Kosmos und Sympathie* (Munich: C.H. Beck).

Robbins, F.E. 1940. ed. and trans. Ptolemy, *Tetrabiblos.* (London: William Heinemann; Cambridge: Harvard University Press).

Rocca-Serra, G. 1980. ed. and trans. Censorinus, *Le jour natal* (Paris: J. Vrin).

Rochberg-Halton, F. 1984. "New Evidence for the History of Astrology," *Journal of Near Eastern Studies* 43, 115–40.

——— 1988. "Elements of the Babylonian Contribution to Hellenistic Astrology," *Journal of the American Oriental Society* 108, 51–62.

Rohr, R.-J. 1980. "A unique Greek sundial recently discovered in central Asia," *Journal of the Royal Astronomical Society of Canada* 74, 271–78.

Rome, A. 1927. "L'astrolabe et le météoroscope d'après le commentaire de Pappus sur le 5e livre de l'Almagest," *Annales de la Société Scientifique de Bruxelles* (Série A) 47, 77–102.

——— 1931. *Commentaires de Pappus et de Théon d'Alexandrie sur l'Almageste.* Tome 1. Pappus d'Alexandrie, *Commentaire sur les livres 5 et 6 de l'Almageste.* Studi e Testi 54 (Rome: Biblioteca Apostolica Vaticana).

Roseman, C.H. 1994. *Pytheas of Massalia: "On the Ocean." Text, Translation and Commentary* (Chicago: Ares).

Rouse, W.H.D. 1982. ed. and trans. Lucretius, *De rerum natura,* 2nd ed., revised by M.F. Smith (Cambridge: Harvard University Press; London: William Heinemann).

Russo, L. 2004. *The Forgotten Revolution: How Science Was Born in 300 B.C. and Why It Had to Be Reborn*, trans. by Silvio Levy (Berlin: Springer).

Sallmann, K. 1988. *Betrachtungen zum Tag der Geburt = De die natali / Censorinus* (Weinheim: VCH).

Sambursky, S. 1959. *Physics of the Stoics* (Princeton: Princeton University Press).

―――― 1962. *The Physical World of Late Antiquity* (Princeton: Princeton University Press).

Samuel, A.E. 1972. *Greek and Roman Chronology: Calendars and Years in Classical Antiquity*. Handbuch der Altertumswissenschaft, ser. 1, pt. 7 (Munich: Beck).

Sandys, J. 1915. *The Odes of Pindar* (London: W. Heinemann; New York: Macmillan).

Sarton, G. 1970. *A History of Science*, 2 vols. (New York: W.W. Norton; first pub. 1959).

Savoie, D., and R. Lehoucq 2001. "Étude gnomonique d'un cadran solaire découvert à Carthage," *Révue d'Archéométrie* 25, 25–34.

Schaefer, B.E. 2005. "The Epoch of the Constellations on the Farnese Atlas and Their Origin in Hipparchus's Lost Star Catalogue," *Journal for the History of Astronomy* 36, 167–96.

Schaldach, K. 2004. "The Arachne of the Amphiareion and the Origin of Gnomonics in Greece," *Journal for the History of Astronomy* 35, 435–45.

Schmidt, M.C.P. 1884a. "Wann schrieb Geminus," *Philologus* 42, 83–110.

―――― 1884b. "Wo schrieb Geminus," *Philologus* 42, 110–18.

Schmidt, O. 1952. "Some Critical Remarks about Autolycus' *On Risings and Settings*," *Den 11. Skandinaviske Matematikerkongress i Trondheim 22–25 August 1949* (Oslo: 1952), pp. 27–32.

Schöne, H. 1903. ed. *Heronis Alexandrini Opera quae supersunt omnia*, vol. 3, *Rationes dimetiendi et commentatio dioptrica. Vermessungslehre und Dioptra* (Leipzig: Teubner).

Schöne, R. 1897. *Damianos Schrift Über Optik mit Auszügen aus Geminos* (Berlin: Reichsdruckerei).

Sélincourt, A. de 1972. trans. Herodotus, *The Histories* (Harmondswoth: Penguin Books).

Sextus Empiricus, *Against the Professors*. See Bury 1987.

Sezgin, F. 1974. *Geschichte des Arabischen Schrifttums. Band V: Mathematik bis ca. 430 H.* (Leiden: E.J. Brill).

Simplikios, *Commentary on Aristotle's* On the Heavens. See Heiberg 1894; Mueller 2004; Mueller 2005.

――――, *Commentary on Aristotle's* Physics. See Diels 1882.

Smart, W.M. 1977. *Textbook on Spherical Astronomy*, 6th ed., revised by R.M. Green (Cambridge: Cambridge University Press).

Smith, A.M. 1981. "Saving the Appearances of the Appearances: The Foundations of Classical Geometrical Optics," *Archive for History of Exact Sciences* 24, 73–99.

―――― 1982. "Ptolemy's Search for a Law of Refraction: A Case-Study in the Classical Methodology of Saving the Appearances," *Archive for History of Exact Sciences* 26, 221–40.

———— 1996. *Ptolemy's Theory of Visual Perception: An English Translation of the* Optics *with Introduction and Commentary*. Transactions of the American Philosophical Society, vol. 86, pt. 2.

———— 1999. *Ptolemy and the Foundations of Ancient Mathematical Optics: A Source Based Guided Study*. Transactions of the American Philosophical Society, vol. 89, pt. 3.

Soubiran, J. 1969. Vitruve, *De l'architecture Livre IX* (Paris: Les Belles Lettres).

Stahl, W.H. 1952. trans. Macrobius, *Commentary on the Dream of Scipio* (New York: Columbia University Press).

Strabo, *Geography*. See H.L. Jones 1959–61.

Swerdlow, N.M. 2004. "Ptolemy's *Harmonics* and the 'Tones of the Universe' in the *Canobic Inscription*," in C. Burnett, J.P. Hogendijk, K. Plofker, M. Yano, eds., *Studies in the History of the Exact Sciences in Honour of David Pingree* (Leiden: Brill).

Tannery, P. 1893. *L'histoire de l'astronomie ancienne* (Paris: Gauthier-Villars & Fils).

———— 1897. *La géométrie grecque* (Paris: Gauthier-Villars & Fils).

Taub, L. 2003. *Ancient Meteorology* (London and New York: Routledge).

Theōn of Smyrna, *Mathematical Knowledge Useful for Reading Plato*. See Dupuis 1892.

Thompson, D.J. 1988. *Memphis under the Ptolemies* (Princeton: Princeton University Press).

Tittel, K. 1910. "Geminos I," in *Paulys Realencyclopädie der classsischen Altertumswissenschaft. Neue Barbeitung*. Halbband XIII (Stuttgart: A. Druckenmüller), cols. 1026–50.

Todd, R. 1976. *Alexander of Aphrodisias on Stoic Physics* (Leiden: E.J. Brill).

———— 1982. "Cleomedes and the Stoic Concept of the Void," *Apeiron* **16**, 129–36.

———— 1985. "The Title of Cleomedes' Treatise," *Philologus* **129**, 250–61.

———— 1990. *Cleomedis Caelestia (ΜΕΤΕΩΡΑ)* (Leipzig: Teubner).

———— 1993. "The Manuscripts of the Pseudo-Proclan *Sphaera*," *Revue d'Histoire des Textes* **23**, 57–71.

———— 2003a. "Damianus (Heliodorus Larissaeus)," in Virginia Brown, ed., *Catalogus Translationum et Commentariorum*, vol. 8 (Washington, D.C.: Catholic University of America Press), pp. 1–5.

———— 2003b. "*Geminus and the Ps.-Proclan Sphaera*," in Virginia Brown, ed., *Catalogus Translationum et Commentariorum*, vol. 8 (Washington, D.C.: Catholic University of America Press), pp. 7–48.

Toomer, G.J. 1984. trans. *Ptolemy's* Almagest (London: Duckworth).

Tredennick, H. 1961–62. ed. and trans. Aristotle, *The Metaphysics* (Cambridge: Harvard University Press; London: W. Heinemann).

Trypanis, C.A. 1978. ed. and trans. Callimachus, *Aetia, Iambi, lyric poems, Hecale, minor epic and elegiac poems, and other fragments* (Cambridge: Harvard University Press; London: Heinemann).

Valerio, V. 1987. "Historiographic and numerical notes on the Atlante Farnese and its celestial sphere," *Der Globusfreund* **35/37**, 97–124.

Van der Waerden, B.L. 1960. "Greek Astronomical Calendars and Their Relation to the Athenian Civil Calendar," *Journal of Hellenic Studies* 80, 68–180.

————— 1974. *Science Awakening II. The Birth of Astronomy* (New York: Oxford University Press).

————— 1984. "Greek Astronomical Calendars I. The Parapegma of Euctemon," *Archive for History of Exact Sciences* 29, 101–14.

Ver Eekke, P. 1933. trans. Pappus d'Alexandrie, *La collection mathématique* (Paris: Desclée De Brouwer).

Vettius Valens, *Anthologies*. See Pingree 1986; Bara 1989.

Vitruvius, *On Architecture, IX.* See Soubiran 1969.

Vlastos, G. 1975. *Plato's Universe* (Seattle: University of Washington Press).

Von Boeselager, D. 1983. *Antike Mosaiken in Sizilien* (Rome: Giorgio Bretschneider).

Wachsmuth, C. 1863. *Ioannis Laurentii Lydi liber de ostentis et Calendaria Graeca omnia* (Leipzig: Teubner). 2nd ed., 1897.

Weinstock, S. 1951. *Catalogus Codicum Astrologorum Graecorum*; vol. 9, *Codices Britannicos*; pt. 1, *Codices Oxonienses* (Brussels: In Aedibus Academiae).

Wicksteed, P.H., and F.M. Cornford 1960–63. eds. and trans. Aristotle, *The Physics*, 2 vols. (Cambridge: Harvard University Press).

Index

www.ingramcontent.com/pod-product-compliance
Ingram Content Group UK Ltd.
Pitfield, Milton Keynes, MK11 3LW, UK
UKHW040942151224
452260UK00002B/14/J